A Geographical Century

Vladimir Kolosov •
Jacobo García-Álvarez •
Michael Heffernan •
Bruno Schelhaas
Editors

A Geographical Century

Essays for the Centenary of the International Geographical Union

Springer

Editors
Vladimir Kolosov
Institute of Geography
Russian Academy of Sciences
Moscow, Russia

Michael Heffernan
School of Geography
University of Nottingham
Nottingham, UK

Jacobo García-Álvarez
Department of Humanities: History,
Geography and Art
Carlos III University of Madrid
Getafe, Madrid, Spain

Bruno Schelhaas
Archive for Geography
Leibniz Institute for Regional Geography
Leipzig, Germany

ISBN 978-3-031-05418-1 ISBN 978-3-031-05419-8 (eBook)
https://doi.org/10.1007/978-3-031-05419-8

This Springer imprint is published by the registered company Springer Nature Switzerland AG
The registered company address is: Gewerbestrasse 11, 6330 Cham, Switzerland

Preface

It is a unique privilege to be so intimately involved with a distinguished international organization such as the International Geographical Union (IGU) as it celebrates one hundred years since its formal establishment. The present volume reflects on this illustrious history in great depth and detail and, in doing so, offers an important viewpoint at a critical juncture—not only on the organization itself, but on the nature and development of the discipline of geography in general. So much has changed in a 100-year period of history that has been punctuated by the cataclysmic World War II, countless national and regional conflicts, political revolutions, sundry economic disasters and stock market crashes—and now even a global pandemic that has taken its toll, one way or the other, on just about everybody on the planet. This has indeed been a tumultuous period. Despite such massive challenges, since the establishment of a scientific union for Geography in 1922 and its acceptance as a member of the recently hatched International Research Council—itself a product of negotiations that took place at the end of the First World War—our beloved discipline has flourished and continues to gain in significance.

Over above the turbulent political and economic events that have characterized the last one hundred years, we now find ourselves facing problems of titanic proportions that have arisen due to the unsustainable utilization of the earth's rich but limited resources expressed. What we now refer to as the climate emergency is the most obvious manifestation of a wider and continually evolving human–environmental crisis. In straddling the natural and social sciences and humanities, geography aims to integrate the study of both natural and human realms and their interactions, focusing on space, places, and regions, addressing and questioning both short-term and longer-term processes and their resultant patterns. In this sense, geography is a science that is critical to understanding the processes, patterns, and trajectories of our future Earth, an understanding that is vital if we are to give future generations any chance at all of an equitable and sustainable future. As an international organization, the IGU oversees a discipline attempting to grapple with some very big issues indeed. It is no mean responsibility.

As this volume surely demonstrates, the IGU has evolved to be a diverse, even heterogeneous, body but one that has consistently demonstrated its effectiveness as a platform for synthesis. It should, therefore, continue to prove to be a highly valued source of intellectual reinforcement for the discipline. One hundred years down the line, we can of course pat ourselves on the back for presiding over an increasingly dynamic organization, but the nature and scale of the current challenges confronting humankind suggests that this is no time to rest on our laurels—there is much work ahead. I express the sincere hope that the IGU President in 2122 will be able to look back with pride on what will then be our bicentennial and feel a sense of satisfaction that this work was indeed well done!

Cape Town, South Africa

Michael E. Meadows
President: International Geographical Union 2020–2024

Acknowledgements

This volume is the result lengthy discussions dating back to February 2016 when an initial meeting was organized in Leipzig involving members of the IGU Centennial and Sesquicentennial Task Force, the IGU Commission on the History of Geography, and the Leibniz Institute for Regional Geography. The general editors wish to thank all those organizations and individuals—especially the section editors and chapter authors—who have generously supported this project, many from positions of responsibility in IGU executive bodies and commissions. Special thanks are extended to past and present members of the Commission on the History of Geography who have been actively involved in this project from the beginning; to Marie-Vic Ozouf-Marignier and Nicolas Verdier for invaluable advice and hospitality; and to Philip Jagessar for editorial assistance.

Contents

Editors and Contributors

About the Volume Editors

Vladimir Kolosov is Deputy Director of the Institute of Geography in the Russian Academy of Sciences, Foreign Member of the Belgian Royal Academy of Sciences, Honorary Member of the French and Polish Geographical Societies and a Doctor Honoris Causa of the University of Le Havre in France. He has lectured and conducted research at universities in France, the USA, Finland, and Germany. His research interests lie in the fields of political geography and geopolitics, social geography, and world cities. From 2006 to 2020, Professor Kolosov served on the Executive Committee of the International Geographical Union (IGU) and was President from 2012 to 2016. He has participated in many international projects.

Jacobo García-Álvarez is Professor of Human Geography at the Carlos III University of Madrid, which he joined in 2003. His research work has focused on the history of geography, historical geography, and political geography, with special attention to político-administrative divisions, boundary-making processes, and territorial identities and ideologies in Spain. He has lectured in several universities and research centres in Europe, Latin America, and the USA. He chaired the IGU Commission on the History of Geography and the International Union of History of Science and Technology from 2008 to 2016. He has coordinated the History of Geography, Historical Geography, and Political Geography Research Group in the Institute of Historiography at the Carlos III University in Madrid since 2012.

Michael Heffernan is Professor of Historical Geography at the University of Nottingham. He has also taught at the universities of Cambridge, Loughborough, and UCLA, and was an Alexander von Humboldt Research Fellow at the University of Heidelberg. He is a Fellow of the British Academy, the Academia Europaea, the Royal Academy of Arts, and both the Royal Geographical and Royal Historical Societies. A former Vice-Chair of the IGU Commission on the History of Geography, he is interested in the histories of geographical and environmental thought from the eighteenth century to the present.

Bruno Schelhaas received his Ph.D. from the University of Leipzig. He is Head of the Archive for Geography at the Leibniz Institute for Regional Geography in Leipzig and the IGU's official archivist. He is member of the Steering Committee of the Commission History of Geography of the International Geographical Union. His interests include the history of geography and cartography, historical geography, and archival science. In 2020, he was corresponding editor of Springer's Historical Geography and Geosciences volume *Decolonising and Internationalising Geography*.

About the Part Editors

Federico Ferretti is full Professor of Geography at the 'Giovanni Maria Bertin' Department of Education Studies in the University of Bologna in Italy. He received his Ph.D. from the universities of Bologna and Paris I (Panthéon-Sorbonne) in 2011. After research and teaching experiences in Italy, France, Switzerland, and Brazil, he was Associate Professor in the School of Geography at University College Dublin in Ireland from 2015 to 2021, working on the history and philosophy of geography and on critical and anarchist geographies, with a special focus on Latin America. Since 2016, he has been Secretary/Treasurer of the IGU Commission on the History of Geography.

Marcella Schmidt di Friedberg is a full Professor of Geography at the 'Riccardo Massa' Department of Human Sciences for Education at the University of Milano-Bicocca in Italy. Since 2016, she has been Chair of the IGU Commission on the History of Geography and Deputy Director of the Marine Research and High Education Center in Faaf-Magoodhoo in the Republic of the Maldives. Her research interests lie in cultural geography, hazards and resilience, gender geography, and the history of geographical thought. She has recently been working on relations between nature, culture, memory, and landscape in different national contexts, from the Mediterranean to Japan.

Alexander B. Murphy is a still-active Professor Emeritus in the Department of Geography at the University of Oregon in the USA, where he long held the Rippey Chair in Liberal Arts and Sciences. A former President of the American Association of Geographers and co-chair of the IGU's Commission on Political Geography, he specializes in political and cultural geography, with regional emphases on Europe and the Middle East. In the early 2000s, he chaired the U.S. National Academy of Sciences—National Research Council Committee charged with identifying 'Strategic Directions for the Geographical Sciences'.

Michael E. Meadows is Emeritus Professor and Visiting Research Scholar in the Department of Environmental and Geographical Science at the University of Cape Town, South Africa, and Professor in the School of Geographic and Ocean Sciences at Nanjing University. He was Visiting Professor in the School of Geographic Sciences at East China Normal University from 2016 to 2020. Professor Meadows obtained his undergraduate degree from the University of Sussex and Ph.D. from the University of Cambridge in the UK. His research interests and publications lie broadly in the field of physical geography and more specifically concern Quaternary environmental change and the geomorphological and biogeographical impacts of natural and anthropogenic changes. Professor Meadows has served as IGU Secretary-General and Treasurer and will be IGU President from 2020 to 2024.

Contributors

Ronald F. Abler Department of Geography, Pennsylvania State University, University Park, PA, United States

Trevor Barnes Department of Geography, University of British Columbia, Vancouver, Canada

Mireia Baylina Universitat Autònoma de Barcelona (UAB), Barcelona, Spain

Ruishan Chen Department of Landscape Architecture, School of Design, Shanghai Jiao Tong University, Shanghai, China

Pascal Clerc Cergy Paris University, Paris, France

Karl Donert European Association of Geographers (EUROGEO), Wardamme, Belgium

Debin Du School of Urban and Regional Science, Institute for Global Innovation and Development, East China Normal University, Shanghai, China

Louis Dupont Sorbonne Université, Paris, France

Federico Ferretti Department of Education Studies, University of Bologna, Italy

Joos Droogleever Fortuijn Department of Geography, Planning and International Development Studies, University of Amsterdam, Amsterdam, Netherlands

Jacobo García-Álvarez Department of Humanities: History, Geography and Art, Carlos III University of Madrid, Getafe, Spain

Maria Dolors Garcia-Ramon Universitat Autònoma de Barcelona (UAB), Barcelona, Spain

Michael F. Goodchild University of California, Santa Barbara, USA

Michael Heffernan School of Geography, University of Nottingham, Nottingham, UK

Heike Jöns School of Social Sciences and Humanities, Loughborough University, Loughborough, UK

Vladimir Kolosov Institute of Geography, Russian Academy of Sciences, Moscow, Russia

Christine Kosmopoulos Chief Editor, Cybergeo, European Journal of Geography, CNRS Laboratory, Géographie-cités, Aubervilliers, France

Nathalie Lemarchand University of Paris, Paris, France

Pengfei Li Haskayne School of Business, University of Calgary, Calgary, AB, Canada

Yingjie Li Department of Fisheries and Wildlife, Center for Systems Integration and Sustainability, Michigan State University, East Lansing, MI, USA

Virginie Mamadouh Department of Geography, Planning and International Development Studies, University of Amsterdam, The Netherlands

Michael E. Meadows Department of Environmental & Geographical Science, University of Cape Town, Rondebosch, South Africa

Rafael de Miguel González EUROGEO & University of Zaragoza, Zaragoza, Spain

Janice Monk SIROW, University of Arizona, Tucson, USA

Alexander B. Murphy Department of Geography, University of Oregon, Eugene, USA

André Reyes Novaes Universidade do Estado do Rio de Janeiro, Rio de Janeiro, Brazil

Jorge Olcina Cantos Department of Regional and Physical Geography, University of Alicante, Alicante, Spain

Denise Pumain Emeritus Professor, University Paris I Panthéon-Sorbonne, Paris, France;
CNRS Laboratory, Géographie-cités, Aubervilliers, France

Xionghe Qin Institute for Global Innovation and Development, East China Normal University, Shanghai, China

Rindra Raharinjanahary Antananarivo University, Antananarivo, Madagascar

Michael Roche School of People Environment and Planning, Massey University, Palmerston North, New Zealand

Bruno Schelhaas Leibniz Institute for Regional Geography, Leipzig, Germany

Marcella Schmidt di Friedberg Department of Human Sciences for Education, University of Milano-Bicocca, Italy

Benno Werlen Department of Geography, UNESCO Chair on Global Understanding for Sustainability, Friedrich Schiller University Jena, Jena, Germany

Marek Więckowski Institute of Geography and Spatial Planning, Polish Academy of Sciences, Warsaw, Poland

Chao Ye School of Geographic Sciences, East Cina Normal University, Shanghai, China

Annah Zhu Environmental Policy Group, Wageningen University, Wageningen, The Netherlands

1 Introduction

Vladimir Kolosov, Jacobo García-Álvarez, Michael Heffernan, and Bruno Schelhaas

The International Geographical Union (IGU) is the only global organisation representing the entire discipline of Geography. The IGU brings together geographers from more than 100 countries, as scholars, teachers, and practitioners. Formally established in Brussels under the auspices of the International Research Council in 1922, the IGU is one of the oldest scientific associations in the world, though it was preceded by a sequence of ten International Geographical Congresses (IGCs), inaugurated in Antwerp in 1871.

In July 2022, the IGU will celebrate its centenary at an extraordinary congress in Paris. This provides an opportunity for the global community of geographers to reflect on the internationalisation of Geography and the IGU's role in fostering this process and the wider cause of scientific international cooperation. The internationalisation of Geography, initiated more than a century before the current, turbulent era of globalisation, has been a remarkably resilient and creative endeavour that has overcome many reversals and diversions during two World Wars, severe economic depressions, Cold War military and geopolitical conflicts, the imposition and collapse of dictatorships, and the end of communism in most parts of the world.

As the IGU's history makes clear, internationalism and the interdisciplinarity are mutually sustaining aspirations. Geography's internationalism has always been closely associated with the discipline's productive diversity and inherent interdisciplinary character. Inspired by traditions of teaching and research inherited from the natural sciences, the social sciences, and the humanities, Geography's internationalisation, revealed by the expanding range and quality of collaborations between geographers in different parts, facilitated by the IGU, has greatly expanded the discipline's conceptual ambition and empirical evidence.

Like most sciences, Geography has continually evolved over the past century. The pace of change has accelerated in recent decades in response to rapidly changing economic, social, political and above all environmental challenges. New research directions have emerged across the discipline, enabled by the rapid development of new information and communication technologies and new

V. Kolosov (✉)
Institute of Geography of the Russian Academy of Sciences, Moscow, Russia
e-mail: vladimirkolossov@gmail.com

J. García-Álvarez
Department of Humanities: History, Geography and Art, Carlos III University of Madrid, Getafe, Spain
e-mail: jacobo.garcia@uc3m.es

M. Heffernan
School of Geography, University of Nottingham, Nottingham, UK
e-mail: mike.heffernan@nottingham.ac.uk

B. Schelhaas
Leibniz Institute for Regional Geography, Leipzig, Germany
e-mail: b_schelhaas@leibniz-ifl.de

V. Kolosov et al. (eds.), *A Geographical Century*,
https://doi.org/10.1007/978-3-031-05419-8_1

methods and theoretical approaches. Modern, twenty-first century Geography, now more diverse and interdisciplinary than ever, is perfectly attuned to a world of complexity. It provides a unique disciplinary perspective on the most difficult global problems, from climate change to the diffusion of pandemics. As an international educational science, modern Geography also deepens popular understanding of the nature-society interface, the widening gap between the global North and global South, the growing geographical inequalities that impede the economic, social, and cultural prosperity within regions and nation-states, and the processes of social fragmentation and polarisation that have arisen in recent years.

The IGU has been critically important to Geography's development as an international, interdisciplinary project. The IGU's many activities are carried out by 43 Commissions, some associated with specific sub-disciplines such as Geomorphology and Political Geography, others organised regionally or thematically such as those concerned with Coastal Systems, Cold and High Latitude Regions, Geographical Education, the Mediterranean Basin, Latin America, and Africa. Each IGU Commission is itself a global association with memberships that range from a few hundred to several thousand.

This volume of newly commissioned essays is not the first to reflect on the IGU's history and significance. Two important volumes reviewed aspects of the IGU's history and current situation in the early 1970s and mid-1990s. The first volume, published to mark the centenary of the first IGC and the 50th anniversary of the IGU in 1972, was entrusted by the IGU's Executive Committee to Philippe Pinchemel, the founding chair of the IGU Commission on the History of Geographical Thought (renamed in 2008 as the Commission on the History of Geography). The volume focused mainly on the history of the IGCs, reviewing in six chapters, co-written by no fewer 15 authors, the organisation and leading participants at these events, as well as their role in the development of different branches of geographical inquiry (Union Géographique Internationale 1972).

The second volume, published as the IGU prepared to celebrate its 75th anniversary in 1997, was edited by Marie-Claire Robic, Anne-Marie Briend and Mechtild Rössler. It was written mostly by five researchers belonging to, or associated with, the EHGO team at CNRS-Université Paris I created by Philippe Pinchemel and directed at that time by Marie-Claire Robic. Organised into five sections and 18 chapters, *Géographes face au monde* was the product of thorough and systematic research on the history and the development of IGU after World War Two, and was illustrated with dozens of maps, statistical tables, and photographs. The editors and authors made use of an extensive bibliography of published work in the history of science and science and technology studies. The volume also included some of the first essays to make systematic use of previously unknown archival documentation, complemented by personal testimonies and recollections from five prominent geographers (Robic et al. 1996).

It is not our intention to replicate, challenge or replace these outstanding commentaries, both of which contain richly diverse and carefully researched essays that will remain fundamental works of reference. That said, a great deal of new unpublished documentation on the IGU's history has emerged in the past 25 years, including congress proceedings, excursion guides, collected volumes by commissions and national committees, newsletters, journals, reports, minutes, keynote lectures and presented papers, as well as unpublished correspondence. These resources, partially reviewed in the substantial body of new published work since 1996, provides an opportunity to reconsider the history of the IGU as it marks its centenary and to reflect on its role within the discipline over an entire century.

The story of the IGU's archival records, recounted in some of the essays in this volume, is an interesting commentary on the complex process of Geography's internationalisation since 1922 (see also Hodder et al. 2021). Like many international organisations, the IGU has never had a permanent office. The organisation's administration, and associated archives, have continually moved from place to place, following each newly elected Secretary, though some IGU documents have been purloined for more sinister political reasons, notably during the German occupation of Paris in World War Two. This constant mobility prevented any systematic attempt to catalogue the collection, or even to preserve the documentary record, until 1982 when attempts were made to establish a more professional archive of materials covering the years from 1956 onwards (Schelhaas and Pietsch 2020). The archive's mobility did not end there, however, as documents were relocated to London and then Rome before finally moving to their permanent home at the Leibniz Institute for Regional Geography in Leipzig where they are now available for consultation by any interested scholar.[1]

Geoffrey Martin, a renowned US-based historian of Geography, established the first catalogue of the IGU archives in 2002. He gathered together 214 boxes of documents and developed a useful organising structure for the collection based on twelve series. Many of these documents were derived from the offices of IGU Secretaries and Presidents. The material focuses mainly on the Executive Committee, IGCs, member countries and commissions. Besides these written sources, the IGU archives also contains visual materials, including a collection of videos, the 'Dialogue Project', that features interviews with leading twentieth century geographers. Many of these are available online on YouTube[2] and the website of University College Dublin.[3] The IGU is always interested in receiving new documents related to its history in any format—paper, digital, or audio-visual. Several essays in this volume, especially in Part One, draw on materials from the IGU archive.

This new volume of essays is divided into four parts, in an attempt to provide both chronological and thematic coherence. The six chapters in Part One, edited by Federico Ferretti, provide a historical perspective focusing on the evolution of Geography as an international science before and after the foundation of the IGU in 1922. The first two chapters, by Michael Heffernan and Bruno Schelhaas, provide new accounts of Geography's institutional internationalisation, focusing on the IGCs, the creation and early history of the IGU, and its relations with other organisations. Both chapters make extensive use of the IGU's archives. The next four chapters explore specific historical and geographical themes within this broad context. Based on the experiences of geographers in the USSR, China and Poland, Vladimir Kolosov, Marek Więckowski, Debin Du, and Xionghe Qin consider how Geography's internationalisation was simultaneously challenged and facilitated by the geopolitical divisions of the Cold War within socialist countries, a story in which the IGU played a fascinating role. In their chapter, Trevor Barnes and Michael Roche provide a critical commentary on the circulation and dissemination of geographical ideas, especially mathematically based theories and concepts associated with the so-called 'quantitative revolution', emphasising once again the importance of the IGU as a forum for international exchange. In the final two chapters in this Part, Heike Jöns and Joos Droogleever Fortuijn explore the complex meaning of scientific internationalism, drawing on recent research in science and technology studies. These two chapters provide detailed, extensively illustrated analyses of recent IGCs and the structures, commissions, and personalities that have shaped the values and practices of the IGU in the post-Cold War era.

The five chapters in Part Two, edited by Marcella Schmidt di Friedberg, address some of the current challenges facing the international community of geographers. The authors of these chapters consider how the IGU has responded to these challenges and what further actions might be necessary in the future. The impact of the Internet, information technologies and open

[1] See https://igu-online.org/igu-archives/. For more information, please contact Bruno Schelhaas, the IGU archivist, at archiv@leibniz-ifl.de.

[2] https://www.youtube.com/channel/UC1WzSi02jYP3QgjseHxKB3g and https://www.youtube.com/channel/UCH2zyPfp_AaPpxZFJfX62BA.

[3] https://www.ucd.ie/geography/research/lifeexperienceascatalystforcross-disciplinarycommunication/.

access journals is addressed in the first chapter by Denise Pumain and Christine Kosmopoulos. In the second chapter, Rindra Raharinhjanahary, Nathalie Lemarchand and Louis Dupont discuss the importance of linguistic diversity in the creation and dissemination of geographical knowledge in a context where English has become the globally hegemonic form of scientific communication. The third chapter, by Rafael de Miguel González and Karl Donert, is a critical overview of the past, present, and future of international geographical education, focusing on the importance of the IGU and its connections with UNESCO and other educational agencies. The fourth chapter, by Pascal Clerc and André Reyes Novaes, provides a novel intellectual history of the most fundamental of geographical divisions, between the 'global South' and the 'global North', which has shaped the development of the modern discipline of Geography, as well as the policies and practices of the IGU. In the fifth chapter, Mireia Baylina, Maria Dolors García-Ramón and Janice Monk, chart the interwoven histories of feminism and internationalism in modern Geography, emphasising the role of the IGU's Commission on Gender and Geography in promoting women's voices in the discipline across all parts of the world and as an institutional space in which previously overlooked questions and perspectives can be formulated and debated.

The five chapters in Part Three, edited by Alexander Murphy and Michael Meadows, offer a more prospective approach and reflect on the twenty-first century challenges confronting international Geography and the IGU in particular. This Part considers Geography's capacity as a 'bridging discipline' between the natural, social and human sciences, and its potential as an integrative, interdisciplinary project. This theme is explored with reference to the human-environmental problems of the Anthropocene in the first chapter, by Ruishan Chen, Annah Zhu, Yingjie Li, Pengfei Li, Chao Ye and Michael Meadows. In the second chapter, Benno Werlen demonstrates how the International Year of Global Understanding, an IGU contribution to the UNESCO International Years programme initiated in 2016, promises a new geographical paradigm and worldview for the twenty-first century. The final three chapters review the potential contribution of Geography and IGU for the analysis and resolution of some of the main questions relating to environmental and territorial sustainability, discussed by Jorge Olcina; the information society, analysed by Michael Goodchild; and a range of social, demographic and educational questions considered by Alexander Murphy and Virginie Mamadouh. As the title of this Part suggests, these chapters consider Geography's potential as a 'dream discipline', the 'science of the twenty-first century', that can make valuable contributions to the challenges of the present and offer a vision of a better and more secure world. The volume closes with a critical concluding essay by IGU Past President Ronald Abler.

This volume includes contributions from 35 authors based in 16 countries from all regions of the world. They include physical geographers, human geographers, and specialists in Geographical Information Science. As such, the volume mirrors the richness and diversity of contemporary international geographical inquiry. We hope these essays will be of interest to all those concerned with the past, present, and future of the IGU and of international Geography.

References

Hodder J, Heffernan M, Legg S (2021) The archival geographies of twentieth-century internationalism: nation, empire and race. J Hist Geogr 71:1–11

Robic MC, Briend AM, Rössler M (eds) (1996) Géographes face au monde. L'Union Géographique Internationale et les Congrès Internaionaux de Géographie/Geographers and the World: The International Geographical Union and the International Geographical Congresses. L'Harmattan, Paris

Schelhaas B, Pietsch S (2020) (Re-)writing the history of IGU? A report from the archives. In: Schelhaas B et al (eds) Decolonising and internationalising geography: essays in the history of contested science. Springer, Cham, pp 127–140

Union Géographique International, Commission d'Histoire de la Pensée Géographique/International Geographical Union. Commission on the History of Geographical Thought (1972) La géographie à travers un siècle de Congrès internationaux/Geography through a Century of International Congresses. Caen, UNESCO

Part I

Geography as International Science: Historical Perspectives on the IGU

Internationalising Geography, 1871–1945

2

Michael Heffernan

Abstract

This chapter provides a critical commentary on attempts to internationalise the discipline of Geography before and after the establishment of the International Geographical Union in 1922. Drawing on previously unused archives, the chapter analyses how a liberal geographical internationalism was created and performed at ten International Geographical Congresses, from Antwerp in 1871 to Rome in 1913, in an ultimately unsuccessful attempt to challenge the intellectual influence of nationalism and imperialism. The early history of the International Geographical Union and the controversies associated with the International Geographical Congresses in Cairo (1925), Cambridge (1928), Paris (1931), Warsaw (1934) and Amsterdam (1938), discussed in the second and third sections of the chapter, illustrate how Geography's fragile internationalism was repeatedly compromised through the interwar period.

Keywords

Internationalism · History of Geography · International Geographical Congresses/Union · Late nineteenth and early twentieth centuries

2.1 Introduction

> *Science is essentially international, and every worker finds, from time to time, the need of freeing himself from the intellectual preoccupations of his fellow-countrymen. This is especially the case with geography, which, of all the branches of knowledge, requires most to be studied from the standpoint of a citizen of the world.*

This quotation, from the personal recollections of British geographer Charles Arden-Close (1947, 144), expressed a widely-shared view about Geography's international affinities after World War Two. Geography's internationalism, measured by the discipline's enthusiasm for international congresses, organisations and research projects, waxed and waned in the decades before Close wrote these words, in productive tension with the competing ideologies of nationalism and imperialism that exerted more obvious and extensively researched influence (Bell et al. 1995; Godlewska and Smith 1995; Driver 2001). This chapter provides a brief history of Geography's internationalism from the 1870s to the 1940s in the belief that the discipline's international future can only be assured by a critical engagement with its contested international past.

M. Heffernan (✉)
School of Geography, University of Nottingham, Nottingham NG7 2RD, UK
e-mail: Mike.Heffernan@nottingham.ac.uk

V. Kolosov et al. (eds.), *A Geographical Century*,
https://doi.org/10.1007/978-3-031-05419-8_2

2.2 Internationalism Performed: International Geographical Congresses, 1871–1913

Geography was one of the first sciences to embrace the spirit of internationalism, barely a generation after the first geographical societies were established in Paris (1821), Berlin (1828) and London (1830). The ten International Geographical Congresses (IGCs) organised between 1871 and 1913 reflected a growing enthusiasm for international scientific conferences whose number increased from 40 in the 1870s to 200 by the 1900s (Table 2.1; Feuerhahn and Rabault-Feuerhahn 2010; Fox 2016, 19; Rasmussen 1990; Robic 2010, 2013; Schroeder-Gudehus 1990).[1]

These early IGCs attracted between c. 400 and c. 1500 delegates, the latter figure reached only at the London and Berlin congresses in 1895 and 1899. The first IGC, organised in Antwerp seven months after the Franco-Prussian war, provided a geopolitical template for subsequent congresses by emphasising how international scientific collaboration could improve otherwise tense relations between rival nation-states. International collaboration was never the unchallenged motivation, however, and most IGCs reflected local and national ambitions, often forcefully expressed during the debates at every congress about the location of the next meeting.

Most early IGCs were organised by geographical societies whose number, size and wealth increased exponentially in the late nineteenth century. Surviving photographs suggest IGCs delegates were mainly though not exclusively male (Robic and Rössler 1996b). Financial support was often provided by the municipal authorities of host cities, anxious to associate themselves with prestigious international conferences. The IGCs in Antwerp (1871), Berne (1891), and Geneva (1908) benefitted from a conviction, equally prevalent in Belgium and Switzerland, that small, politically neutral countries were the natural custodians of modern internationalism (Herren 2017; Herren and Zala 2002; Laqua 2013). These three IGCs attracted fewer than 600 delegates each but were among the most international in terms of nationalities represented.

Delegates at the 1871 Antwerp IGC were invited to consider 87 questions ranging from the general ('What are the best ways of teaching geography?') to the highly specific ('What would be the consequences of creating a man-made inland sea in the northern Sahara?') (Briend 1996, pls I-VIII). Subsequent IGCs adopted a similar approach, with questions linked to continually changing congress themes. The c. 50 questions circulated for the 1881 Venice IGC included inquiries about the geography of nervous ailments and the nature of geographical causality, the latter prompted by post-Darwinian debates about the influence of the physical environment on human activity (Arden-Close 1947, 151).

As the IGC series became better established, local organising committees arranged increasingly elaborate social events, partly to encourage media interest. The tradition of inviting heads of state, royal patrons and aristocratic grandees to gala events began at the 1881 Venice IGC where the opening ceremony was attended by King Umberto I and Queen Consort Margherita. At the 1889 Paris IGC, delegates were entertained at the private residence of Prince Roland Bonaparte, traveller, anthropologist and photographer whose fabulous wealth derived not from his famous family but from his wife's controlling stake in the Monte Carlo casinos. For the 1895 London IGC, Queen Victoria and the Prince of Wales were the official patrons and the Duke of York (later George V) presided at the opening ceremony. Prince Albrecht of Prussia, cousin of Kaiser Wilhelm II, performed a similar role at the 1899 Berlin IGC.

Several IGCs were associated with public commemorations, anniversaries and urban expositions. The 1871 Antwerp IGC, the brainchild of Charles Ruelens, keeper of the Royal Library in Brussels, was part of a campaign to promote the importance of Flanders in the history of cartography which focused initially on two proposals, conceived separately in 1869, to erect statues of Abraham Ortelius in his native

[1] The proceedings of the IGCs discussed in this chapter were published in various formats and with titles too lengthy to repeat. Bibliographical details are listed in Briend 1996, 321-323.

Table 2.1 International Geographical Congresses 1871–1938. The varying durations reflected differing approaches to field excursions, some of which took place during the proceedings, especially in earlier congresses, though most were arranged afterwards and are excluded from these dates

Year	Host city
1871 (14–22 August)	Antwerp
1875 (1–11 August)	Paris
1881 (15–22 September)	Venice
1889 (5–10 August)	Paris
1891 (10–14 August)	Berne
1895 (26 July–3 August)	London
1899 (28 September–4 October)	Berlin
1904 (8–22 September)	Washington DC, and other US cities
1908 (27 July–6 August)	Geneva
1913 (27 March–4 April)	Rome
1925 (1–9 April)	Cairo
1928 (18–25 July)	Cambridge
1931 (16–24 September)	Paris
1934 (23–31 August)	Warsaw
1938 (18–28 July)	Amsterdam

Antwerp and Gerardus Mercator at his birthplace in nearby Rupelmonde (Shimazu 2015).

The 1889 Paris IGC, overseen by Ferdinand de Lesseps, architect of the Suez Canal and president of the Paris Geographical Society (SGP), was one of hundreds of international scientific conferences that took place in the French capital during the Exposition Universelle, a six-month celebration of the French Revolution's centenary that attracted more than 32 million visitors (Rasmussen 1989). The Paris Exposition 'served as an unfailing source of amusement and recreation' for the 530 IGC delegates whose programme included visits to exhibitions of Bonaparte's disturbing photographs of indigenous peoples (Morgan 1889, 552).

Delegates at the 1904 IGC in the United States assembled at Hubbard Memorial Hall in Washington DC, the National Geographic Society's newly-built headquarters, before travelling by chartered train to Philadelphia, New York, Niagara Falls, Chicago, and Saint Louis, where their programme was absorbed into the International Congress of Arts and Sciences, the intellectual centrepiece of the 1904 World's Fair to mark the centenary of the Louisiana Purchase. The 1904 IGC was intended to symbolise the westward advance of the American frontier and some delegates continued their journey to Santa Fe, the Grand Canyon, the Rocky Mountains, San Francisco and Mexico.

Most IGCs made their own contributions to urban expositions in the form of public exhibitions of maps, globes and instruments of survey and exploration. The tradition began at the 1871 Antwerp IGC with a modest exhibition celebrating the Flemish origins of modern cartography (Briend 1996, pl. III). For the 1875 Paris IGC, a more ambitious and politically contentious exhibition was organised in the Palais de l'Industrie, created for the 1855 Exposition Universelle. This featured hundreds of 'objects connected… with the science and practice of Geography' supplied by commercial and educational companies in the hope of winning official prizes.[2] Pride of place was given to a huge wall-map of France, roughly 10 m square, on which the 'lost provinces' of Alsace-Lorraine, ceded to the new German Empire four years earlier, were defiantly represented as part of 'la mère patrie', albeit with a dark shaded border (Dunlop 2015, 152–3, pl. 8).

[2]Quoted in a parliamentary question by Richard Monckton Milnes, Baron Houghton, to Foreign Secretary Edward Stanley, Earl of Derby, asking the British government to fund a representative to take charge of the British exhibits in Paris. See *Hansard (House of Lords)* 28th May 1875, vol. 224, cc 1005-7.

The 1875 map exhibition opened two weeks before the IGC and continued for a month afterwards, a tactic repeated at the 1881 Venice IGC. The success of these exhibitions, which were widely reported in newspapers, prompted mutterings of discontent that the scientific work of both congresses had been overshadowed by commercial ventures (Arden-Close 1947, 151). In response to these concerns, the cartographic exhibitions at the 1889 Paris IGC were self-consciously educational and scientific. Historical maps from French and European libraries were displayed in the Palais des Arts Libéraux, one of the cavernous Exposition buildings, alongside richly coloured demographic and medical maps of France prepared by Victor Turquan, the new head of the French Statistical Bureau.

The desire to incorporate international scientific conferences into urban expositions partly explains the irregular sequence of IGCs before 1914. The 1891 Berne IGC took place just two years after the 1889 Paris IGC under pressure from the Swiss government, determined to attract international conferences to its federal capital during the 700th anniversary celebrations of its foundation. Conversely, the 1913 Rome IGC, originally scheduled to take place during the city's Esposizione Internazionale d'Arte in 1911 to mark the 50th anniversary of Italian unification, was repeatedly postponed, thereby undermining the original rationale to locate the congress in the Italian capital.

Only 400 delegates attended the 1913 congress, confirming a growing scepticism among serious-minded university geographers that IGCs prioritised social events and commercial ventures over scientific discussion. Some argued that congresses should avoid large capital cities where social and cultural distractions were many and varied. Recalling the 1913 IGC, Close noted that 'even those who knew Rome fairly well could not resist the temptation of prowling about the city when they should have been attending to the affairs of the Congress' (Arden-Close 1947, 169).

The failure of early IGCs to convert high-sounding resolutions into sustainable long-term projects prompted an important organisational change at the 1891 Berne IGC when commissions were created to develop international research projects between congresses (Collignon 1996, 117). One of the most important commissions was established in response to a lecture at the Berne congress by Albrecht Penck, professor of Geography at the University of Vienna and later the University of Berlin, who challenged national cartographic agencies to collaborate in the compilation of a new 1:1 million International Map of the World (IMW) based on common conventions and symbols (Penck 1892).

The IMW was endlessly discussed at successive IGCs but little was achieved before two important conferences, in London in 1909 and Paris in 1913, secured the necessary international agreements, despite the withdrawal of the United States on the eve of the second conference (Robic 1996b; Pearson et al. 2006; Pearson and Heffernan 2015; Rankin 2017). The outbreak of World War One halted progress on the IMW and other IGC initiatives, including a proposal at the 1913 Rome IGC to establish a World Union of Geographical Societies, a recognition that Geography's internationalisation required a permanent co-ordinating organisation (Arden-Close 1947, 169).

2.3 Internationalism Interrupted: World War One, the International Research Council, and the International Geographical Union

Geography became a science of war in most belligerent nation-states between 1914 and 1918. Dozens of Germany's most distinguished scientists, including Fritz Haber, Ernst Haeckel and Max Planck, signed the 'Manifesto of the 93' in October 1914 expressing full support for the German cause (MacLeod 2018). As Haber later wrote, 'the duty of the scientist is to humanity in peacetime but the fatherland in war' (Harris 1992).

The geographical societies in London, Paris and New York, home to substantial map collections and specialist cartographic facilities, were partially converted into military and strategic research centres linked to government departments (Heffernan 2000a). Existing geographical projects, including those originally conceived as international collaborations, were made to serve narrower national objectives associated with the war effort. British cartographers in the Royal Geographical Society (RGS), led by the society's waspish secretary, Arthur Hinks, re-configured their hesitant pre-war work on the IMW into a new project to prepare an outline 1:1 million base-map of Europe and the Middle East for future peace negotiations (Heffernan 1996).

Longstanding personal friendships and professional collaborations were broken, sometimes permanently. Sven Hedin, the Swedish explorer of Central Asia, previously feted by the RGS and the SGP and awarded an honorary knighthood by the British government, was expelled from both geographical societies early in the war following his vocal support for the German cause. Shortly after he was awarded the RGS Founders' Medal on the eve of World War One, Penck was arrested as an enemy alien and placed under house arrest in London until January 1915 (Heffernan 2000b).

Geography's mobilisation continued into the post-war period when some of the discipline's most distinguished figures were recruited as scientific experts to advise national delegations at the Paris Peace Conferences. Isaiah Bowman, Director of the American Geographical Society (AGS) in New York, was Chief Territorial Specialist to the US delegation, and Emmanuel de Martonne, professor of Geography at the Sorbonne, played a similar role for the French delegation. Both men were prominently involved in wartime American and French geopolitical 'think-tanks', the House Inquiry and the Comité d'Études, based in the AGS and the SGP respectively (Heffernan 1995, 2001, 2002; Lowczyk 2010; Prott 2014; Reisser 2012; Smith 2003, 113–80; Souton and Davion 2015).

Delegations from aspiring nations seeking independence from old European empires also relied on geographical experts at the Peace Conferences. Jovan Cvijić was an eloquent advocate of Serbia-Yugoslavia and Eugeniusz Romer played a similar role for the Polish delegation. Countries hoping to avoid dismemberment also drew on geographical experts. The Budapest geographer Pál Teleki, later the country's Prime Minister, was a prominent member of the Hungarian delegation (Crampton 2006; Győri and Withers 2019; Seegel 2018; Warmoes et al. 2016). The small, hastily assembled German delegation, to whom peace terms were dictated, had no scientific advisers to contest putative territorial changes. The German academic advisers, sociologist Max Weber and historian Hans Delbrück, focused on challenging the 'war guilt clause' in the Versailles Treaty that insisted Germany and its allies bore sole responsibility for the war.

The Peace Conferences confirmed the wartime impression of Geography as a science of territory, land and power. Some geographers celebrated this new-found political importance but others were concerned by the discipline's renewed association with nationalism and imperialism. From their perspective, a reformed international Geography was necessary to reflect the 'esprit de Genève' embodied in the League of Nations and other international organisations. The International Committee on Intellectual Cooperation (ICIC), established in 1922 and directed by luminaries such as Henri Bergson, Albert Einstein, Marie Curie, Thomas Mann and Gilbert Murray, sought to encourage international collaboration in the arts and sciences, a project later co-ordinated from the Paris-based International Institute of Intellectual Cooperation (IIIC) (Renoliet 1999). This included programmes for a new university science of International Studies and reviews of school textbooks in History and Geography to persuade reluctant education ministries to emphasise international peace and cooperation in national curricula (Fleure 1934; Unstead 1934; McCarthy 2011).

Unfortunately, the International Research Council (IRC), created by the Versailles Treaty to facilitate international scientific collaboration, undermined these ambitions by limiting the geographical scope of international scientific renewal. The IRC emerged from meetings in late 1918 and early 1919 involving senior representatives from the Royal Society in London, the Academy of Sciences in Paris, and the National Academy of Sciences in Washington DC (Greenaway 1996, 1–18). Formally constituted in Brussels on 28th July 1919, the IRC sought to establish a network of international scientific unions across all disciplines that would act in a coordinated manner according to agreed rules (Greenaway 1996, 19–32).

Despite their rhetorical support for scientific internationalism, the IRC's Executive Committee was determined to destroy Germany's scientific pre-eminence.[3] Scientists from Germany and its war time allies, including Austria, Hungary and Bulgaria, were to be excluded from international collaborations, including scientific congresses organised by affiliated IRC associations. Even scientists from neutral countries were viewed with suspicion and only welcomed into IRC associations to prevent their potential recruitment by an alternative, German-led international coalition (Cock 1983; Fox 2016, 45–73; Irish 2015; Reinbothe 2019; Schroeder-Gudenhus 1966, 1973, 1986, 2014). The IRC policy was absurdly self-defeating. In the decade before the outbreak of war in 1914, more than 40% of scientific papers in Chemistry and Physics were published in German, 33% in English, and 12% in French. By excluding German scientists from international collaboration, the IRC was cutting itself off from almost half the world's experts in many scientific disciplines (Richard 1990, 401; see also Forman 1973; Iaria et al. 2018).

The IRC initially comprised four international unions for Astronomy (IAU), Chemistry (IUPAC), Geodesy and Geophysics (IUGG), and Radio Science (known by its French acronym URSI). Two additional unions were established for the Biological Sciences (IUBS) and Mathematics (IMU) in 1919 and 1920. At the second IRC general assembly in Brussels in July 1922, new international unions were created for Physics (IUPAP) and Geography. The International Geographical Union (IGU) was primarily a French proposal, initiated by Guillaume Grandidier, the SGP secretary, with support from Charles Lallemand, the French president of the IUGG, the French Service Géographique de l'Armée, and several Belgian and Italian geographers (Robic 1996a). The explicit objective was to eradicate Germany's previously dominant influence from the discipline.

The RGS initially opposed this idea. Hinks had no objection to the document's anti-German objectives but was innately hostile to all international ventures, including the IMW. 'International scientific congresses of the non-official kind have been useful in the past for making acquaintances and exchanging ideas', he claimed, 'but it is doubtful if they have contributed to science any positive results commensurate with the labour they have absorbed' (Hinks 1920, 142).

Grandidier's proposal was welcomed by the IRC's Executive Committee, however, and a full constitution was formulated at the 1922 General Assembly, with input from previously sceptical British geographers. Bonaparte, by then SGP president, was elected as the first IGU president, with Close, recently retired as director of the Ordnance Survey (OS), as Secretary. A chastened Hinks reluctantly accepted the new arrangements (Hinks 1922).

An IRC-compliant IGU was viewed with dismay by the German geographical community, justifiably proud of its leading role in the discipline's history. Relations between German geographers and their colleagues in other countries, notably France, had been difficult in the past but the IRC's systematic anti-German policy caused deep and lasting resentment (Rössler 1990). Penck, who became Rector of the University of Berlin during the war, was especially aggrieved. As the IMW's main progenitor, Penck had now to accept the humiliating prospect of exclusion from the project's delayed

[3]The committee included Arthur Schuster, a German-born British physicist and secretary at the Royal Society; Émile Picard, a French mathematician and permanent secretary at the Academy of Sciences; George Ellery Hale, an American astronomer and foreign secretary at the National Academy of Sciences; and Georges Lecointe, director of the Belgian Royal Observatory.

post-war development. Rejecting his pre-war internationalism, Penck spent most of the 1920s challenging the geopolitical injustices of the Versailles treaties, especially the allocation of the Danzig 'corridor' to Poland, intensifying his acrimonious dispute with Romer, a former student (Hagen 2009; Labbé 2018; Seegel 2018).

2.4 Internationalism Re-launched: From Cairo to Cambridge

The IGU's first task was to decide where the next IGC should be located. A decision taken at the 1913 Rome IGC to hold the next congress in St Petersburg (then Petrograd and soon to be Leningrad) was no longer viable as the Soviet Union was excluded from the IRC. Bonaparte, supported by some Italian geographers, agreed to a hastily prepared proposal from the Egyptian government to organise an IGC in Cairo in 1925 to mark the 50th anniversary of the Royal Egyptian Geographical Society (EGS) (Reid 1993).[4]

Close was angered by this decision, partly because he was not consulted but mainly because his contacts in the British Foreign Office suggested the new Egyptian King, the Italian-educated Fuad I who had served as EGS president from 1915 to 1918, intended to use the IGC to promote an anti-British message at a time when the Foreign Office was blocking attempts by the Egyptian government to join the League of Nations. Close also suspected Bonaparte's motives. The EGS, which Fuad proposed to refurbish at enormous expense, was immediately adjacent to the Institut d'Égypte, the headquarters of the Napoleonic survey of Egypt established by Bonaparte's great uncle in 1798. Close feared a Cairo IGC, located in this historically significant setting, would draw an unhelpful contrast between Britain's imperial legacy in independent Egypt, revealed by its continuing military presence, and France's sophisticated cultural influence, suggested by an international scientific congress likely to be conducted in French.

Close's anxieties worsened when he was informed that the congress organisers, unaware that the IGU was bound by IRC rules, had sent invitations to geographical societies in Germany, Austria, Hungary, Bulgaria and Turkey. Special invitations had even been issued to the family of the German explorer Georg August Schweinfurth, the EGS's inaugural president. Faced with this embarrassing situation, Close decided to resign as IGU Secretary and was only persuaded to remain, following appeals from de Martonne and Bowman, on condition that invitations to 'ineligible' organisations were rescinded, apparently unmoved by the indignation this confusion generated in Germany.

Hinks refused to accept the validity of the Cairo proposal, repeating his opposition to congresses in capital cities, 'where hotels are expensive', and his preference for 'small, less well-known, and geographically interesting places' (Hinks 1922, 294). IGCs should have 'no banquets, receptions, gala performances at the Opera, or other costly impediments to work', he insisted, and should be located 'in a University town, or a quiet health resort, or in any pleasant place with comfortable but simple accommodation, and a sufficiency of meeting rooms that could be borrowed or hired… Such a Congress would not attract a crowd, and would be all the better for that; but if it did not attract the right kind of serious geographers, then perhaps the present international congresses are better left uncalled' (Hinks 1923, 443).

Despite this blustering, preparations continued in Cairo. Bonaparte died, at the age of 65, a few months before the congress began and was replaced as IGU president by Nicola Vacchelli, director of the Istituto Geografico Militare, president of the Italian Geographical Society, and the official geographer of Mussolini's Fascist regime. Close overcame his reservations and played a prominent role in the Cairo congress, later describing it as an 'interesting gathering' (Arden-Close 1947, 172). The congress generated five volumes of published proceedings and

[4]The following discussion draws on unpublished materials in the papers of Charles Arden-Close in the archives of the RGS-IBG, CFA1-6.

was less bacchanalian than Hinks feared, though there were several languid social interludes, including a tea party beside the Pyramids, hosted by Fuad, a visit to the Opera, and lengthy excursions along the Nile to Luxor and Howard Carter's recently excavated archaeological sites in the Valley of the Kings (Martonne 1925).

It was agreed in Cairo that the next IGC should take place in Britain in 1928, away from the bright lights of London. Close's self-serving suggestion that the congress should be organised in his home town of Winchester was sensibly rejected in favour of a proposal from Cambridge, where Hinks taught surveying, with a preliminary IMW meeting at the RGS in London. Preparations began in earnest, overseen by a local committee chaired by Frank Debenham, co-founder of the Scott Polar Research Institute in Cambridge and a veteran of Robert Falcon Scott's ill-fated Terra Nova expedition to the Antarctic in 1910–13. Close, by now RGS president, produced a colourful conference logo and persuaded the OS to prepare a new map of Cambridge highlighting the conference buildings.

As the IRC's anti-German policy had long since unravelled, especially after the 1925 Locarno Treaties opened the door to German membership of the League of Nations, Close expected full German participation at the Cambridge congress (Robic and Rössler 1996a). A few months before the congress was due to begin, however, Debenham learned that German geographers were planning a 'counter-boycott' and would refuse to accept their invitations. Close, aware that the RGS's international reputation had been tarnished by an unfortunate dispute with the Norwegian explorer Roald Amundsen, asked Bowman and a senior Foreign Office official to write to the Berlin Geographical Society reinforcing the importance of German re-engagement (Hinks 1927).

The German geographers stood firm, however, as did most of their colleagues from previously excluded countries. Their absence was widely lamented, not least by the British Foreign Secretary Austen Chamberlain, one of the architects of the Locarno Treaties and guest of honour at the congress dinner, hosted at Trinity College by the Nobel Prize winning physicist J. J. Thomson. Chamberlain noted that 'German statesmen' had re-joined the League of Nations two years earlier and were 'welcomed to that great society as colleagues and friends'. It was a matter of deep regret, he concluded, that German geographers 'would not do what the German government had done, and accept the welcome that was waiting for them' (quoted in Hinks 1928, 264).

The Cambridge IGC attracted more than 450 delegates from 30 countries and improved the previously marginal position of Geography in the city's ancient university. Shortly after the congress, the university established a professorial position in Geography, awarded to Debenham in 1931 (Ogilvie 1928). Four separate IGC map exhibitions were arranged at the British Museum and the RGS in London, and at the University Library and the Geography Department in Cambridge. Five new IGC commissions were established, in addition to the existing three, including a spin-off IMW project, devised by the socialist archaeologist O. G. S. Crawford, whom Close appointed at the OS, to compile an international 1:1 million map of the Roman Empire, the Tabula Imperii Romani (Hauser 2008; Robic 1996c).

2.5 Internationalism Compromised: International Geographical Congresses 1931–1941

These projects were reaffirmed at the next three IGCs in Paris, Warsaw and Amsterdam. The 1931 Paris IGC was chaired by de Martonne and took place during the Exposition coloniale internationale, a bizarre six-month celebration of colonialism, presented by the organisers as a progressive, modernising and international process inaugurated by the French occupation of Algiers a century earlier. Close to nine million people visited the Exposition's pavilions in the Bois de Vincennes to marvel at exhibits that included carefully recreated African villages and a perfect replica of the Angkor Wat temple (Morton 2000).

The 750 IGC delegates in Paris included representatives from 41 countries, though the majority were French (Clout 2005, 2012). More than 100 British geographers attended, the largest foreign delegation, though some senior figures, notably Close, stayed away. The continuing absence of German geographers was by now a serious embarrassment, made worse by the symbolic decision to organise a German-language geographical conference at the same time in the 'international' city of Danzig (Clout 2005, 16).

As Marie-Claire Robic (1996d, 2009) has shown, the 1931 Paris IGC reinforced the main trends of geographical research at the time, including population growth, migration, urbanisation, and urban and regional planning, all enhanced by more experimental forms of thematic cartography (Arrault 2007; Bashford 2014). Bowman was installed as the new IGU president in 1931 and his grand proposals for large-scale population resettlement across the world's 'pioneer fringe' exemplified these themes (Bowman 1931; Smith 2003, 211–234).

The large Polish delegation in Paris, led by Romer and generously supported by the Polish government, successfully campaigned for the 1934 IGC to be organised in Warsaw, further alienating German geographers still resentful at the loss of territory to the new Polish state. The Warsaw IGC was a highly-charged event overseen with characteristic energy by Romer who organised, among other things, a major cartographic exhibition to reinforce the legitimacy of Poland's international borders. Bowman, who played an important role in securing these borders at the Paris Peace Conference, was feted as a national hero (Smith 2003, 281).

More than 700 delegates attended, less than half from Poland. Forty-four countries were represented. The 87 French geographers, the largest foreign delegation, far outnumbered their colleagues from Britain, Italy and the United States. Thanks to Bowman's tireless efforts and skilful diplomacy, a 50-strong German delegation also attended, alongside smaller groups from Austria, Hungary and Bulgaria. A handful of Soviet geographers, enthusiastically courted by Romer, also took part (Clout 2013; Górny 2018; Jackowski et al. 2014).

Whatever satisfaction Bowman felt at Germany's long delayed re-engagement with the IGU was tempered by the obvious irony that their involvement had been facilitated by the newly installed Nazi regime in Berlin, ostensibly hostile to internationalism yet determined to re-assert a German presence in each and every international arena. The disciplined German delegation in Warsaw, led by Ludwig Mecking from the University of Münster, was dominated by younger academics, including several Nazi party members, whose presentations were carefully vetted by the new Deutsche Kongress-Zentrale (DKZ) to ensure consistency with Nazi ideology (Herren 2002; on Mecking, see Fahlbusch et al. 1989; Mecking 1934).[5]

Close, who was disturbed by the political content of German presentations in Warsaw, replaced Bowman as IGU President and worked diligently to prepare for the next IGC in Amsterdam in 1938. This was organised by anthropologist Johannes Pieter Kleiweg de Zwaan and Edward John Voûte, secretary of the Royal Dutch Geographical Society. Lectures took place in the Koloniaal Instituut where Queen Wilhelmina, the congress patron, was a regular visitor. Eleven volumes of proceedings were printed in advance of the congress in an attempt to shorten verbal presentations (Arden-Close 1947, 174–9).

More than 1230 delegates attended the Amsterdam congress, two-thirds from outside the Netherlands. The French and German delegations (147 and 137) were of comparable size and easily outnumbered representatives from the UK (94), the USA (73), Poland (61) and Italy (46). The German delegation included Mecking, by then at the University of Hamburg, but was officially overseen by Wolfgang Panzer from the University of Heidelberg. Voûte's careful scheduling ensured Panzer and his German

[5]Close's diary of his visit to the Warsaw IGC includes fascinating photographs and oblique comments on the political activities of German delegates (RGS-IBG archives, CFA1).

colleagues were protected from critical scrutiny by other delegates. As a direct result of his obliging administration, Voûte became mayor of Amsterdam during the Nazi occupation and was later condemned to a lengthy prison sentence for collaboration (Meershoek 2013). On the eve of World War Two, international Geography had become an ideological contest played out with strained politesse in IGC lecture halls, exhibition spaces, gala dinners and field excursions.

The political significance of international geographical organisations and congresses was highlighted during World War Two, often in tragic ways. Geographers in some occupied countries, notably Poland, were identified as dangerous intellectuals and enemies of the Reich. Romer survived the war under an assumed identity as a humble labourer but many Polish geographers were arrested, imprisoned and even killed. Stanisław Pawłowski, professor of Geography at Poznań University, where he also served as rector, and IGU vice-president from the Amsterdam congress in 1938, was arrested by the Gestapo in October 1939 and murdered on 6th January 1940. His children, Przemysław and Wanda, later died in Auschwitz. Following the Sonderaktion Krakau of November 1939, which targeted academics at Jagiellonian University, the geographer Jerzy Smoleński was imprisoned in Sachsenhausen and died there two months later (Jackowski et al. 2019).

In order to monitor and control scientific internationalism, the SS officer in charge of the DKZ, Karl Franz Schweig, personally directed a 'special action' in Paris on 30th July 1940 to seize IGU archives from de Martonne's offices in the Sorbonne. These documents were relocated to Berlin and subsequently lost or destroyed, as were other papers stolen the following month from the offices of Paul Michotte, the IGU's recently appointed secretary-general, at the University of Louvain (Herren 2002, 81–82). In December 1941, a new geographical society was established in Berlin, the Deutsche Geographische Gesellschaft (DGG), to co-ordinate the existing geographical organisations and ensure more efficient control by the Nazi regime. The DGG organised an international geographical 'congress' in Würzburg in March 1942, attended by a handful of compliant European delegates from Italy, Spain, Bulgaria and Finland whose deliberations were published in an expensively bound volume, edited by the Berlin geographer Norbert Krebs (Krebs 1943; see also Praesent 1942; Rössler 1990, 196; Troll and Fischer 1949, 119–120).

The minimal international presence in Würzburg prompted attempts to entice scholars in occupied countries to join forces with German geographers on projects deemed valuable to the Nazi authorities. In May 1942, the Kiel geographer Oskar Schmieder, who had previously worked with Carl Sauer in Berkeley, was dispatched to France in a futile attempt to 'recruit' leading geographers, including Roger Dion.[6] French geographers trained in the liberal, republican tradition of Paul Vidal de la Blache resisted these inducements even though their own government in Vichy was no less determined than the Nazis to develop ideologically amenable forms of geographical inquiry. Vichy educational reforms paved the way for a new agrégration in Geography that enhanced the discipline's independent status in higher education, separate from History under which it was previously subsumed (Clout 2015; Ginsburger 2017; Ginsburger et al. 2021; Heffernan 2005).

2.6 Conclusion

Several different forms of geographical internationalism jostled for pre-eminence in the decades considered in this chapter (Robic 1996a, 27–30; Meadows 2020). During the late nineteenth and early twentieth centuries, many leading geographers advocated a liberal scientific internationalism in order to challenge the aggressive nationalism and imperialism of the time. Their efforts to internationalise Geography were continually undermined by entrenched national rivalries, however, most obviously during and after World War One. Internationalism was also defined in this period in geographically restricted

[6]DKZ Records, Hoover Institution Archives (Stanford University), 126/G3.

terms, as an aspiration relevant only to Europe, the United States and a handful of mainly British 'settler' colonies. Attempts to include perspectives from 'the global south', notably at the Cairo IGC in 1925, proved contentious.

During the 1930s, geographers associated with authoritarian, ultra-nationalist regimes in Italy, Germany and elsewhere proposed alternative geographical internationalisms that rejected liberal democracy and promoted instead various kinds of racial theorising and distinctive interpretations of human-environment interactions. This intellectual detour had tragic moral and political consequences during World War Two.

As the next chapter shows, a liberal geographical internationalism was re-established after 1945 under the auspices of a re-launched IGU. International collaborations were constrained, however, by the intensifying Cold War that limited interaction between geographers on either side of the East-West divide. The end of the Cold War raised the enticing prospect of a new era of international intellectual and scientific cooperation but this opportunity has only partially been realised. It is difficult to predict what forms of geographical internationalism are likely to emerge in the twenty-first century, though one thing seems certain: as we confront deepening environmental crises, the inexorable growth of China's intellectual and geopolitical power, intensifying nationalism and populism in Europe and North America, and a global pandemic whose long-term consequences remain unclear, international geographical perspectives are more urgent than ever.

References

Arden-Close C (1947) Geographical byways and some other geographical essays. Edward Arnold, London

Arrault J-B (2007) Géographie et mondialisation: les géographes français face à la crise des années 1930. Annales de Géographie 657:493–512. https://doi.org/10.3917/ag.657.0493

Bashford A (2014) Global population: history, geopolitics, and life on earth. Columbia University Press, New York

Bell M, Butlin R, Heffernan M (eds) (1995) Geography and imperialism, 1820–1940. Manchester University Press, Manchester

Bowman I (1931) The pioneer fringe. American Geographical Society, New York

Briend A-M (1996) Bibliographie: les publications de l'UGI. In: Robic M-C, Briend A-M, Rössler M (eds) Géographes face au monde. L'Union géographique internationale et les Congrès internationaux de géographie. L'Harmattan, Paris, pp 317–431

Clout H (2005) Geographers in their ivory tower: academic geography and popular geography in Paris 1931. Geogr Ann B 87(1):15–29. https://doi.org/10.1111/j.0435-3684.2005.00175.x

Clout H (2012) French geographers under international gaze: regional excursions for the XIIIth International Geographical Congress, 1931. Belgeo: Revue Belge de Géographie 1–2:1–15 https://doi.org/10.4000/belgeo.6248

Clout H (2013) France, Poland and Europe: the experience of the XIVth International Geographical Congress, Warsaw 1934. Belgeo: Revue Belge de Géographie 4:1–10 https://doi.org/10.4000/belgeo.12138

Clout H (2015) French geographers during wartime and German occupation. J Hist Geogr 47:16–28. https://doi.org/10.1016/j.jhg.2014.04.003

Cock AG (1983) Chauvinism and internationalism in science: the International Research Council, 1919–1926. Notes Rec R Soc Lond 37(2):249–288. https://doi.org/10.1098/rsnr.1983.0013

Collignon B (1996) Les commissions, "l'organe le plus important de l'UGI." In: Robic M-C, Briend A-M, Rössler M (eds) Géographes face au monde. L'Union géographique internationale et les Congrès internationaux de géographie. L'Harmattan, Paris, pp 156–177

Crampton JW (2006) The cartographic calculation of space: race mapping and the Balkans at the Paris Peace Conference of 1919. Soc Cult Geogr 7(5):731–752. https://doi.org/10.1080/14649360600974733

Driver F (2001) Geography militant: cultures of exploration and empire. Blackwell, Oxford

Dunlop CT (2015) Cartophilia: maps and the search for identity in the French-German Borderland. University of Chicago Press, Chicago

Fahlbusch M, Rössler M, Siegrist D (1989) Conservatism, ideology and geography in Germany 1920–1950. Polit Geogr Q 8(4):353–367. https://doi.org/10.1016/0260-9827(89)90032-3

Feuerhahn W, Rabault-Feuerhahn P (eds) (2010) La fabrique internationale de la science: les congrès scientifiques de 1865 à 1945. Revue Germanique Internationale (Special Issue) 12:5–258 https://doi.org/10.4000/rgi.236

Fleure HJ (ed) (1934) The teaching of geography in relation to the world community. Cambridge University Press, Cambridge

Forman P (1973) Scientific internationalism and the Weimar physicists: the ideology and its manipulation in Germany after World War I. Isis 64(2):150–180. https://doi.org/10.1086/351079

Fox R (2016) Science without frontiers: cosmopolitanism and national interests in the world of learning, 1870–1940. Oregon State University Press, Corvallis

Ginsburger N (2017) Historiens et géographes au scalpel de Vichy. Revue d'Histoire des Sciences Humaines 31:163–185. https://doi.org/10.4000/rhsh.446

Ginsburger N, Robic M-C, Tissier J-L (eds) (2021) Géographes français en Seconde Guerre Mondiale. Éditions de la Sorbonne, Paris

Godlewska A, Smith N (eds) (1995) Geography and empire. Blackwell, Oxford

Górny M (2018) A vacuum to be filled: Central and Eastern Europe in the times of 'geography without the Germans.' Studia Historiae Scientiarum 17:253–272. https://doi.org/10.4467/2543702XSHS.18.010.9330

Greenaway F (1996) Science international: a history of the International Council of Scientific Unions. Cambridge University Press, Cambridge

Győri R, Withers C (2019) Trianon and its aftermath: British geographers and the dismemberment of Hungary, c. 1915-c. 1922. Scottish Geographical Journal 135(1–2):68–97. https://doi.org/10.1080/14702541.2019.1668049

Hagen J (2009) Mapping the Polish corridor: ethnicity, economics and geopolitics. Imago Mundi 62(1):63–82. https://doi.org/10.1080/03085690903319325

Harris H (1992) To serve mankind in peace and the fatherland in war: the case of Fritz Haber. Ger Hist 10 (1):24–38. https://doi.org/10.1093/gh/10.1.24

Hauser K (2008) Bloody old Britain: O. G. S. Crawford and the archaeology of modern life. Granta, London

Heffernan M (1995) The spoils of war: the Société de Géographie de Paris and the French empire, 1914–1919. In: Bell M, Butlin R, Heffernan M (eds) Geography and imperialism 1820–1940. Manchester University Press, Manchester, pp 221–264

Heffernan M (1996) Geography, cartography and military intelligence: the Royal Geographical Society and the First World War. Trans Inst Br Geogr 21(3):504–533. https://doi.org/10.2307/622594

Heffernan M (2000a) Mars and Minerva: centres of geographical calculation in an age of total war. Erdkunde 54:320–333. https://doi.org/10.3112/erdkunde.2000.04.03

Heffernan M (2000b) Professor Penck's bluff: geography, espionage and hysteria in World War I. Scottish Geographical Journal 116:267–282. https://doi.org/10.1080/00369220018737100

Heffernan M (2001) History, geography and the French national space: the question of Alsace-Lorraine, 1914–1918. Space and Polity 5:27–48. https://doi.org/10.1080/13562570120049645

Heffernan M (2002) The politics of the map in the early twentieth century. Cartogr Geogr Inf Sci 29(3):207–226. https://doi.org/10.1559/152304002782008512

Heffernan M (2005) Geography, empire and National Revolution in Vichy France. Polit Geogr 24:731–758. https://doi.org/10.1016/j.polgeo.2005.02.001

Herren M (2002) 'Outwardly… an innocuous conference authority': National Socialism and the logistics of international information management. Ger Hist 20 (1):67–92. https://doi.org/10.1191/0266355402gh245oa

Herren M (2017) Geneva, 1919–1945: the spatialities of public internationalism and global networks. In: Jöns H, Meusburger P, Heffernan M (eds) Mobilities of knowledge. Springer, Heidelberg, pp 211–226

Herren M, Zala S (2002) Netzwerk Aussenpolitik. Internationale Kongresse und Organisationen als Instrumente der schweizerischen Aussenpolitik 1914–1950. Zürich: Schweizer Beitrage zur internationalen Geschichte

Hinks AH (Anon.) (1920) A proposal for an International Geographical Union. Geogr J 55(2):139–142. https://doi.org/10.2307/1781587

Hinks AH (1922) The International Geographical Union. Geogr J 60(4):291–294. https://doi.org/10.2307/1781315

Hinks AH [Anon.] (1923) The future of international congresses. Geogr J 61(6):440–443. https://doi.org/10.2307/1780821

Hinks AH [Anon.] (1927) Captain Roald Amundsen and the Society. Geogr J 70(6):572–575. https://doi.org/10.2307/1782920

Hinks, AH [Anon.] (1928) International Geographical Congress, Cambridge, 1928. Geogr J 72(3):259–267. http://www.jstor.org/stable/1781976

Iaria A, Schwarz C, Waldinger F (2018) Frontier knowledge and scientific production: evidence from the collapse of international science. Quart J Econ 133 (2):927–991. https://doi.org/10.1093/qje/qjx046

Irish T (2015) The university at war, 1914–25: Britain, France, and the United States. Palgrave-Macmillan, London

Jackowski A, Bilska-Wodecka E, Sołjan I (2014) 80 years after the 14th Congress of the International Geographical Union in Warsaw, 23–31 August 1934. Geogr Pol 87(2):295–308. https://doi.org/10.7163/GPol.2014.18

Jackowski A et al. (eds) (2019) Geografia polska i geografowie w latach II wojny światowej. IGIGP UJ, Kraków

Krebs N (1943) Sitzungsberichte der Zusammenkunft europäischer Geographen in Würzburg 16. bis 19. März 1942. Verlag von Quelle & Meyer, Leipzig

Labbé M (2018) Eugene Romer's 1916 Atlas of Poland: creating a new nation state. Imago Mundi 70(1):94–113. https://doi.org/10.1080/03085694.2018.1382118

Laqua D (2013) The age of internationalism and Belgium, 1880–1930: peace, progress and prestige. Manchester University Press, Manchester

Lowczyk O (2010) La fabrique de la paix: du Comité d'Études à la Conférence de la Paix, l'élaboration par la France des traités de la Première Guerre mondiale. Économica, Paris

MacLeod R (2018) The mobilisation of minds and the crisis of international science: the Krieg der Geister and the Manifesto of the 93. J War Cult Stud 11 (1):58–78. https://doi.org/10.1080/17526272.2016.1264753

Martonne E de (1925) Le Congrès du Caire et l'avenir des Congrès géographiques internationaux. Annales de

Géographie 190:289–300. https://doi.org/10.3406/geo.1925.8294
McCarthy H (2011) The British people and the League of Nations: democracy, citizenship and internationalism. Manchester University Press, Manchester
Meadows M (2020) Perspectives on internationalism in geography. Geogr Pol 93(4):493–503. https://doi.org/10.7163/GPol.0189
Mecking L (1934) Blut und Boden – Erdkundliche Bildung im neuen Staat! Erdkunde in der neuen deutschen Bildung und Erziehung. Geographische Anzeiger (blätter Für Den Geographischen Unterricht) 35:1–6
Meershoek AJJ (2013) Voûte, Edward John (1887–1950), in Biografisch Woordenboek van Nederland http://resources.huygens.knaw.nl/bwm1880-2000/lemmata/bwm5/vote
Morgan ED (1889) The geographical congress in Paris. Proceedings of the Royal Geographical Society and Monthly Record of Geography 11(9):552–559 http://www.jstor.org/stable/1801337
Morton PA (2000) Hybrid Modernities: representation and architecture at the 1931 Colonial Exhibition. MIT Press, Cambridge, Mass, Paris
Ogilvie AG (1928) Great Britain: essays in regional geography (published on the occasion of the twelfth International Geographical Congress at Cambridge). Cambridge University Press, Cambridge
Pearson AW et al (2006) Cartographic ideals and geopolitical realities: international maps of the world from the 1890s to the present. Can Geogr 50:149–176. https://doi.org/10.1111/j.0008-3658.2006.00133.x
Pearson AW, Heffernan M (2015) Globalizing cartography? the International Map of the World, the International Geographical Union, and the United Nations. Imago Mundi 67:58–80. https://doi.org/10.1080/03085694.2015.974956
Penck A (1892) Die Herstellung einer einheitlichen Erdkarte em Maßtabe von 1:1,000,0000. In: Compte Rendu du Vème Congrès International des Sciences Géographiques tenu à Berne du 10 au 14 août 1891. Schmid & Franke, Berne, pp 191–198
Praesent H (1942) Die Arbeitssitzung europäischer Geographen in Würzburg. Geographische Zeitschrift 48(4–5):177–185 http://www.jstor.org/stable/27816164
Prott V (2014) Tying up the loose ends of national self-determination: British, French, and American experts in peace planning, 1917–1919. Hist J 57(3):727–750. https://doi.org/10.1017/S0018246X14000041
Rankin W (2017) Zombie projects, negative networks, and multigenerational science: the temporality of the International Map of the World. Soc Stud Sci 47 (3):353–373 https://doi.org/10.1177/0306312716658138
Rasmussen A (1989) Les congrès internationaux liés aux Expositions Universelles de Paris (1867–1900). Mil Neuf Cent 7:23–44. https://doi.org/10.3406/mcm.1989.976
Rasmussen A (1990) Jalons pour une histoire des congrès internationaux au XIXe siècle: régulation scientifique et propagande intellectuelle. Relations Internationales 62:115–133 http://www.jstor.org/stable/45344366
Reid DM (1993) The Egyptian Geographical Society: from foreign layman's society to indigenous professional association. Poetics Today 14(3):539–572. https://doi.org/10.2307/1773284
Reinbothe R (2019) Deutsch als internationale Wissenschaftssprache und der Boykott nach dem Ersten Weltkrieg. De Gruyter, Berlin
Reisser WJ (2012) The black book: Woodrow Wilson's secret plan for peace. Lanham, MD, Lexington
Renoliet J-J (1999) L'UNESCO oubliée: la Société des Nations et la coopération intellectuelle (1919–1946). Publications de la Sorbonne, Paris
Robic M-C (1996a) La naissance de l'Union géographique internationale. In: Robic M-C, Briend A-M, Rössler M (eds) Géographes face au monde. L'Union géographique internationale et les Congrès internationaux de géographie. L'Harmattan, Paris, pp 23–40
Robic M-C (1996b) Les vœux des premiers Congrès: dresser la carte du monde. In: Robic M-C, Briend A-M, Rössler M (eds) Géographes face au monde. L'Union géographique internationale et les Congrès internationaux de géographie. L'Harmattan, Paris, pp 149–178
Robic M-C (1996c) Résolutions et irrésolutions d'une Cité scientifique: l'entre-deux-guerres. In: Robic M-C, Briend A-M, Rössler M (eds) Géographes face au monde. L'Union géographique internationale et les Congrès internationaux de géographie. L'Harmattan, Paris, pp 179–226
Robic M-C (1996d) L'investissement cartographique des années trente. Cartographie thématique et nationalismes. In: Robic M-C, Briend A-M, Rössler M (eds) Géographes face au monde. L'Union géographique internationale et les Congrès internationaux de géographie. L'Harmattan, Paris, pp 227–240
Robic M-C (2009) La crise des années trente et la tension vers l'expertise géographique: expériences françaises et internationales – une nouvelle frontière? Confins: Revue Franco-Brésilienne de Géographie 5: online https://doi.org/10.4000/confins.5652
Robic M-C (2010) À propos de transferts culturels: les congrès internationaux de géographie et leurs spatialités. In: Feuerhahn W, Rabault-Feuerhahn P (eds) La fabrique internationale de la science: les congrès scientifiques de 1865 à 1945. Revue Germanique Internationale (Special Issue) 12:33–47 https://doi.org/10.4000/rgi.257
Robic M-C (2013) Connaître son monde: les géographes et les savoirs géographiques en congrès internationaux: spatialité et géographismes. Terra Brasilis: Revista da Rede Brasileira de História da Geografia e Geografia Historica 2: online https://doi.org/10.4000/terrabrasilis.863
Robic M-C, Rössler M (1996a) L'UGI, enjeu des relations internationales dans la première moitié du XXe. siècle. In: Robic M-C, Briend A-M, Rössler M (eds) Géographes face au monde. L'Union géographique internationale et les Congrès internationaux de géographie. L'Harmattan, Paris, pp 241–252

Robic M-C, Rössler M (1996b) Sirens within the IGU – an analysis of the role of women at International Geographical Congresses (1871–1996b). Cybergéo: European Journal of Geography 14: online https://doi.org/10.4000/cybergeo.5257
Rössler M (1990) La géographie aux congrès internationaux: échanges scientifiques et conflits politiques. Relations Internationales 62:183–199 http://www.jstor.org/stable/45344370
Schroeder-Gudehus B (1966) Deutsche Wissenschaft und internationale Zusammenarbeit 1914–1928. Dumaret & Golay, Geneva
Schroeder-Gudehus B (1973) Challenge to transnational loyalties: international scientific organizations after the First World War. Sci Stud 3:93–118. https://doi.org/10.1177/2F030631277300300201
Schroeder-Gudehus B (1986) Pas de Locarno pour la science: la coopération scientifique internationale et la politique étrangère des États pendant l'entre-deux-guerres. Relations Internationales 46:173–194 http://www.jstor.org/stable/45344021
Schroeder-Gudehus B (1990) Les congrès scientifiques et la politique de coopération internationale des académies des sciences. Relations Internationales 62:135–148 http://www.jstor.org/stable/45344367
Schroeder-Gudehus B (2014) Les scientifiques et la paix: la communauté scientifique internationale au cours des années 20. Presses de l'Université de Montréal, Montreal
Seegel S (2018) Map men: transnational lives and deaths of geographers in the making of East Central Europe. University of Chicago Press, Chicago
Shimazu T (2015) War, peace, and a geographical internationalism: the 1871 Antwerp International Geographical Congress. Geogr Rep Tokyo Metrop Univ 50:97–105
Smith N (2003) American empire: Roosevelt's geographer and the prelude to globalization. University of California Press, Berkeley and Los Angeles
Souton G-H, Davion I (eds) (2015) Les experts français et les frontières d'après-guerre: les procès-verbaux du Comité d'Études 1917–1919. Bulletin hors-série, Société de Géographie, Paris
Troll C, Fischer E (1949) Geographic science in Germany during the period 1933–1945: a critique and justification. Ann Assoc Am Geogr 39(2):99–137. https://doi.org/10.2307/2560953
Unstead JF (ed) (1934) Geography teaching in relation to world citizenship. League of Nations Union, London
Warmoes I, Hofmann C, Nathan I (eds) (2016) La carte dans les traités de paix de la guerre de 14/18. Cartes & Géomatique 228:111-182

Michael Heffernan is Professor of Historical Geography at the University of Nottingham. He has also taught at the universities of Cambridge, Loughborough and UCLA, and was an Alexander von Humboldt Research Fellow at the University of Heidelberg. He is a Fellow of the British Academy, the Academia Europaea, the Royal Academy of Arts, and both the Royal Geographical and Royal Historical Societies. He is interested in the histories of geographical and environmental thought.

The International Geographical Union in the Post-war Period

3

Bruno Schelhaas

Abstract

This chapter introduces into the history of the International Geographical Union after the Second World War and the development in the 1950s and 60s. After a short period of re-organisation, the Union's life re-started with the International Geographical Congress at Lisbon 1949. The international and often complicated political circumstances formed the basis for the Union's work. IGU was still a Western European and North American dominated union, but with several attempts of integration and cooperation within the international community of geographers. Many communities from Africa, Latin America and Asia played only a very marginal role, with only some exceptions. A special focus is on the Union within the East–West Conflict, with the complicated integration of China and the Eastern bloc members, and with the special case of German geography with its double structures between 1960 and 1990. Besides the political background, the chapter provides also insights into biographies of some leading figures within twentieth century IGU history. Among others, the Swiss geographer Hans Boesch was one of those celebrities, with extraordinary devotion for international geography, often under complicated circumstances and faced with financial, diplomatic and interpersonal problems.

Keywords

History of geography · International Geographical Union · International science organisations · International relations

3.1 Introduction

The century since the foundation of the International Geographical Union (IGU) in 1922 is a long time in modern academic as well as political history, and encompasses many different experiences of several generations of geographers. It is only possible in this chapter to present selected facets of the IGU's past, with a focus on the post-war years and subsequent Cold War period, and a few important figures and events from these years. Fortunately, we can benefit from some standard works on the IGU's history and a larger number of IGU publications (see Pinchemel 1972; Robic, Briend and Rössler 1996). The often complicated international circumstances that informed the IGU's work before World War Two, outlined in the preceding chapter, continued after 1945, characterised by diverse attempts at integration and cooperation within the international community of geographers.

B. Schelhaas (✉)
Leibniz Institute for Regional Geography, Schongauerstrasse 9, 04328 Leipzig, Germany
e-mail: B_Schelhaas@leibniz-ifl.de

V. Kolosov et al. (eds.), *A Geographical Century*,
https://doi.org/10.1007/978-3-031-05419-8_3

3.2 Restart and Continuation After 1945

The International Geographical Congress (IGC) at Amsterdam in 1938 was the last official IGU event before the World War Two. The newly elected Executive Committee (EC) included Emmanuel de Martonne, from France, previously the IGU's Secretary, as President (Delfosse 2001; Dresch 1988) and Paul Michotte, from Leuven in Belgium, as Secretary General (Dept 1945; Ghellinck 1945; Hegenscheidt 1943; Lefèvre 1941).[1] The Paris-Leuven axis was henceforth fundamental for the IGU's administration, and French became the preferred language for most of the IGU's communication.

These arrangements reflected the excellent personal relations between the geography departments in Paris and Leuven, and the strong influence of Emmanuel de Martonne on geographers in Belgium and across Europe. The new EC also included, as Vice Presidents, Charles Close from the UK, who had been closely connected with the IGU from the very beginning; Claude Hale Birdseye from the USA; Willem Everhard Boerman from the Netherlands; Ludwig Mecking from Germany; Stanisław Pawłowsky from Poland; and Antonio Renato Toniolo from Italy. After Michotte's sudden death in 1940, his assistant Marguerite Alice Lefèvre replaced him as secretary (Denis 1986; Polspoel 1964, 1965, 1968). Lefèvre, the first female geographer in the IGU's inner circle, was a member of the Paris geography group from 1922 to 1927 and greatly influenced by both de Martonne and Albert Demangeon. She was very active in several IGU commissions, especially those devoted to Rural Habitats, Erosion Surfaces, and Pliocene and Pleistocene Terraces, and attended every IGC from Cambridge in 1928 to London in 1964 (Droogleever Fortuijn 2019; Robic and Rössler 1996).

We have very little information about the activities of the IGU from 1938 to the end of the war (Schelhaas and Pietsch 2020). The German occupation in Belgium and France in 1940 prevented most scientific activities, including those of the IGU, though de Martonne noted immediately after the war that: "Cependant les hostilités éclatèrent et en novembre 1940 le Bureau décidait de réduire l'activité du Comité Exécutif de l'Union aux questions administratives, sans exclure toute action scientifique qui pourrait rester possible dans chaque pays" (Martonne 1946: 40).

The IGU's activities did not restart immediately after the liberation of Belgium and France in the summer of 1944, nor indeed after end of the war a year later. The Gestapo had dismantled the IGU's offices in Paris and Leuven, in July and August 1940 respectively, and transported the records to Berlin, whence they were partially returned to Belgium two years later a state of considerable disorder (Herren 2002: 81–82; Martonne 1946: 40; see also Schelhaas and Pietsch 2020: 129–131, and Chap. 1.1).

These wartime arrangements were retained until the 1949 Lisbon IGC, the first after World War Two. In her report to the Lisbon General Assembly on the IGU's activities since the 1938, Lefèvre noted that: "Lorsque en 1946 on songea à remettre l'Union en marche, il fallut commencer par reconstituer le Comité Exécutif" (Lefèvre 1950: 86). A new committee was duly elected, led by Martonne and Lefèvre as President and Secretary respectively, assisted by a new team of Vice Presidents that included Herbert John Fleure from the UK; Roberto Almagià from Italy; Willem Everhard Boerman from the Netherlands; George B. Cressey and John K. Wright from the USA); and Eugeniusz Romer from Poland. This group was a combination of longstanding IGU officers such as Fleure, Boerman and Romer (who had each served as Vice Presidents from 1931 to 1938) and new but already established authorities such as Almagià, Cressey and Wright. Romer was elected to replace his Polish colleague Stanisław Pawłowsky, murdered by German forces in 1940, though Lefèvre admitted that Romer would be unable to attend meetings from war-shattered Poland due to the difficulties of travel across the newly created Iron Curtain (Lefèvre 1950: 86; see also Olszewicz 1968; Wilcynski 1992).

[1]For all EC members 1922–2022, see Chap. 7 and Volle 1996b.

The IGU had 24 member states in 1946, provisionally without representatives from the losing nations in World War Two, Germany and Japan, and six commissions (Tables 3.1 and 3.2; Stratton 1946: 83–84; see also Volle 1996a; Collignon 1996; Chap. 7). The status of some countries that had been members of the IGU before the war was at the time still unclear, especially Brazil, Bulgaria and Chile.

The first meeting of the EC took place in July 1946 in London and was followed by meetings in Paris (July and November 1947), Brussels (September 1948) and Lisbon (April 1949) (Lefèvre 1950: 87). It seems that the meetings in 1946–48 were combined with general assemblies or other events of the International Council of Scientific Unions (ICSU) for which de Martonne, Boerman and Lefèvre were representatives for geography (Lefèvre 1950: 93; Stratton 1946: v and 45).

The affiliation with ICSU and UNESCO was very strong in the first post-war decade, mainly because of financial support (Rössler 1996). As Secretary Kimble reported at the General Assembly in Washington 1952: "Needless to say, this substantial program could not have been carried out without the assistance of UNESCO. Since Lisbon, the IGU has received nearly $40,000 from this organization, or approximately twice the income over the same period from dues" (Kimble 1957: 16).

3.3 International Geographical Congresses, 1949–56

The IGU General Assembly at Amsterdam in 1938 agreed to organise the next IGC four years later at Lisbon, but this decision was impossible to implement in 1942. After the first period of re-organising the IGU, the 1949 IGC at Lisbon was the 'official' re-start of IGU after the war. The main figure behind the organisation of the meeting was Orlando Ribeiro (Belo 1999; Daveau 2011, 2012; Ribeiro 2003), an internationally well-known scholar with very close connections to Emmanuel de Martonne. Ribeiro served as Vice President (1949–52) after the congress and then as first Vice President (1952–56). Ribeiro and Ernesto de Vasconcelos (VP 1924–28) are the only delegates from Portugal, an IGU founding member, to have worked in the IGU's inner circle to date. The Lisbon congress proceedings show about 700 participants were at the congress, coming from 37 countries, but half of them in absentia. Nevertheless, the meeting was a first success for the re-established IGU, only a few years after the end of the war.

The first post-war General Assembly saw the election of four new members, all of them of special interest: China, India, Hungary and Turkey. Remarkably, in the official list of congress participants we can find only one Chinese delegate,[2] three from Turkey,[3] but no-one from Hungary or India. The Eastern bloc was only represented by four delegates from IGU member Poland (including former Vice President Eugeniusz Romer) and four from full member Czechoslovakia, but nobody from the Soviet Union, which was not affiliated with the IGU at the time.[4] The actual participation of the Polish and Czechoslovak geographers is unclear, possibly only joining in absentia as John K. Wright reported: "Geographers from the U.S.S.R., Czechoslovakia, and other countries behind the 'iron curtain' did not participate in the sessions, though Poland was unofficially represented by a Polish resident in Great Britain" (Wright 1949: 484).

The delegates also elected a new EC for the period 1949–52 at the Lisbon General Assembly. George B. Cressey became the new President while George Kimble became Secretary (DeVivo 2015: 61–66; Herman 1965; James 1964; James and Perejda 1981). Both men were well established within international science and political affairs. Cressey was a well-travelled expert on Asia, especially China and the Soviet Union, and had already served as IGU Vice President since

[2] Lin Chao, the director of the Institute of Geography at Nanjing in the PR of China.

[3] Hakki Akyol from Istanbul University, together with Suzan Akyol, and Cemal Arif Alagöz from Ankara University.

[4] The third IGU full member in 1949 from the Eastern bloc was Romania, also not present at the Lisbon IGC.

Table 3.1 IGU Member Countries, 1946

Argentina	France	New Zealand	Sweden
Belgium	(Germany)	Norway	Switzerland
Canada	Greece	Poland	United Kingdom
Cuba	Italy	Portugal	USA
Czechoslovakia	(Japan)	Romania	Yugoslavia
Denmark	Morocco	South Africa	
Egypt	Netherlands	Spain	

Source: Stratton 1946: 84

Table 3.2 IGU Commissions, 1946

Study of population problems	Climatic variations	Aerial photographs
Pliocene and Pleistocene Terraces	Early maps	Cartography of Tertiary levelling surfaces

Source Stratton 1946: 83–84

1946. His academic home since the 1930s was Syracuse University, in upstate New York, and during the war he was also engaged in military and governmental affairs. Originally British, Kimble was the first head of the geography department at McGill University in Montreal from 1945 (Bird 1995) and had close connections to the American Geographical Society (AGS), where the IGU office was hosted between 1949 and 1956. He was later closely involved with the influential think-tank, The Twentieth Century Fund, in New York City.

The IGU was still a Western European and North American dominated organisation but for the first time elected two Vice Presidents from the southern hemisphere: Christovam Leite de Castro from Brazil and George Kuriyan from India. Officials from those two countries would play a role in the IGU for the following years, specifically Hilgard O'Reilly Sternberg (VP 1952–60), Speridiao Faissol (VP 1976–84) and Bertha Becker (VP 1996–2000) from Brazil, and Shiba P. Chatterjee (President 1964–68, VP 1968–72), Mohammed Shafi (VP 1984–92) and Ram Babu Singh (VP 2012–18, Secretary 2018–21) from India.

The 1949–52 term of office and the institutional re-organisation of the IGU was mainly based on two activities: the publication of the Newsletter, the IGU's official periodical, and several visits around the globe by Cressey and Kimble. Cressey's idea to publish an English-French newsletter was very successful and the series was soon established as the official series of the Union providing information, announcements, minutes and reports of commissions and national committees. For more than a decade after 1956, all printed IGU works were printed and distributed by a central institution, Geographical Publications Ltd., based in the small town of Bude in Cornwall, England, close to where the IGU President Dudley Stamp lived, in some splendour, at Ebbingford Manor. After Stamp's sudden death during the Latin American Regional Conference in Mexico City in August 1966 (Kimble 1967: 246), the IGU Publication Office was transferred to the headquarters of the British Geographical Association in Sheffield and Alice Garnett became the new publication officer, a position she held for several years.[5]

The second strategy from the IGU's re-organisation in the early 1950s were official but usually informal visits by IGU officers to several countries. As Cressey reported: "During the past

[5]Besides Lefèvre, Garnett was another female protagonist within this generation of European geographers. She was Professor of Geography at the University of Sheffield, Vice President of the Royal Geographical Society and the first woman President of the Institute of British Geographers (Maddrell 2009: 189–195).

two years [1949/50] I have visited 38 countries on behalf of the International Geographical Union" (Cressey 1951: 3). The enormous volume of travel was a new development in the IGU's history under Cressey and Kimble and replicated by later presidents, especially Stamp and Chatterjee. As the official record of the 1952 IGC noted: "Professor Stamp expressed the hope that he would continue the work of his distinguished predecessor, though he did not expect to be able to compete with him in the matter of mileage—or miles per hour!" (Closing General Assembly, August 15, 1952. In: Proceeding IGC Washington 1952: 11). After Stamp's term of office, he finally eclipsed Cressey by visiting 41 countries (Proceedings IGC Rio de Janeiro 1956: 183–184).

One output from the personal visits and data collected about member countries was the volume "The status of geography in countries adhering to the International Geographical Union" (Moorman 1952), with brief descriptions on the state of geography in member countries, excluding France and Eastern Europe. This volume was part of the publications prepared for the 1952 IGC in Washington which was connected with the centenary of the American Geographical Society. This was both a consequence of, and an arena for, the growing influence of American geography, and the event was certainly a personal success for Cressey and Kimble.

The model of organising the IGC at the President's home was continued several times in the following years, at Stockholm (1960), New Delhi (1968), Sydney (1988), Washington (1992) and The Hague (1996), and there was always a connection between the EC and the host country. Around 1200 participants from 62 countries attended the 1952 meeting but many of them, including the delegates from Czechoslovakia and Poland, were registered in absentia. China was listed with ten participants present (six as associated) and one in absentia (Proceedings IGC Washington 1952: 68–76). The General Assembly elected new members, from Europe: Austria and Finland; from Latin America: Colombia, Dominican Republic, Mexico, Uruguay and Venezuela (associated); and from Asia: Pakistan, Indonesia, Israel and Sri Lanka (associated). The internationalisation of the IGU progressed, but not in Africa. The Washington Assembly decided also to exclude the partly long-time members Argentina, Bulgaria, Chile, China (PR), Hungary, Poland, Romania and South Africa, because of missing payments for three years. The problem of continually changing membership payments would be a cause of concern for the IGU until recently.

The reconstruction period of the IGU ended after the Washington meeting, with a growing number of members and commissions, but still only representing a part of international geography without geographers from the Eastern bloc and China, and so many communities from Africa, Latin America and Asia. The following IGC at Rio de Janeiro in 1956 was a highly international event in IGU history, as the first congress in Latin America, with 1084 participants from 59 countries and with six official languages (Lamego 2020). A closer look at the list of participants shows a different picture, with 639 participants from the host country Brazil and 103 from the USA; the French delegation (63) followed by some distance (Proceedings IGC Rio de Janeiro 1956: 4–5). After the further election of new members, the IGU had 42 full member countries and six associated members in 1956. European dominance was still substantial with 23 countries, while Latin America (8), Africa (8) and Asia (7) followed by some distance.

3.4 The IGU and the East–West Conflict

With the election of the Soviet Union as a full IGU member at the Rio General Assembly in 1956, a new chapter started, influenced by political developments and the Cold War (Chap. 4). Soviet participation had been thoroughly prepared, especially through the organisation of the influential Soviet geographer Innokenti Gerasimov and his secretary, Mikhail Gournung. The USSR delegation, comprising 12 participants, was able to present a comprehensive volume in French and Russian on the progress of Soviet geography at the Rio congress (Essais de géographie 1956). On the other hand, the Soviet proposal to have Russian accepted as an official IGU language was not successful at the Rio General Assembly (Proceedings IGC Rio de Janeiro 1956: 187–188).

Two members from the Eastern bloc, the Soviet Union and Poland, would play an important role in the following decades. Geographers from both countries had been highly connected with IGU activities, especially Innokenti Gerasimov (VP 1960–68), Stanislav Kalesnik (VP 1968–72), Feofan F. Davitaia (VP 1972–80) and Vladimir Kotlyakov (VP 1988–96) from the Soviet Union and Eugeniusz Romer (VP 1931–38, 1946–49), Stanisław Leszczycki (VP 1964–68, 1972–76, President 1968–72) and Jerzy Kostrowicki (VP 1976–84) from Poland, which had been an IGU member from 1924.[6]

The leading figure of the Eastern bloc geographers in the 1950s and 60s was undoubtedly Innokenti Gerasimov (Dresch 1986; Zimina and Mashbits 1988), who was a special character as former IGU Secretary Leszek Kosiński, who had moved from socialist Poland to Canada in 1968, pointed out: "Gerasimov… believed he had the right to represent the world of socialist states. Before each congress he would convene a meeting in Moscow, a meeting during which a joint course of action was decided, for instance, whom to vote for or whom to admit to the Union…. In 1964 Gerasimov was a candidate for the Presidency of the IGU, but shortly before the vote a telegram came from Moscow saying that he could not stand; no reason was given. It was too late for him to appeal from this decision and he had to withdraw. He was very displeased and worried, but others were happy, even showing this quite clearly—like the Poles for example—or more 'diplomatically', like the Czechs. In the end, at the last moment Professor Shiba Chatterjee from India was put forward as a candidate; he was elected for the 1964–1968 term" (Kosiński 2014: 290; see also Harris 1996: 294–295).

3.5 A Special Case: Germany

A very difficult and highly political point was the question of re-integrating German geography into the IGU, and connected to that, the independence and international acceptance of GDR geography (Schelhaas 2005). Here we are faced deeply with Cold War politics and see the process of the separation of German geography. Germany's position within the IGU after World War II was confusing. At the first post-war IGC in Lisbon (in 1949) no German delegation was present, although the official congress proceedings listed a group of nine German participants. Hermann Lautensach, who was very close to Orlando Ribeiro, only attended as a private scholar. On his visit to Europe in summer 1950, President George B. Cressey met with German geographers in East and West. In the GDR he visited Harry Waldbaur and Joachim Heinrich Schultze, and in the FRG, Hermann Lautensach, Emil Meynen, Herbert Louis and, later IGU President, Carl Troll. There was strong interest, mainly among the US community, to integrate Germany in the IGU as soon as possible. As Cressey pointed out very briefly: "Germany. A member, but inactive. Visited by Professor Cressey, August, 1950" (IGU Newsletter 2, 1951, 2: 27), and in more detail: "It is hoped that Germany may soon rejoin the Union, but the situation is complicated by the de facto division of the country into East and West. At present no single scholarly organization is able to serve as the adhering body for Germany as a whole. Nor are there suitable academies or research councils in the two zones. As soon as the proper adhering organization exists, Germany will automatically resume her membership, either as one unit or as two" (Cressey 1951: 3).

[6] Aside from the representatives from the Soviet Union and Poland, there was only one EC member from another Eastern bloc between 1946 and 1990: György Enyedi from Hungary who served as IGU Vice President from 1984 to 1992. There were no representatives from Bulgaria, Czechoslovakia, GDR and Romania, and none have been elected from the post-communist regimes in these countries. For more detailed information, see Chap. 4; and for more on Poland, see Degórski (2015); Kosiński (2008) and (2014).

According to the IGU statutes, a full membership was only possible through national academies or research organisations, which delegate National Committees. In West Germany, the German Research Foundation (Deutsche Forschungsgemeinschaft) offered the institutional frame for a National Committee. The situation in the GDR was more complicated. While the German Academy of Science was an eligible organisation, geography was under-represented within the Academy at that time. Günther Köhler was the first geographer to be elected to the GDR Academy in 1956, followed by Edgar Lehmann and Heinz Sanke some years later.

At the next IGC at Washington DC in 1952, a delegation of 20 German members participated but no-one attended from the GDR. At the General Assembly, the Federal Republic of Germany was elected as a full IGU member and re-integration was officially finished. At this point, there existed one National Committee for all German geographers. As Cressey explained: "Due to the division of Germany there was no such single body to continue the long-standing membership which Germany has held in the Union. Following extensive consultation with many geographers and with the International Council of Scientific Unions, the Executive Committee at its meeting in London in 1950 voted to extend a simultaneous invitation to the Deutsche Forschungsgemeinschaft of the Federal Republic and the Deutsche Akademie der Wissenschaft at People's Republic. I am happy to report that the former organization has accepted our proposal, so that Western Germany resumes its adherence. Unfortunately, no reply has been forthcoming from Eastern Germany. Since the statutes make no provision for dropping a country except for non-payment of dues, it is our position that Germany has continued to be a member of the Union but was inactive during the period when it lacked a proper adhering body. No action is called for by the Union with regard to the participation of the Deutsche Forschungsgemeinschaft or with regard to the Deutsche Akademie should it care to adhere on behalf of its respective area" (Cressey 1952: 12–13).

The IGC in Rio de Janeiro in 1956 was the first congress at which geographers from the GDR were invited to participate though one two —Harry Waldbaur and Ernst Neef—attended. At the General Assembly, delegates voted by a majority of 15 to accept GDR geographers as a full and independent member, though the initiative eventually failed due to some formal procedural problems and because the GDR lacked its own national committee of geographers. President Ahlmann ruled that the election of the GDR had to be regarded as "*sub judice*", and devolved the case to the new Executive Committee. The decision to postpone the acceptance of GDR geography was a failure for the officials involved in GDR science policy. In the following months, geographers in both West and East Germany were preoccupied with many difficult diplomatic debates.

A special German-German committee for geography was installed, comprising Johannes F. Gellert, Theodor Hurtig, Julius Büdel and Horst Mensching, but was unable to establish a single German National Committee representing both the GDR and the FRG. Attempts by IGU officials, including Hans W. Ahlmann (President), Hans Boesch (Secretary), Jan Peter Bakker, and Georges Chabot, to mediate between the two German communities ended in failure. Consequently, the National Committee of the GDR was founded in 1959 within the East German Academy of Science, and Johannes F. Gellert was its first chair (followed by Horst Kohl, Heinz Lüdemann and Fritz Hönsch). At the 1960 IGC in Stockholm, the GDR was finally accepted as a full IGU member, with 17 votes in favour, eight votes against and seven abstentions. Bulgaria and Romania were also elected which completed the Eastern bloc's IGU representation. The whole context was highly political, in both East and West Germany, but according to GDR policy the involvement of the country's geographers in the IGU was viewed as a great success. On the other hand, the West German delegation and their chair Erich Otremba were under massive pressure and urged by several ministries and authorities to explain why the IGU was so willing to accept geographers from the GDR. This conflict eventually calmed down but the German division within the ranks of the IGU remained a curious geopolitical fact for many years (Schelhaas 2005: 117–125).

From the 1960s, West German influence within IGU increased in several commissions, especially after the election of Carl Troll as President from 1960 to 1964 and when Walther Manshard served as Secretary from 1976 to 1984. GDR geographers were unable to participate in a similar way though some East Germans became commission members. One prominent GDR representative was Edgar Lehmann, the long-time chair of the Commission for National Atlases and very active in several fields of managing geography under complicated political conditions. In comparison to West Germany, however, geographers from the GDR had limited opportunities to travel to western countries and their involvement was always dependent on political circumstances. Finally, we have to bear another detail in mind: IGU meetings allowed East–West- and German-German encounters, a very important way to keep in contact during the Cold War. Until 1990, two German National Committees were part of the IGU. At an IGU meeting in Beijing on 16th August 1990, the two German chairmen Eckart Ehlers (FRG) and Fritz Hönsch (GDR) presented their decision to merge both committees. This came into existence on 3rd October 1990, the day of German reunification.

3.6 Hans Boesch and the IGU Family

Organisations—and international scientific networks—are only as successful as the people within these communities. The history of IGU was, and partly still is, dominated by a group of male geographers who were mostly, but not always, trained in a European or North American scientific and cultural background. We can easily find influential figures with a deep devotion to the IGU in the list of officers, but the simple schedules of Presidents and Secretaries say nothing about the staff, the ordinary IGU members and the many supporters behind the more important officials. The tradition starts with Charles Close, then Emmanuel de Martonne, Marguerite Alice Lefèvre and, after the war, with George Kimble, Hans Boesch, Chauncy D. Harris, Walther Manshard, Lesek Antoni Kosiński, Eckart Ehlers, Ronald Abler, Vladimir Kolosov and Michael Meadows. They all spent much of their lifetime on the administration and advancement of the IGU, often under complicated political conditions.

One influential and long-time member of the IGU family was Hans Boesch (Kishimoto 1980; Spiess 1978). The Swiss geographer was a remarkable and certainly interesting figure in twentieth century geography, especially within the German-speaking community but also in international geography. He was an example of a global scientist and science manager, multilingual and with extensive experience in applied and academic geography, especially from a European and North American background. We know him as the long-time Secretary and organiser of the IGU but also an acknowledged expert in global economic geography and land use studies. His native town of Zurich, with its two universities and the Geographical Society, was Boesch's academic and private home for his whole career. The Swiss metropolis was a centre for academic, applied and popular geography, but also an excellent place to travel from for scientific and business purposes around the world.

The special cultural and political situation of Switzerland in the twentieth century, with its neutrality, wealth and different cultures and languages, allowed Hans Boesch to develop an impressive and accelerated career based on huge ambition, tremendous work, some coincidences and influential contacts. After Boesch's years as a geologist, he started teaching geography in 1939 at the University of Zurich. In his early 30s, and under difficult circumstances during the war, Boesch already obtained influential academic positions, notably as director of the Geography Department at the University of Zurich in 1942, which he retained until 1978. He was the undisputed and generally accepted director. After

World War Two, Boesch was able to widen his international relationships through travel, especially across the western hemisphere. Obviously, Boesch focused first on the German-speaking community. Boesch was the first foreign geographer to lecture in Germany after World War Two and was internationally significant too when he accepted an invitation to a guest lectureship at the University of Zurich in 1948–49 from future IGU President Carl Troll. This was a very important step for the re-integration of German geography into the international scientific community, occurring only four years after the end of war.

With regard to Switzerland's position within international geography in the past, we have to certainly mention the two International Geographical Congresses at Berne, in 1891, and at Geneva, in 1908. Switzerland has been a full IGU member since 1928. The IGU was a significant arena for Boesch to realise his vision of a global science. At the Lisbon meeting in 1949, he assumed the position of Vice President. At the ninth General Assembly of the 1956 Rio de Janeiro IGC, Boesch was elected Secretary and Treasurer. He succeeded George Kimble, and remained in those positions for three periods of office from 1956 to 1968. This was only comparable with Charles Close, Emmanuel de Martonne, Ronald Abler and Michael Meadows, who served for long terms of office as Secretary or later as President.

Zurich was the headquarters of the IGU for many years. Boesch also benefitted from two personal relationships within the IGU family: Dudley Stamp, President from 1952 to 56, and Carl Troll, President from 1960 to 64. He was close to both since at least the 1940s. It was apparent that Boesch was the first choice for the position of Secretary and Treasurer. His Swiss origin and the country's political neutrality were certainly helpful in garnering a majority within the General Assembly; his ability to speak different languages fluently and his knowledge of the inner workings of the IGU were also of great advantage.

One statement from the newly elected President Carl Troll, at the closing ceremony of the IGC at Stockholm in 1960, underlines their satisfaction with the Secretary and Treasurer: "As president of the Union you were lucky to have the experienced cooperation of our Secretary-Treasurer, Prof. Hans Boesch. In the IGU-correspondence the collegiality between president and Secretary-Treasurer is clearly documented: 'Dear Hans' from one side, 'dear Hans' from the other, and 'Dear Hans and Hans' in letters from Executive members to both. We all, during these days, were witnesses of Prof. Boesch's astonishing ability in everything, his masterly clarity, his familiarity with statutes, languages and organization techniques. The activity of the different commissions and the budget document Boesch's eagerness, care and skill in obtaining funds for the benefit of our extensive work" (IGC Stockholm 1960, Final Congress Report: 15).

Such exorbitant praise was certainly part of the closing ceremony genre, but we can also interpret Troll's description of Hans Boesch as a perfectly genuine desire to give credit for his exemplary organisation. Boesch served under three IGU presidents—Hans W. Ahlmann, Carl Troll and Shiba Prasad Chaterjee—and clearly found professional and practicable solutions to administrative and financial challenges, supported by his colleagues from the Geography Department at Zurich. In his farewell address at the New Delhi IGC in 1968, Boesch presented a sentimental review of his time at IGU: "one day the first President whom I had the privilege to serve, George Cressey, swivelled around in his chair and suddenly asked me, 'Would you like to become a Vice-President in 1949.' And I said, 'Why not.' I had no idea at that time what the IGU was and what the Vice-President would have to do. I accepted and I was elected almost unanimously. I think procedures were not exactly democratic" (IGC New Delhi 1968 Congress Proceeding: 113).

Another detail from his review is also telling: Boesch had no deep interest in becoming IGU President after a long period as Secretary. As he explained: "they asked me, would I like to become President of the Union; I said, 'No, I want to continue to work'" (IGC New Delhi

1968 Congress Proceeding: 113). The move from Secretary to President was one possible career trajectory within IGU, however, realised by Charles Close and Emmanuel de Martonne in the past and by Ronald Abler and Michael Meadows more recently.

3.7 The IGU as Global and Professional Scientific Organisation

Hans Boesch's successors continued the IGU's established and well-organised administration. These were Chauncy D. Harris (1968–76), Walther Manshard (1976–84), Leszek Antoni Kosiński (1984–92), Eckart Ehlers (1992–2000), Ronald Abler (2000–08) and Michael Meadows (2010–18), all enormously devoted to international geography, who often served under complicated circumstances and faced financial, diplomatic and interpersonal problems.

Besides focusing on the Secretaries, a simple view at the list of Presidents offers several kinds of interpretation too. After George B. Cressey and Dudley Stamp, the General Assembly unanimously elected Hans Wilhelmsson Ahlmann from Sweden (President 1956–60) and Carl Troll from Western Germany (1960–64). Western European dominance was clear in this period. With the election of Shiba P. Chatterjee from India (1964–68), the IGU had for the first time a geographer from the southern hemisphere in its highest position, but who also happened to be very familiar with French and British geography (Mookerjee 1998). As mentioned before, he was only a substitute choice because the dedicated President, Innokenti Gerasimov, withdrew some hours before the London General Assembly in 1964. The Polish geographer Stanisław Leszczycki (1968–72) was the first and only from the Eastern bloc to occupy this position, although coming from Europe again. Jean Dresch (1972–76) and Michael John Wise (1976–80) followed, which showed that the President was still rooted in Europe, although both characters also had a strong international background.

Akin Mabogunje from Nigeria was the first African President (1980–1984), with experience as Vice President from 1972 onwards. The African representation within the IGU was, and partly still is, marginal with some exceptions. The Kingdom of Egypt was an IGU founding member, Morocco and South Africa followed in 1926, but both with a strong colonial background. We also have to mention two officers from Egypt, Ahmed Hassanein Bey (VP 1924–31) and Hassan Awad (VP 1956–64).[7] Since the 1960s, several African countries have joined but with varying times of participation.

Peter Scott, Roland J. Fuchs, Herman Theodoor Verstappen, Bruno Messerli, Anne Buttimer, Adalberto Vallega, José Luis Palacio Prieto, Ronald Abler, Vladimir Kolosov and Yukio Himiyama would continue the IGU's agenda. But we still have to keep in mind that the IGU was and is much more than a long list of famous Presidents, Secretaries, and Vice Presidents. The IGU commissions and events, some of them dating back to even before 1922, are essential for any international scientific exchange.[8] Finally, the IGU was and remains a fruitful arena for geographers and their allies. It provides a context for scientific and social encounters, for the formulation and promotion of policy in science and society, and for building friendships without cultural or political barriers. The following chapters expand on the past, present and future of international geography, with many additional details about the IGU, its members, commissions and events.

Acknowledgements I wish to thank my co-editors and Philip Jagessar for their support and advice during the final elaboration of this chapter.

[7]It was not until 1992 that Folasade Iyun (Nigeria) was elected Vice President from Africa again, followed by Lindisizwe M. Magi (South Africa) 2004.

[8]For the development of the commissions between 1922 and 2022, see Chap. 1.6.

References

Belo D (1999) Orlando Ribeiro seguido de uma viagem breve à Serra da Estrela. Assírio & Alvim, Lisboa

Bird BJ (1995) Geography at McGill University: a 50 year perspective, 1945–95. McGill University, Montreal

Collignon B (1996) Les commissions, 'l'organe les plus important de l'UGI.' In: Robic MC, Briend AM, Rössler M (eds) Géographes face au monde. L'Union Géographique Internationale et les Congrès Internationaux de Géographie. Éd. L'Harmattan, Paris, pp 117–146

Cressey G (1951) The State of the Union. The IGU Newsletter 2(2):3–4

Cressey G (1952) Report of the President. In: Proceedings of the 8th General Assembly and the 27th International Geographical Congress, Washington D.C., August 8–15, 1952. Washington 1957, p 11–14

Daveau S (2011) Orlando Ribeiro (1911–1997). Ponto de partida, lugar de econtro. Biblioteca Nacional de Portugal, Lisboa

Daveau S (2012) Orlando Ribeiro. 1911–1997. Geogr Biobibliographical Stud 31:30–55

Degórski M (2015) Polacy w Międzynarodowej Unii Geograficznej. Przegląd Geograficzny 87(1):163–166

Delfosse C (2001) Emmanuel de Martonne, tisseur de réseaux internationaux de géographes. In: Baudelle G, Ozouf-Marignier MV, Robic MC (eds) Géographes en pratiques (1870–1945). Le terrain, le livre, la cité. Presses Université de Rennes, Rennes, p 189–206

Denis J (1986) Marguerite Alice Lefèvre 1894–1967. Geogr Biobibliographical Stud 10:106–110

Dept GG (1945) Hulde aan Prof. Paul Michotte. Bulletin de la Société Belge d'Études Géographiques 14 (1):44–49

DeVivo M (2015) Leadership in American academic geography: the twentieth century. Lexington Books, Lanham

Dresch J (1986) Innokententiy Petrovich Gerasimov (1905–1985). Annales De Géographie 95(528):233–236

Dresch J (1988): Emmanuel de Martonne. 1873–1955. Geographers: Biobibliographical Studies 12:73–82

Droogleever Fortuijn J (2019) Women geographers and the International Geographical Union. Boletín De La Asociación De Geógrafos Españoles 81:1–16

Essais de géographie (1956). Recueil des articles pour le XVIII. Congrès International Géographique. Académie des Sciences de l'URSS-Société de Géographie de l'URSS, Moscou, Leningrad

de Ghellinck A (1945) Hommage à Paul Michotte. Bulletin De La Société Belge D'études Géographiques 14(1):36–44

Harris C (1996) "On political issues ...": Chauncy Harris interviewed by Mechtild Rössler. In: Robic MC, Briend AM, Rössler M (eds) Géographes face au monde. L'Union Géographique Internationale et les Congrès Internationaux de Géographie. L'Harmattan, Paris, pp 291–304

Hegenscheidt A (1943) Le géographe Paul Michotte. Université catholique de Louvain, Louvain

Herman T (1965) George Babcock Cressey, 1896–1963. Ann Assoc Am Geogr 55(2):360–364

Herren M (2002) 'Outwardly... an innocuous conference authority': National Socialism and the logistics of international information management. Ger Hist 20 (1):67–92

James P (1964) George Babcock Cressey 1896–1963. Geogr Rev 54(2):254–257

James P, Perejda AD (1981) George Babcock Cressey 1896–1963. Geogr Biobibliographical Stud 5:21–26

Kimble GHT (1957) Report of the Secretary-Treasurer. In: Proceedings of the 8th General Assembly and the 27th International Geographical Congress, Washington D. C., 8–15 Aug 1952. Washington 1957, p 15–17

Kimble GHT (1967) Obituary: Laurence Dudley Stamp, 1898–1966. Geogr Rev 57(2):246–249

Kishimoto H (1980) Hans Boesch 1911–1978. In: Kishimoto H, Kümmerly H (eds) Geography and Its Boundaries: A Publication to Commemorate the Work of Prof. Dr. Hans Boesch. Kümmerly und Frey, Bern, p 127–129

Kosiński L (2008) Międzynarodowa Unia Geograficzna. In: Jackowski A (ed): Historia geografii polskiej. Wydawnictwo Naukowe PWN, Warszawa, p 477–488

Kosiński L (2014) Geographica Polonica: a window onto the world—an interview with Professor Lesek Antoni Kosiński. Geographica Polonica 87(2):277–294

Lamego M (2020) How international was the International Geographical Congress in Rio de Janeiro 1956? On location and language politics. In: Schelhaas B et al (eds) Decolonising and Internationalising geography: essays in the history of contested science. Springer, Cham, pp 113–126

Lefèvre MA (1941) In memoriam Paul Lambert Michotte 1876–1940. L. Durbeck, Louvain

Lefèvre MA (1950) Rapport sur l'activité de l'Union Géographique Internationale depuis l'Assemblée Générale d'Amsterdam en 1938. In: Comptes rendus du Congrès Internationale de Géographie, Lisbonne 1949. Tome premier. Lisbonne, p 85–93

Maddrell A (2009) Complex locations: women's geographical work in the UK 1850–1970. Wiley-Blackwell, Chichester

Martonne E de (1946) Rapport sur l'activité de l'Union Géographique Internationale depuis la dernière Assemblée du Conseil International des Union Scientifiques. In: Stratton FJM (ed) (1946) The fourth General Assembly of the International Council of Scientific Unions. Reports of proceedings. Cambridge University Press, Cambridge, p 39–41

Mookerjee S (1998) Shiba P. Chatterjee. Geogr Biobibliographical Stud 18:21–35

Moorman MA (ed) (1952) The status of geography in countries adhering to the International Geographical Union. International Geographical Union, Washington D. C.

Olszewicz B (ed) (1968) Stanisław Pawłowsky. W dwudziestą piątą rocznicę (1940–1965). Państwowe Wydawn. Naukowe, Warszawa
Pinchemel P (ed) (1972) Geography through a century of International Congresses. L'Imprimer Ozanne, Caen
Polspoel GL (1964) La carrière et l'activité scientifique de Mademoiselle M. A. Lefèvre. Professeur à l'Université de Louvain. Acta Geographica Lovaniensa 3:5–21
Polspoel GL (1965) L'hommage au Professeur M. A. Lefèvre. Petit histoire d'un grand jubilé. Bulletin De La Société Belge D'études Géographiques 34:217–223
Polspoel GL (1968) In memoriam Mademoiselle Marguerite Lefèvre. Bulletin De La Société Belge D'études Géographiques 37:27–31
Ribeiro O (2003) Memórias de um géografo. João Sá da Costa, Lisbon
Robic MC, Rössler M (1996) Sirens within the IGU: an analysis of the role of women at International Geographical Congresses (1871–1996). Cybergeo. https://doi.org/10.4000/cybergeo.5257
Robic MC, Briend AM, Rössler M (eds) (1996) Géographes face au monde. L'Union Géographique Internationale et les Congrès Internationaux de Géographie. L'Harmattan, Paris
Rössler M (1996) Double bind and global change: the IGU between ISSC and ICSU. In: Robic MC, Briend AM, Rössler M (eds) Géographes face au monde. L'Union Géographique Internationale et les Congrès Internationaux de Géographie. L'Harmattan, Paris, pp 253–258
Schelhaas B (2005) Institutionelle Geographie auf dem Weg in die wissenschaftspolitische Systemspaltung. Die Geographische Gesellschaft der DDR bis zur III. Hochschul- und Akademiereform 1968/69. Leibniz-Institut für Länderkunde, Leipzig
Schelhaas B, Pietsch S (2020) (Re-)Writing the history of IGU? A report from the archives. In: Schelhaas B et al (eds) Decolonising and internationalising geography: essays in the history of contested science. Springer, Cham, pp 127–140
Spiess E (1978) Professor Dr. Dr. h.c. Hans Boesch: 24. März 1911 bis 16. August 1978. Geographica Helvetica 33(4):169–172
Stratton F J M (ed) (1946) The fourth General Assembly of the International Council of Scientific Unions. Reports of proceedings. Cambridge University Press, Cambridge
Volle D (1996a) La carte des États : vers une couverture du monde ? In: Robic MC, Briend AM, Rössler M (eds) Géographes face au monde. L'Union Géographique Internationale et les Congrès Internationaux de Géographie. L'Harmattan, Paris, pp 41–62
Volle D (1996b) L'universalité et ses limites. In: Robic MC, Briend AM, Rössler M (eds) Géographes face au monde. L'Union Géographique Internationale et les Congrès Internationaux de Géographie. L'Harmattan, Paris, pp 63–81
Wilczynski W (1992) Stanisław Pawłowsky, 1882–1940. Geogr Biobibliographical Stud 14:69–82
Wright JK (1949) The sixteenth international geographical congress, Lisbon, 1949. Geogr Rev 39(3):482–487
Zimina P, Mashbits AG (1988) Innokentii Petrovich Gerasimov. Geogr Biobibliographical Stud 12:83–94

Bruno Schelhaas received his Ph.D. from the University of Leipzig. He is Head of the Archive for Geography at Leibniz Institute for Regional Geography in Leipzig and is the archivist of the International Geographical Union. His interests include the history of geography and cartography, historical geography and archival science. He is member of the Steering Committee of the Commission History of Geography of the International Geographical Union. 2020 he was the corresponding editor of Historical Geography and Geosciences volume "Decolonising and Internationalising Geography".

4 Internationalization of Geography in the Bipolar World: Socialist Countries During the Cold War

Vladimir Kolosov, Marek Więckowski,
Debin Du, and Xionghe Qin

Abstract

The chapter's purpose is to examine interactions between geographers of socialist countries with IGU and the world geographical community at large in 1945–1990. The authors consider some specific national trends in the development of geography in the former USSR, Poland, and China under the conditions of ideological constraints and geopolitical tensions. A special attention is paid to the forms and impact of internationalization on geography in these countries and the ways of the dissemination of scientific information. The authors show that participation of geographers from their countries in the activities of IGU was of particular importance in the extension of international contacts. It improved the positions of geography in the country and at the same time stimulated the use of new methods and approaches in geographical studies, and allowed spreading of national geographical concepts abroad. A particular role in internationalization belonged to academic leaders. In general, the development of geographical science in socialist countries follows the global paradigm.

Keywords

Internationalization · Geography · Cold war · Russia · China · Poland

4.1 Introduction

The internationalization of geography, like other disciplines, developed unevenly in time and space and depended not just on the size, geographical location, and level of development of each country and the scale of research undertaken but also, of course, on the political situation in the world. World War Two and the material difficulties of the restoration period, the confrontation between the military-political blocs in the era of the bipolar world, local wars and armed conflicts, and ideological constraints together limited interactions between scientists (Robic et al. 1996). In socialist countries, a

V. Kolosov (✉)
Institute of Geography, Russian Academy of Sciences, Moscow, Russia
e-mail: vladimirkolossov@gmail.com

M. Więckowski
Institute of Geography and Spatial Planning, Polish Academy of Sciences, Warsaw, Poland

D. Du
School of Urban and Regional Science, Institute for Global Innovation and Development, East China Normal University, Shanghai, China

X. Qin
Institute for Global Innovation and Development, East China Normal University, Shanghai, China

V. Kolosov et al. (eds.), *A Geographical Century*,
https://doi.org/10.1007/978-3-031-05419-8_4

planned state economy and the centralization of power largely determined the differences in the directions and methodology of geographical research, the nature and territorial pattern of international contacts of geographers. Discrimination and persecution of individual geographers for political reasons provoked a sharp reaction from the scientific community, which led to boycotts and new restrictions on contacts. As a result of fundamental shifts that began in the 1980s and led to radical changes in the economic and political map of the world, most of the political factors that limited international scientific cooperation seemed to have disappeared into oblivion (Paasi 2015; Meadows 2020). However, the deterioration of the political situation in the 2010s has reminded us once again of the ghosts of the past.

The usual indicators of the internationalization of research are the number of joint papers; the number of references to foreign authors; double and multiple affiliations; the number and composition of participants in international projects; the number of international conferences and their participants, academic exchanges, etc. The use of most of them, however, became possible only recently. The authors addressed earlier written sources, in particular, reports about IGU events and the participation in them of geographers from their countries, and documents of national geographical societies and associations.

The objective of this chapter is to analyze the impact of the split of the world on two political systems in the twentieth century on interactions between geographers of socialist countries and the rest of the world. What forms did it take, what role did IGU play in it? To what extent did the development of geographical science in socialist countries follow the global paradigm? The authors of this chapter try to answer these questions using the cases of their countries—the USSR, Poland, and China. The degree of openness of these countries and the general situation, history, structure, and organization of geographical science differed significantly, but they also had common features.

4.2 Despite Geopolitical and Ideological Constraints: The Internationalization of Soviet Geography

4.2.1 Political Context: Geography Under the Socialist System

There is a widespread opinion about the closedness of the Soviet Union and the limited possibilities of Soviet scientists to participate in international collaborations. This statement, however, can only be fully attributed to a relatively short period of the late 1930s—early 1950s, which covered the years of World War II and the apogee of the Stalin dictatorship. The USSR officially joined the IGU only in 1956, after the death of Stalin.

However, even in the early 1930s small Soviet delegations took part in the International Geographical Congresses in Paris (1931) and Warsaw (1938). But at most of the post-war IGCs, Soviet participants made up one of the largest groups. Despite these obstacles, Soviet geography was developing in a wide international context. Firstly, foreign publications were accessible. The abstract monthly journal *Geography* consisted of separate thematic and regional issues. They contained not only bibliographic information, but also detailed one to three-page summaries of the most valuable articles from foreign journals. The All-Union Institute of Scientific and Technical Information in Moscow (*VINITI*), which ran these journals, also published reviews of the state of the art in different areas of geography. Secondly, the leading libraries were subscribed to almost all international and national geographical journals. Thirdly, a special Moscow publishing house *Progress* translated the most interesting books into Russian. Fourthly, since the 1970s geographers of the academic institutes and leading universities had the opportunity to implement international projects—first of all, with colleagues from socialist countries, but also from Western countries. These were, for example, projects of the UNESCO program "Man and Biosphere", and later, in the 1980s—International Geosphere-Biosphere Program, "Human Dimension of Global Change", and others. In the late 1970s, Soviet and French geographers initiated a unique interdisciplinary comparative project "Alps—Caucasus". Soviet-Polish seminars on population geography held alternately in two countries were notable milestones in the development of Soviet human geography.

The largest event in the history of Soviet geography was the XXIII IGC in Moscow (1976). It remained for a long time unsurpassed in terms of the number of participants (about 4000) and scope. Within the framework of the congress, 28 symposia were held in the capitals of all union republics and in many large cities (Gerasimov 1976; Gerasimov et al. 1976) (Figs. 4.1 and 4.2).

Finally, many foreign scientists visited the Soviet Union on an individual basis. In many foreign countries, there was an undoubted interest in Soviet geography, a sign of which was, for example, the establishment in the USA of the journal "Soviet Geography: Review and Translations" (1960) later transformed into "Eurasian Geography and Economy".

Historically, Russian geography was closely connected with German anthropogeography and relied on a developed landscape analysis. Geography has always been considered in Russia and the USSR as part of Earth sciences. The number of scholars in human geography is significantly less than in physical geography, which partly includes some disciplines that belong to geophysics in the West (glaciology, etc.). This situation had its own advantages: Soviet human geography traditionally paid much attention to the use of natural resources and environmental protection. Many ambitious projects (e.g., atlases) have been implemented in close collaboration between physical and human geographers. At the same time, paradoxically, the concept of integrated geography as a science that studies the geographic envelope of the Earth in the integrity of its natural and human elements made its way with great difficulties and became generally recognized only in the 1960s, since dogmatic Soviet Marxism strictly divided "natural laws" and "laws of the development of

Fig. 4.1 The opening ceremony of the XXIII International Geographical Congress was held in the Palace of Congresses in the Moscow Kremlin. Photo: Tamara Galkina, 1976

Fig. 4.2 Preparing the XXIII International Geographical Congress. From left to right: Secretary General of the congress Dr. Vladimir Annenkov, Chair of the Local Organizing Committee Academician Innokenty Gerasimov, Academician Vladimir Kotlyakov, and Professor Valery Chichagov. Photo: Tamara Galkina, 1976

society" (Kolosov and Treivish 2009). In 1966 Soviet geographer David Armand, a brilliant scholar, who spoke five languages fluently, published the book *For Us and Grandchildren* grounding many principles of the concept called later "sustainable development" (Armand 1966). The unity of geography became the basis of the concept of constructive geography aimed at transforming the environment, promoted by Academician Innokenty Gerasimov, the leader of Soviet geography in 1950–1985.

The historical centralization of the Russian state and its high role in society contributed to the development of original economic and geographical theories. They had a significant impact on human geography and the practice of territorial planning in the socialist countries of Central and Eastern Europe and developing countries. According to Ivan Aleksandrov and Nikolay Baransky, who founded in the 1930s the so-called regional school of economic geography, the task of economic geography is to ground a new efficient territorial division of labor between different parts of the country. The objective was to optimize within the boundaries of a certain territory (economic region) the use of natural resources, historically accumulated material values, and skills of the population, together with new technologies and a rational combination of economic branches. The boundaries of this territory should be delimited in the course of a geographical study.

The concept of territorial production complexes (TPCs) was another well-known Soviet geographical and economic theory developed in the early 1950s. TPC was understood as a territorial combination of industrial enterprises, infrastructure and services created for the development and maintenance of the main branch(es) of industry on a given territory thanks to favorable conditions. A TPC included enterprises that supplied the main industries with raw materials, energy, equipment, and processed by-products, and infrastructure and industries destined to meet the internal needs of the region's population. Some premises of the TPC theory are similar to the concept of industrial cluster proposed much later by M. Porter, which provoked in post-Soviet Russia a series of publications comparing these approaches (Kolosov 2017).

4.2.2 Processes and Channels of Internationalization

Perhaps the most regular and comparable data are found in the reports on International Geographical Congresses. Forty articles and other publications in the leading Soviet geographical journals before and after each IGC since 1956 are a mirror of the attitude of Soviet geographers to international cooperation and the IGU. Extensive reports on the results of the IGCs contain the assessments of Soviet geography against a global background and a look at the state of world geographical science.

Soviet geographers paid much attention to the participation in the activities of IGU. Various institutions and departments published before each monographs in English and Russian summarizing the results of research on individual sub-disciplines and areas. On the one hand, the IGU and its congresses were one of the main channels for informing foreign colleagues about geographical research in the Soviet Union and increasing its influence through, for example, disseminating the experience of Soviet territorial planning. The reports always included detailed information on the election of Soviet geographers as commissions' members and Chairs and Vice-Presidents of IGU. Soviet geographers were usually well represented in the IGU Executive Committee: after Gerasimov, a Georgian geographer Feofan Davitaya served from 1972 until his death in 1979 and Academician Vladimir Kotlyakov was also a EC member from 1988 to 1996.

Any Soviet publications passed ideological filters, the "density" of which changed over time. In articles on international cooperation, a tribute to the official ideology was paid, especially until the 1980s, by including phrases about the advantages of Soviet (Marxist) geography, information on the ratio of reports submitted to IGCs by scientists from socialist, developing and capitalist countries, and complaints about the petty topics of their presentations in comparison with the Soviet ones.

Presentations at the IGCs were divided into "progressive" and "apologetic". "Progressive" presentations contained, for example, criticism of former colonial powers or large companies and transnational corporations not interested in the development of territories in developing countries, or an analysis of depressed areas and unemployment (Gerasimov and Annenkov 1975). The geographical journal of Moscow University noted that IGCs helped "to define new lines of ideological struggle between progressive and conservative forces among geographers in capitalist and developing countries" (Gerasimov 1969; Saushkin 1969, 1977). The "indisputable ideological advantage" of Soviet geography was seen in its constructive approach, that is, the hypothetical possibility to implement the ideas proposed by geographers in a centralized planned economy.

Radical socio-economic differences between the USSR and its allies, and other countries resulted in the fact that some sub-disciplines could not be developed in socialist countries in principle not only because of ideological taboos, but also the absence of the subject of study itself —for example, electoral geography. The geography of retail trade, based on market principles, was noted by Saushkin as a new field of research.

The convictions or illusions of the possibility and expediency of complete controllability and centralization of social processes were indirectly reflected in the positions of Soviet representatives in the IGU. Gerasimov advocated reducing the number of commissions and "eliminating spontaneity" in their emergence. These proposals were approved by the XX IGC Program Committee but rejected by the IGU General Assembly.

The leadership of the Soviet National Committee was particularly wary of the meetings on political geography since in the USSR there was a risk of its association with geopolitics, which in turn had long been identified with Nazi geopolitics. The official Soviet delegation and Soviet members of the IGU Executive Committee were obliged to note and, if necessary, to fight "the use of congresses for political declarations" (for example, "in the report on national identity of French-speaking population in Quebec" at the 22nd IGC in Montreal). However, after the

XXV IGC in Paris, Gerasimov changed his mind. Upon his return, he urged the development of political geography in the USSR. However, at his insistence, the term "political geography" in the name of the IGU working group created at the congress was replaced by the euphemism "world political map".

By the mid-1980s, after Gorbachev came to power, ideological accents disappeared altogether. The report on the Sydney IGC in 1988 contains no ideological motives at all and ends with complaints about the insufficient representation of Soviet geographers in the IGU, especially young scholars, the lack of funds for the extension of remote sensing, and the acquisition of scientific equipment abroad (Kotlyakov and Annenkov 1989).

The journal reports allow to identify the topics of the greatest interest to the Soviet official delegation—new methods of collecting, analyzing, and mapping information, new theoretical concepts, areas, and trends in the development of geographical research. For Soviet geographers, particularly important were the papers on the holism of geography, the protection of environment and rational use of natural resources, cartography, geographical education, and other "integrating" topics.

A great attention was paid to the "methodological rearmament of science" (in the expression of Gerasimov): new methods of remote sensing and study of the Earth from space, sources of geographic information, and mathematical and statistical methods of its analysis. At the London IGC in 1964, Soviet geographers actively supported the creation of a commission on quantitative methods in geography (Gerasimov 1965). Interest in them evolved in the Soviet Union almost simultaneously with the "quantitative revolution" in the West. Since the mid-1960s, for about 10–15 years, the application of mathematical methods was one of the most "fashionable" trends in Soviet geography. Yulian Saushkin and Konstantin Salishchev, the leaders of Soviet human geography and cartography, respectively, were among the most prominent advocates of new methods in Moscow State University. At the Institute of Geography, Yuri Medvedkov contributed much to the development of mathematical methods (later, he emigrated to the USA). In the late 1970s, as in other countries, there was a certain disappointment with the limited heuristic potential of mathematical methods. The fascination with them gave way to a new trend characteristic also of Soviet geography: humanization.

They were related, in particular, to the accelerating processes of the emergence of new areas and sub-disciplines. In accordance with the dogma on the primacy of material production, not by chance, human geography has long been called economic geography. However, since the 1950s population geography gradually gained strength. It quickly contributed to the emergence of urban geography. In the 1970s, rural geography and the geography of services, and then social and political geography appeared. These processes resulted from the complication of the structure and needs of society, but also undoubtedly from the influence of the tendencies in world geography and the creation of the respective IGU commissions.

"Environmentalization" was another important trend in world geography in the 1970s. The leaders of the "integral" Soviet geography based on the focus on the nature-society interface, claimed leadership in this area. Characteristically, the central event at the 1976 Moscow IGC was the symposium "Man and Environment", which brought together the full color of world geography and took place aboard a specially rented vessel moving along the Don and Volga. The topics of this symposium were environmental changes and monitoring, the quality of information, and natural disasters. Considerable attention was paid at the Moscow IGC to anthropogenic climate changes and fluctuations of glaciers as their indicator, and the consequences of the greenhouse effect.

Soviet geographers maintained close ties with their Polish colleagues. In Poland, long before the Second World War, original and well-known schools emerged in many branches of geography. Polish geographers were keeping more contact with Western scholars than in other socialist countries. Soviet-Polish conferences and joint participation in international research projects served as another channel for internationalization of geography.

4.3 Polish Geography's Way to the West

4.3.1 Political Context: Geography Under the Socialist System

Under communism, human geography in Central and Eastern Europe was forced to adapt to a Soviet model of geography; and to communist decision-makers, the old human geography, with its achievements and rich tradition, reeked of "bourgeois" ideas (Jackowski et al. 2014). Instead, it was recommended that geography should adopt Marxist methodology. Luckily, these decisions were never fully implemented and the Marxist methodology never became obligatory. Indeed, Polish geographers found the theoretical and methodological studies of their Western colleagues very appealing. Although contacts with Soviet science were encouraged, if from the position of humble "younger brothers"; while contacts with "fraternal countries of people's democracy" were tolerated, these contacts were anyway regulated quite closely. On the other hand, all attempts at contact with the "bourgeois West" were treated with great suspicion (Kosiński 2008).

There was a dispute between specialists in regional geography and so-called applied geography to explain to what extent geography should serve society. There was a very serious dispute at the time over political geography, whether it should exist, and if so, to what extent and in which direction. Of course, disputes also concerned geopolitics, but at this time the "climate" for these kinds of research was unsuitable. In the 1960s and the 1970s, many publications could appear only under censorship. Motivations should come from the USSR, and comparisons with the Soviet experience should be made. A typical example of a motivation of the country's division for the needs of planning and administration by the reference to the Soviet concept of region can be found in a work by K. Dziewoński.

> A similar problem, although in an even more acute form, arose in Russia after the October Revolution, when the demands of socialist planned economy created the necessity of setting up a new network of basic administrative units (which ex definition was to be at the same time units of planning and of economic management) as well as of an additional network of larger units of planning (grouping together several administrative areas of the highest rank — the Gosplan regions), In this way the establishment of the optimal regional administrative and political division of the federal state became, right from the beginning of the planned economy, one of the leading tasks facing the state authorities. (Dziewoński et al. 1964)

However, in the post-war period, Polish geographers made significant and original contributions to the development of such fields as the theory of economic regions and settlement systems (K. Dziewoński and J. Kondracki), the agricultural geography of the world (J. Kostrowicki), and urban planning (B. Malisz). The most often quoted works include publications by S. Leszczycki, A. Kukliński, P. Korcelli, Z. Chojnicki and R. Domański. Scholars who were not Communist Party members, but wanted to make a serious career, did not have much of a chance. But in the Institute of Geography of the Polish Academy of Sciences, the situation was different (Śleszyński and Kosiński 2014). Professor J. Kondracki became a member of the IGU Commission on regionalization and organized its symposium in Jabłonna (1963), one of the most important conferences in Polish geography at that time. Reports on the activity of this Commission, which had existed for several years, were published in Poland. Professors Andrzej Wróbel (1928–1999) and Kazimierz Dziewoński (1910–1994) played an active role in these conferences and publications. They concluded that it was necessary to distinguish between different types of economic regions: those defined for practical planning purposes and economic regions considered as phenomena to be studied by geographical and other sciences. The authors believed that efforts should be made to further clarify their mutual relations and to avoid misunderstandings in using these terms correctly. The need to develop an international system of terms in this field, recognized by all principal languages, was emphasized (Dziewoński et al. 1964).

One of the leading Polish geographers of the 1970s, J. Kostrowicki chaired the IGU Commission on Agricultural Typology. He saw its major achievements in discussing and adopting a general concept of agricultural typology, the criteria, methods, and for identifying the types of agriculture worldwide. This concept was applied to modeling the spatial organization of agriculture and planning agricultural development (Kostrowicki 1977). Kostrowicki believed that classification is fundamental for the progress of any discipline and any valid classification is to be based on the commonly agreed principles. According to him, the scientific objectives of an agricultural typology could be summarized as follows: (1) developing agricultural geography as a scientific discipline; (2) putting into order the existing knowledge on world agriculture and its areal similarities, differences, and interrelationships; (3) contributing to a better understanding of agriculture as a complex phenomenon on a world, continental, and national scale (Kostrowicki 1970).

4.3.2 Processes and Channels of Internationalization

The leading human geographers (S. Leszczycki, K. Dziewoński, and J. Kostrowicki) published in foreign journals though finding a way into and place in Western periodicals was a near impossibility for Poles throughout the communist era. As early as in the 1950s it was decided that it was necessary to inform foreign geographers about the achievements of Polish geography. First, abstracts in English and Russian were added to the papers in the revived *Czasopismo Geograficzne* (Geographical Journal) and *Przeglad Geograficzny* (Geographical Review. There was also a plan to issue special publications to mark international conferences, in particular, a special issue of *Przeglad Geograficzny* which was eventually published in 1956 before the Congress in Rio de Janeiro). Ultimately, there was a plan to publish a continuous series of publications in a Western language.

International congresses and conferences have always served as pretexts for publishing special issues of Polish academic journals. They regularly reported about the most relevant international events. *Geographia Polonica*—Poland's oldest scientific journal in the field of geography published in English—prepared many special issues devoted to the IGU congresses. Particularly detailed was the report on the 25th Congress held in Paris in 1984. So, information circulated the other way as well.

The desire to organize bilateral encounters can also be listed as one of the reasons behind the founding of *Geographia Polonica.* The first of these was a Polish-British seminar devoted to human geography held in September 1959 in Nieborów. Meetings with British geographers were followed by the workshops with American, French, Czech, Hungarian, and other scholars. There were about a dozen such meetings over two decades, which contributed to the establishment of many friendly contacts maintained for years. Thanks to these seminars, Poland played a key role in academic exchanges in this part of Europe where the West and the East met (Śleszyński and Kosiński 2014). Looking back, we can see that the collaboration of Polish geographers with their Western colleagues was unique, not only in comparison with other countries of the "Soviet Bloc", but also when compared with other academic disciplines in Poland. Foreign literature, too, was more easily accessible in Poland than in other countries of the Eastern bloc.

There were no Polish participants at the first post-War IGC in Lisbon in 1949 nor at the Washington IGC in 1952, though a five-person delegation attended the 1956 IGC in Rio de Janeiro. An important breakthrough came with the 1960 IGC in Stockholm which paved the way for Stanisław Leszczycki's election to the IGU Vice Presidency at the London IGC in (1964) and the President at the New Delhi IGC in 1968. This marked a new high point in the history of Polish geography. Professor Leszczycki was one of Polish scholars who perfectly understood the importance of international cooperation and, thanks to his position, his colleagues in Poland enjoyed a period when foreign contacts were easier (Kosiński 2008).

In 1953 Leszczycki became co-chair of the Polish Committee for the IGU and then took over the chairmanship for the twenty years from 1956 to 1976. He represented Polish geography in the IGU, managed preparation for its congresses and conferences, and usually was the head of the Polish delegation. He was the most visible Polish geographer at international fora, which reflected his dominant position in Poland, too, thanks not only to his organizational skills, great energy, and diligence, but also diplomatic talent and valuable relations with the authorities (Kosiński 2008). He concluded his presidential term at the 22nd IGC in Montreal in 1972 by delivering a speech in which he presented his views on geography and its future (Leszczycki 1972, 1973). Speaking about the numerous attempts to establish a definition of its object, he concluded that the search for a too precise definition did not make much sense, especially in a situation where various disciplines and branches of science overlap on many fronts. Geography was defined by a spatial approach, whereby phenomena and processes occurring on the Earth's surface, including the relationship between humans and the environment, were studied. In recognizing Professor Leszczycki's merits, the IGU awarded him an honorific *Laureat d'Honneur* in 1988. It was stressed that he had played an important role in rebuilding scientific life in Poland after the war, and in developing scientific cooperation between scientists from different countries, especially those with different political systems at a time when this cooperation was particularly difficult.

The participation in the International Geographical Olympiad (*iGeo*) was also a sign of the internationalization of Polish geography. Together with the Netherlands, Poland was one of the Olympiad's initiators and the Polish team has an outstanding record of achievements in this competition (Barwiński et al. 2014). Since 1974, a national Geography Olympiad has been held annually in Poland.

The evolution of geographical science in China proceeded in a completely different way than in the Soviet Union or Poland, but at the same time, it also had features in common with them, associated with the extreme difficulty of personal contacts with foreign scientists, a great contribution to the process of internationalization of scientific leaders, institutions and academic journals headed by them.

4.4 The Impact of Geopolitical Issues on the Internationalization of Geography in China

4.4.1 Political Context: Geography Under the Socialist System

Modern geography in Chinese universities is undoubtedly a field of knowledge imported from the West. In the early period before the Xinhai Revolution of 1911, Chinese geography developed under the impetus of western modern scientific and democratic revolutionary ideas. In the period 1927–1937, tireless efforts were made by renowned scientists to ensure the rapid development of modern Chinese geography (Hu 1936). Kezhen Zhu, who graduated in the United States founded in 1921 the first department of geography in China at the Southeast University (the predecessor of Nanjing University), and in cooperation with other geographers—the Journal of Human Geography. Before the foundation of new China, the circulation of western geographical thought and the emergence of a large number of geographical departments and research institutions led to the creation of contemporary Chinese geography.

World War Two had a major impact on geographical research. During the war, it was difficult to obtain sustainable national funding, and geographic research stalled. For example, the department of geography of Nanjing National Central University trained no more than 100 graduates (Ren 1943). Under the slogan of scientific salvation, the Chinese Anti-Japanese War inspired the patriotism of Chinese geographers. After the war, the long-term goal of geography was to use its achievements for building New China (Gu 1936). About 1200 geoengineers participated in the country's reconstruction and economic recovery. The war against Japan strengthened patriotic sentiments in Chinese society which sustained the use of geography in the construction of new China (Gu 1936).

During the Cold War era, Chinese geography went through a period of differentiation and established closer contacts with other disciplines, especially ecology, economics, and sociology. The theoretical level and the spectrum of methods used in geographical studies significantly improved (Wu and Zhang 1999; Zhu 2004). Progress in Chinese geography was backed by learning from the experience of Soviet geography, in particular the theories of the regional school and the concept of the TPCs.

In Soviet science, the value of economic geography was determined by its contribution to the study of factors of production and the optimization of industrial development. During the Cold War, this branch of geography was dominated by economic utilitarianism, which was actively supported by state funding. Other subdisciplines of human geography (such as urban geography, political geography, etc.) were viewed as bourgeois fabrications (Li 1981). For example, geographic theories related to geopolitics have been considered as justification for the expansion of Western empires (Qian and Zhang 2021).

The dogmatic model of the development of physical and human geography in the Soviet Union weakened the integrity of geography. In many areas of human geography research ceased for almost 30 years (Wu et al. 2002). The situation changed only after the beginning of reforms and the policy of openness. Combined with the theories and methods of Western geography, Chinese research in human geography began to enrich itself. There has been a shift from cultural and relational turns in human geography to quantitative paradigms (from central place theory to Krugman's geographical economics). Thus, geopolitical factors significantly influenced the assimilation and application of the theories and practices of Soviet and Western science in Chinese geography.

4.4.2 Processes and Channels of Internationalization

Though in the early twentieth century some geographical research institutions were created in Beijing, Shanghai, and other cities, there was no coordination of their activity (Weng et al. 1934) and no organization that could cooperate with IGU and foreign geographical societies (Zhang et al. 2019). From 1949 onwards, Chinese geographers began to establish international contacts, mainly with Soviet geographers. Some books published in the Soviet Union were translated into Chinese (Vitver 1956; Baransky 1956; Kalesnik et al. 1956). Although China was formally admitted into the IGU in 1949, prior to 1978 the Geographical Society of China (GSC) hardly ever communicated with Western geographical societies. Chinese geographers were almost unknown and seldom cited in the West.

During the reforms, China established diplomatic and cultural relations with a large number of countries around the world. The GSC maintained close cooperation with the IGU and promoted the internationalization of geography with its assistance. In 1984, the GSC resumed its full membership in IGU and formed the National IGU Committee. Chinese geographers began to attend the IGU congresses and other associated events. Since 1988, the GSC has shaped Chinese delegations at the General Assembly. Chinese geographers are usually the largest group of participants at IGCs. With the help of the IGU, Chinese geographers strengthened academic exchanges and disseminated the results of their studies around the world.

China's most well-known geographers, such as Chuanjun Wu, Changming Liu, Dahe Qin, Chenghu Zhou, and Bojie Fu, were successively elected as Vice Presidents of the IGU. They set up the Commissions on Geography for Future Earth, Agricultural Geography, and Land Engineering. The IGU enables Chinese geographers to enhance cooperation with other commissions and international geographical organizations.

In 1983 and 1989, respectively, the GSC joined the International Frozen Soil Association and the International Geomorphology Association. The GSC established bilateral exchanges with geographical communities in Japan, France, South Korea, the United States, Russia, India, the United Kingdom, and Kazakhstan, and hosted the China-Japan-Korea Joint Conference on Geography and the Asian Geography Conference. The 33rd International Geographical Congress in Beijing (2016) was one of the largest in IGU history by the number of participants. The success of these fora reflects the contribution of Chinese geography to the IGU and contributes to the internationalization of geography.

4.4.3 Scientific Journals and Publications

The international activities of geography resulted in internationalization of geographical publications in terms of their structure and research topics. The language and the presentation of geographical publications much more often meet the requirements of foreign readers. In the journal of the Geographical Society of China founded in 1934, each article has always had an English abstract. Publications in this journal have also changed accordingly and are complemented by English references to meet the needs of international readers (Geographical Society of China 1954).

The study of geopolitical issues was also one of the important indicators of the internationalization of geography. In the early 1970s, under the dual pressure from the United States and the Soviet Union, China's diplomatic strategy had gradually turned to unite with the third world to fight against hegemonism. In this context, considerable attention was paid to maritime law and national interests of different countries. Chinese geographers advocated Peru's claim of a 200-mile maritime economic zone, which directly threatened the so-called navigation freedom and fishing rights of the United States, the Soviet Union, and the other maritime powers in the waters around the Latin America (Li 1978). The geopolitical problems of the war in the Middle East drove Professor Jinxi Qian and his team to publish many influential papers in the field of the world energy geopolitics, which still has important implications for the geography of international energy (Qian 1984).

4.4.4 International Exchanges

International exchanges also affected the process of geography internationalization. In the early twentieth century, the Chinese government received the Boxer Indemnity voluntarily returned by the Eight Power Allied forces. This was used to fund Chinese students to study in Western countries (Tang 1989). For example, Kezhen Zhu was admitted to the United States for the second term of the Boxer Indemnity program and studied in the agricultural college of the University of Illinois, and then returned home to found modern Chinese geography.

In the early Cold War, because of good relations between China and the Soviet Union, many Chinese studied geography and theories of industrialization in the Soviet Union. In its turn, the Soviet Union sent many geographers and experts to China to help in the reconstruction of China's national economy. After the start of reforms and opening up, Chinese geographers were more prone to go to study in Europe and the United States. Coming back home, these geographers became the backbone of the development of Chinese geography and made a great contribution to internationalization of geography. Some world-renowned Chinese geographers have also worked in Western countries for many years and became important intermediaries for international academic exchanges with foreign countries.

4.5 Conclusion

The economic capabilities of the country, the needs of the national economy and the principles of its functioning, the role of the state in research, and historical traditions of national geographical schools determine the development of geographical science: the relationship between disciplines, theoretical and methodological equipment, connection with practice. But the political regime, the geopolitical situation, and the country's place in the geopolitical world order were ultimately the strongest factors. Two world wars and the arrival to power of communist regimes significantly influenced geography in many countries. In the most difficult years, the interaction of Soviet, Chinese, and Polish geographers with their foreign colleagues practically ceased or was extremely limited. The USSR and post-war Poland joined the IGU only in 1956, and China renewed its membership only in 1984.

However, thanks to leading scholars who studied at foreign universities or kept close contacts with partners abroad, even in these years, geography in these countries were by no means isolated from the rest of the world. The access to foreign publications played a significant role. As soon as shifts in the international situation improved geographers immediately profited from these new opportunities. The participation of geographers from socialist countries in the activities of IGU played an invaluable role and for some time became almost the main form of internationalization of geographical studies. On the one hand, it helped to strengthen the prestige and institutional positions of geographers in their country, and on the other, and most importantly, it gave impetus to the development of new directions, theoretical approaches, and methods. New ideas made their way, despite ideological prohibitions that only slowed down the emergence of pioneering topics. The activity of IGU created a platform for professional contacts and networks. The number of publications abroad was increasing, including in cooperation with foreign co-authors, academic exchanges were gradually expanding, and joint projects began to be developed. In general, the development of geography on the other side of the "Iron Curtain" followed global patterns: further differentiation, "quantitative revolution", followed by powerful trends toward humanization and "environmentalization"—the integration of physical and social geography on a new theoretical and methodological basis.

References

Armand DL (1966) For us and grandchildren. Mysl', Moscow (in Russian)

Baransky NN (1956) Economic geography of the U.S.S.R. Foreign Languages Publishing House, Moscow

Barwiński M, Sawicki T, Uroda J (2014) Poles in the international geography olympiad (iGeo). Geogr Pol 87(2):309–316. https://doi.org/10.7163/GPol.2014.19

Geographical Society of China (1935) The history of geographical society of China. Acta Geogr Sin 2 (1):149–153

Geographical Society of China (1954) The afterword of acta geographica sinica. Acta Geogr Sin 20(1):112–115

Gerasimov IP (1965) General results of the XX international geographical congress in London. Izvestia of the Academy of Sciences of the USSR. Ser Geogr 1:3–10 (in Russian)

Gerasimov IP (1976) XXIII international geographical congress. Geoforum 7(2):87–88; 7(3):159–160

Gerasimov IP et al (1975) XXII international geographical congress in Canada and the participation of Soviet scientists in it. In: XXII international geographical congress in Canada (Montreal, August 1972) Nauka, Moscow, pp 5–18 (in Russian)

Gerasimov IP et al (1976) XXIII international geographical congress. Scientific activity (a review). Izvestia of the USSR Academy of Sciences. Geogr Ser 6:5–24 (in Russian)

Gu Y (1936) The significance and mission for the joint annual conference of seven scientific groups. Pop Sci 3(10):20

Hu X (1936) Prospects for scientific development of China. Sci Time 3(10):22

Jackowski A, Bilska-Wodecka E, Sołjan I (2014) 80 years after the 14th congress of the international geographical union in Warsaw, 23–31 August 1934. Geogr Pol 87(2):295–308. https://doi.org/10.7163/GPol.2014.18

Jackowski A et al (2018) Tradycja geograficznych kontaktów polsko-francuskich. In: Więckowski M et al (eds) Geografia—wspólne idee, wspólne wizje, Sto lat współpracy polsko-francuskich. WGiSR UW, Warsaw, pp 11–32

Kalesnik SV, Yongluan T, Zhengxian W (1956) Principles of general geography. Geological Press, Beijing

Kish G (1992) International geographical union: a brief history. Geo J 26(2):224–228

Kosiński L, Stanisław L (2007) Międzynarodowa Unia Geograficzna. Przegl Geogr 79(3–4):381–387

Kolosov V (2017) Post-socialist and post-Soviet geography. In: The international encyclopedia of geography: people, the earth, environment, and technology. Wiley-Blackwell, Hoboken, NJ. https://doi.org/10.1002/9781118786352

Kolosov V, Treivish A (2009) A Russian-language geography. In: Kitchin R, Thrift N (eds) International encyclopedia of human geography, vol 1. Elsevier, Oxford, pp 457–462

Kosiński LA (2008) Międzynarodowa Unia Geograficzna. In: Jackowski A, Liszewski S, Richling A (eds), Historia Geografii Polskiej. Wydawnictwo Naukowe PWN, Warsaw, pp 477–488
Kosiński LA (2014) Geogr Pol 87:2:277–294. https://doi.org/10.7163/GPol.2014.21
Kosiński LA, Śleszyński P (2014), Geographia Polonica: a window onto the world—an interview with Professor Jackowski—will geography remain geography? Pondering the state of geography. Geogr Pol 87 (2):251–265. https://doi.org/10.7163/GPol.2014.16
Kotlyakov VM, Annenkov VV (1989) The XXVI international geographical congress. Izvestia of the academy of sciences of the USSR. Ser Geogr 1:7–19
Leszczycki S (1964) Research problems in Polish geography. Geogr Pol 37(1):7–22
Li C (1978) Geographical analysis of 200 mile maritime rights in Peru. J Shanghai Normal Univ 1:61–64
Li Z (1981) A geography conference of historical significance. Acta Geogr Sin 36:345–348
Meadows ME (2020) Perspectives of internationalism in geography. Geogr Pol 93(4):493–503. https://doi.org/10.7163/GPol.0189
Paasi A (2015). Hot spots, dark-side dots, tin pots': the uneven internationalism of the global academic market. In: Meusburger P, Gregory D, Suarsana L (eds) Geographies of knowledge and power. Springer, Dordrecht, pp 247–262. https://doi.org/10.1007/978-94-017-9960-7_12
Robic M-C, Briend A-M, Rössler M (eds) (1996) Géographes face au monde. L'Union géographique internationale et les Congrès internationaux de géographie. L'Harmattan, Paris
Qian J (1984) The economy of developed capitalist countries under the impact of oil: impact and countermeasures. World Oil Probl 1:1–29
Que W (1995) Historical geography in China. J Hist Geogr 21(4):361
Ren M (1943) Geographical research and economic development. Orient Mag 39(12):23
Saushkin YG (1969) The XI international geographical congress (Delhi, December 1968). Geogr Sch 3:81–85 (in Russian)
Saushkin YG (1977) The XXIII international geographical congress and higher school. Bulletin of Moscow State University. Series 5. Geography 4:3–8
Sen L (1936) People in the evolution of Chinese geography semi-monthly magazine (trans Yiliang Z). Chinese Hist Geogr 5(10):66
Tang J (1989) The Boxer indemnity returned by the United States and Tsinghua University. Educational Research of Tsinghua University 2:87–92
Vitver IA (1956) World economic geography. The Fifties Press, Shanghai
Wu C, Zhang J (1999) Review the development of geography in the 20th century and prospect for geography in the 21st century: congratulation on the 90th anniversary of the founding of the geographical society of China. Acta Geogr Sin 5:385–390
Wu C, Liu S, Yang Q (2002) Geography of China in the 20th century. Fujian Education Press, Fuzhou
Zhang G, Huang J, Zhu H (2019) The geographical society of China and the development of the Acta Geographica Sinica. Acta Geogr Sin 74(11):2197–2208
Zhu K (2004) The responsibility of geographer in China, vol 1. Complete works of Zhu Kezhen. Shanghai Science and Technology Education Press, Shanghai

Vladimir Kolosov is Deputy Director of the Institute of Geography, Russian Academy of Sciences, Foreign Member of Belgian Royal Academy of Sciences, Honorary Member of French and Polish Geographical Societies and Doctor Honoris Causa of the University of Le Havre, France. He lectured and conducted research at the universities of France, USA, Finland and Germany. His research interests lie in the fields of political geography and geopolitics, social geography and world cities. In 2006–2020 Professor Kolosov served in the IGU Executive Committee and in 2012–2016—as President of the IGU. He participated in many international projects.

Marek Więckowski is Professor and Head of Department at the Institute of Geography and Spatial Planning of the Polish Academy of Sciences (PAS), Warsaw. His studies are focused on political boundaries and cross-border cooperation, tourism, regional development and other fields. Since 2012 he is Editor in Chief of the *Geographia Polonica* journal. In 2014–2017 he served as Director of the PAS Scientific Centre in Paris. Professor Więckowski has a number of academic awards and is also known as journalist, writer, photographer and traveler.

Debin Du is Professor and the Dean of the School of Urban and Regional Science and the Institute for Global Innovation and Development in the East China Normal University, Shanghai. He had been a Fulbright Visiting Scholar at the Department of Geography of the University of California, Berkeley. His research focuses on the world geography and geopolitics. He also holds many important academic positions, such as Director of World Geography Committee of the Geographical Society of China and the Editor-in-Chief of the Journal of World Geographic Research.

Xionghe Qin is a post-doc at the Institute for Global Innovation and Development in the East China Normal University. He received his Ph.D. in human geography from East China Normal University and has been a visiting Ph.D. student in the Department of Geography and Geo-Information Science at University of Illinois Urbana-Champaign. His research interests are world economic geography and regional development.

The International Circulation and Dissemination of Geographical Concepts and Ideas

5

Trevor Barnes and Michael Roche

Abstract

Using the examples of both regional geography and the new geography, this chapter's purpose is to examine how IGU conferences and symposia for the fifty-year period 1920–1970 were part of a larger apparatus that allowed the international circulation of geographical ideas along with their development, amendment and reproduction. The chapter is divided into three sections. First, we discuss the role of the conference as a site in the circulation of disciplinary ideas. We stress the importance of both travelling scholarship and conferences as sites for performing new knowledge. Second, we discuss the role of the IGU in formalising and disseminating regional geography. While the region was an object of geographical research from the late nineteenth century, it was not formally recognised by the IGU as a separate thematic interest until 1938. Third, we turn to the new geography that emphasised abstract theory and quantification, which received its first IGU exposure in 1938 with Walter Christaller's presentation on central place theory. It was not until after the War, however, that the new geography became a general methodological approach. That was the significance of the 1960 Lund IGU Symposium in Urban Geography and from 1964 furthered by the IGU Commission on Mathematical Geography.

Keywords

Conferences · Knowledge circulation · New geography · Regional geography

T. Barnes (✉)
Department of Geography, University of British Columbia, Vancouver, Canada
e-mail: trevor.barnes@geog.ubc.ca

M. Roche
School of People Environment and Planning, Massey University, Palmerston North, New Zealand

5.1 Introduction

On 21st July 1938, at the 15th International Geographical Congress (IGC) in Amsterdam, the German geographer, Walter Christaller (1938: 21–22), presented *Rapports fonctionnels entre les agglomérations urbaines et les campagnes.*[1] It was a summary of his central place theory developed five years earlier in his doctoral dissertation at the University of Erlangen, *Die zentralen Orte in Süddeutschland* (Christaller 1933). Christaller's paper based on a stark hexagonal geometry, numerical data drawn from many Southern German telephone users, and a belief in a law-governed geographical order was all a bit

[1]The title in the congress programme was in French, but the printed paper was in German. It is highly likely that the paper was presented in German (Christaller 1938).

V. Kolosov et al. (eds.), *A Geographical Century*,
https://doi.org/10.1007/978-3-031-05419-8_5

too much for several regional geographers who attended and returned "communications".[2] Even before those comments were aired, however, the session's *président*, the Swiss regional geographer, Charles Biermann, felt compelled to weigh-in, chiding Christaller for a presentation that:

> seems too rigid to me…. Man becomes a machine which obeys the laws of natural science and in which urban agglomerations are arranged according to a geometric series. I do not believe that the human trial takes place in this mathematical order. Besides … I do not find a word about life and society and the problems of the countryside. (quoted in Djament and Covindassamy 2005, Section 47, our translation)

So it was that the dispute began between an older regional geography and what came to be known as "the new geography", or "spatial science", or "the quantitative revolution"—theoretical, mathematical, scientific, law-like. Twenty-two years later, Christaller gave a second paper at another International Geographical Union (IGU) event, this one in Lund, Sweden in 1960. Linked to the main IGC in that year in Stockholm, the IGU Symposium in Urban Geography was a full-throated celebration of what the new geography had become since 1938 in Amsterdam. Now Christaller was accorded pride of place: the opening speaker of the symposium. Many of the subsequent papers at the symposium were also devoted to his central place theory, given by European and especially American acolytes.[3] The new geography had arrived. Moreover, on this occasion, despite the presence of notable regional geographers including perhaps the then most well-known of all, Richard Hartshorne, there was no open criticism.

By using the examples of both regional geography and the new geography, the purpose of this chapter is to examine how IGU conferences and symposia are part of a larger apparatus allowing the international circulation and dissemination of geographical ideas along with their development, amendment and reproduction. Staying with our two exemplar cases, regional and the new geography, though, means we have restricted our study of the IGU and its role in the circulation of geographical ideas to roughly only a 50-year period in the middle of the twentieth century (ca. 1920–1970). Nonetheless, we believe that our larger conceptual framework developed in the next section is robust enough to be applied to other eras within the history of the IGU.

The chapter is divided into three sections. First, we discuss the role of the conference as a site in the circulation of disciplinary ideas. We stress the importance of both travelling scholarship and conferences as platforms for performing new knowledge. Second, we discuss the role of the IGU in formalising and disseminating regional geography. Its academic origins were in the nineteenth century associated with geographers in France and Germany. Strangely, though, the region was not formally recognised as a separate thematic interest by the IGU until 1938, although de facto it was an object of interest much earlier (Kish 1972b: 86). Thirdly, we turn to the new geography that as already discussed received IGU exposure in 1938 with Christaller's presentation. While the use of mathematics in geography was longstanding, linked to surveying, geodesy and cartography, it was not until the post-Second World War period that mathematical geography was transformed from an analytical technique to a general epistemological approach. That was the significance of the 1960 Lund IGU Symposium in Urban Geography and from 1964 furthered by the IGU Commission on Mathematical Geography.

5.2 Conferences, Performance and Circulation

No conferences, even IGU congresses and symposia, are innocent. They are always political events, creatures of their time and place, with power coursing through them at various scales. They serve and maintain, and on occasion frustrate, assorted social interests that *inter alia* are

[2] The paper was written and circulated before the congress, receiving 15 written "communications" (discussed and quoted in Djament and Covindassamy 2005: Section 48).

[3] The participant list and programme are in Norborg (1962: vi–vii).

personal, disciplinary, institutional, national, epistemological and ideological.

The most immediate function of the earliest International Geographical Congresses (IGCs) was to legitimate, shape and institutionalise primarily European geography. In 1871, the first IGC in Antwerp attracted some 600 subscribers —400 attendees—from 20 countries and lasted eight days (98% of participants were from Europe; Kish 1972a: 49). Martin (1992: 5) writes, it "was the first time that so many people interested in things geographic were assembled in one place". They included academics, diplomats, government administrators, educationists, military officers, cartographers and assorted geographical "savants" (Martin 1992: 5). They gave the IGC the provenance to advertise the virtues of the discipline especially to European governments. In the terms of the sociologist of science, Latour (1987), the early IGCs demonstrated that geography commanded strong and diverse allies, stabilising the discipline, fostering at selected sites its institutionalisation and the gathering of what Martin (1992: 6) calls its "paraphernalia".

The early version of the discipline at the IGCs was a geography "in the wild" (Callon 2008). It was unruly, lacked formalisation, made up of a set of radically varied practices derived from diverse users of geographical knowledge who included travellers and explorers, amateur local geographical societies, military strategists, diplomats, colonial civil servants, makers of gazetteers and atlases, business owners and commerce students, and academics (see Driver 2001: Chapters 1 and 2). In this sense, geography was a "boundary object", meaning different things to different communities, but nonetheless garnering sufficient overall support to be sustained (Star and Griesemer 1989). Increasingly over the first half-century of the IGC's existence there were calls to domesticate geography, to tame it, to discipline the discipline. But it was not until 21st July 1922 in Brussels that those efforts were realised following the establishment of the International Geographical Union. It became the governing body for the IGC, emphasising above all geography as an academic science along with associated rigorous forms of communication and discussion (Kish 1972b).

It was those communications and discussions that circulated and travelled, that shaped the discipline often far from where those ideas were first articulated. As Raj (2013: 344) writes, the term circulation.

> serves as a strong counterpoint to the unidirectionality of "diffusion" or even of "dissemination" or "transmission"… which all imply a producer and an end user. "Circulation" suggests a more open flow—and especially the possibility of the mutations and reconfigurations coming back to the point of origin. Moreover, the circulatory perspective confers agency on all involved in the interactive processes of knowledge construction.

Recognition that ideas travel is critical. From the 1960s sociologists of science argued that scientific ideas are produced within local historical contexts rather than the product of a universal rationality (Secord 2004: 657). This begs the question, though, of how locally produced ideas become generally accepted, seemingly true all over? The consensus now is that it is through travel, the movement of ideas from one local context to another. As Ophir and Shapin (1991: 16) put it, the belief that "ideas float free in the air … [as] heavenly knowledge" has been abandoned, replaced by an emphasis on "lateral movements between mundane places". Consequently, as Secord (2004: 664) asserts, even more important than "science in context" is "knowledge in transit". Circulation becomes a central concept, emphasising the continual flow of knowledge into and out of particular sites, which as it does is amended and revised.

Large conferences such as the IGC are one of those sites around which geography's knowledge circulates. In the case of the IGC, the actual geographical site changes for each meeting affecting who attends and from where, as well as the kinds of issues discussed and how they are discussed. At those conferences, knowledge is produced through the performances of those attending. Participants dress up, rise to their feet, go on stage often with a script in hand and with various props perform knowledge in front of an audience (Craggs and Mahoney 2014). In many cases, that knowledge ends there, on the conference floor. It stops because the presenter runs out of time, or poorly executes, or is in a language not widely understood,[4] or is too specialised, or jargon-laden, or is achingly dull, or regurgitates existing knowledge.[5] But just sometimes the performance takes on an extended life, it circulates, moving beyond the micro-geography of the conference hall to scale up, travelling to outside spaces, to regions, countries, the world. As it does, it can be changed and modified, even returning to future IGCs in revised forms and subject to yet further revision.

That is what happened to Christaller's (1938) Amsterdam presentation. Performed in German, subsequently it circulated elsewhere through the journeys of the conference "Records" and perhaps even more importantly through the corporeal travels of those who were there, or even those who heard of the event second- or third-hand.[6] While there were criticisms of Christaller by the session chair, there were also several discussants who praised him, calling his approach "fundamentally significant" (Djament and Covindassamy 2005: Section 48, our translation). The Dutch geographer Willem Boermann, who organised the Amsterdam meeting, was a fan and at least indirectly likely behind his invitation (van Meeteren forthcoming). Consequently, it was not so surprising that in 1960, having languished for some time in a post-War wilderness, Christaller circulated back onto the IGU conference stage at Lund with his central place theory but this time he and it were lionised (Norborg 1962: 157–165). In the interim his central place theory had been reinterpreted, changed and modified by a group of especially young North American trained urban and economic geographers. They then did in Lund what Christaller had done in 1938 in Amsterdam: they performed central place theory at the lectern, propelling its yet further circulation including to future IGU conferences, consolidating the hitherto inchoate new geography. The twin processes of performance and circulation at the IGC were of course not peculiar to Christaller and his central place theory. They were part of the choreography of any international circulation of geographical ideas that we now discuss using our two case studies.

5.3 Regional Geography and the IGC

Our first example is regional geography (Livingstone 1992: Chapter 8). While there were papers about the region at early IGCs, it was only after the establishment of the IGU in 1922 that they became a staple. The importance of the region was readily apparent at the 1928 IGC at Cambridge. Despite IGU President General Nicola Vacchelli, Director of the Instituto Geografico Militare of Florence, saying that "the period of discoveries is far from finished", academic discussion about the region was prominent (Anon 1928: 259). There were 17 presentations listed under "Regional Geography". Probably, however, regional geography's greatest impact at that meeting was through the publication for attendees of the volume, *Great Britain: Essays in Regional Geography*. Edited by Alan Ogilvie and eminent agricultural scientist Russell (1930), it comprised 24 chapters, mainly by senior geographers. Even Vacchelli, in his closing address, gave "high commendation to the 'Regional Studies of Great Britain'" (Anon 1928: 266).

[4] Until 1960, the IGU allowed presentations in six languages: French, English, Italian, German, Spanish and Portuguese; Harris (2001: 675). Afterwards it was two, French and English.

[5] The diary of the New Zealand geographer, George Jobberns, who attended the 1952 IGC in Washington DC, recorded many of these performative failings: poor time management, poor chairing, the challenges of multilingual sessions even when translation services were offered, and overly specialised papers (Roche 2019: 148).

[6] It is not known how many conference attendees understood German, but before World War One German academics dominated the social sciences and many non-Germans academics, including Americans, necessarily learned the language. Even during the interwar period, the American geographers Carl Sauer, Richard Hartshorne, Edward Ullman, Chauncy Harris and Edward Ackerman could read and speak German. This became less common after World War Two with implications for IGU events (discussed further below). But with English becoming the global academic lingua franca over the last 25 years or so the issue is no longer so acute, although it clearly puts non-native English speakers at a disadvantage.

The 1931 congress in Paris was even more of a stage for regionalism, particularly that of the French school derived primarily from writings of Paul Vidal de la Blache (Clout 2005a, 2012). Vidal was already dead, but his influence remained through the conference performances of his many students. In 1931, the number of French university geographers was still small—a total of 18 geography professors and five other teaching staff (Clout 2012: Paragraph 8). Because of Vidal's earlier work, however, rather than being seen as "a poor relative of history", French geography was "self confident … developing its own identity" (Clout 2005a: 15). Especially important was the regional monograph that offered meticulous studies of typically French rural *pays* that seamlessly integrated physical and cultural landscape processes. Apart from featuring papers about the French regional method and its execution, the Paris congress included several pre- and post-congress field trips. Vidal's son-in-law, Emmanuel de Martonne, wrote and organised an accompanying set of matchless guidebooks. Led by doctoral students, experienced professors and physical scientists, the field-trips "introduced congress-goers to the practice of academic geography in France" (Clout 2012: 3). In turn, such performances, according to Clout (2005a: 15), "provided inspiration for teachers of geography in the UK, the USA and elsewhere who were seeking to set their subject on a firm footing". More generally, the 1931 IGC established French regional geography as dominant, eclipsing its long-time rival from Germany. Indeed, German geographers had been banned from IGC events as punishment for Germany's involvement in the First World War. That was rescinded by the Paris IGC, but the Nazi party took umbrage and although still not in power urged German geographers not to attend—only seven came (Fig. 5.1).

Fig. 5.1 Emmanuel de Martonne (1873–1955) 70 years old. *Source* Archives privées de la famille Birot

That changed significantly at the 1934 IGC in Warsaw. Poland was viewed by the Nazis, by then the German government, as Germany's lost *lebensraum*. Fifty German geographers attended, the same number that came from the UK, and twice as many as from America even though the IGU President was the American Isaiah Bowman (Clout 2005b: Paragraph 3). According to some observers, French regional geographers at this conference were treading water, stuck in rural studies, lacking their earlier "cohesion and vigour" (Clout 2005b: Paragraph 9). "Section V Regional Geography" included only 11 presentations, with, as one reviewer noted, "in nearly all the cases regional studies methodology receiv[ing] incidental treatment" (Anon 1935: 143; Kish 1979).[7] The action was with the Polish geographers who were exploring the geography of urbanisation and industrialisation, deploying quantification and referencing even laws of location (Clout 2005b: Paragraph 9; see also Jackowski et al. 2014). That trend continued at the 1938 Amsterdam congress, except the new work was carried forward by a German, Christaller.[8] Regionalists were still well represented at that

[7]Some presenters listed did not attend—a conspicuous example being New Zealand geomorphologist Sir Charles Cotton who had papers printed in both the Lisbon and Washington congresses but participated in neither.

[8]Tragically, many of the Polish geographers who presented at Warsaw and who in some ways prepared the ground for Christaller were murdered by the Nazis after they invaded that country in September 1939 (Clout 2005b, Paragraph 19).

meeting, however, with 35 regional geography presentations, reacting so strongly against Christaller's work.

The first post-war IGC was in Lisbon, 1949. For Freeman (1972), its structure and concerns were more closely linked to the past than the post-war world. There were only seven presentations in the Regional Geography Section, but related issues were discussed in "section VII Methodology, Teaching and Bibliography". According to one British participant: "Attendances were largest and discussions most vigorous in sessions devoted to papers dealing with theory and practice in the delimitations of geographical regions" (Anon 1949: 80). The leading discussants were Yann-Moruvan Globet, Orlando Ribeiro and Fábio de Macedo Soares Guimarães. Globet spoke on *La formation et l'évolution des régions anthropo-géographiques.* His Ph.D. was on the political geography of Ireland and although Professor of Commerce at the École des Hautes Études Sociales in Paris, he was not part of the French geographical establishment. Orlando Ribeiro was the leading figure in Portuguese geography in the 1940s and 1950s. He studied under de Martonne at the Sorbonne and also grasped German approaches to regional geography. Fábio de Macedo Soares Guimarães was Director of the Brazilian Geography National Council. His initial geographical training was guided by French regional geographers teaching in Brazil and shaping his major work, a Vidalian inflected study of Brazilian regions. In 1945, he pursued further geographical training in regional planning in the United States, later helping to consolidate geography at the Brazilian Institute of Geography and Statistics.

In sum, while immediate post-war French regional geography may have lost some cohesion and vigour, it remained in circulation in Lisbon. But, as a sign of things to come, as we will discuss in the next section, also presenting at Lisbon was Edgar Kant of Lund University who gave a paper on geographical terms (Anon 1949: 80).

The 1952 Washington IGC gave voice to an American version of regional geography. There were 25 presentations in the IGC section "Regional Geography". That showing was in part a result of the influence of Richard Hartshorne's tome, *The Nature of Geography* (1939), which celebrated regional geography. Significantly, Hartshorne drew on German rather than French regionalism, especially as articulated by Alfred Hettner. Congresses had always provided an opportunity for the hosts to present national points of view. Given Hartshorne's influence in American geography, it was unsurprising that regional geography was so prominent at the Washington IGC.

Regions also remained an important organising principle for geographers outside America. Carl Troll and Borivoje Z. Milojević, professors of geography at the universities of Bonn and Belgrade respectively, jointly presided over Section V, Regional Geography. They believed the section "offered an opportunity for the consideration of what must always be regarded as one of the central objectives of geographical study" (Edwards 1953: 86). Troll worked on landscape ecology, climatology and mountain lands, while Milojević developed regional geographical methods and published on regions in Yugoslavia (Stanković 2006). George Chabot (Sorbonne), an active participant at the 1931 congress spoke in Washington on *La valeur scientifique de la géographie régionale*. Never one to hide under a bushel the merits of French regionalism, he believed that American geography was 50 years behind French thinking and familiar "only with the generalizations of the French school" (Freeman 1972: 207). In contrast "advancing an American view of regional study, Professor Derwent Whittlesey (Harvard), while [following Hartshorne in] denying the existence of the region as a self-given entity, emphasized its importance as a tool for geographical work" (Edwards 1953: 86).[9]

George Jobberns, who in 1937 was the inaugural university geography appointment in New Zealand, kept a diary of the Washington congress. Jobberns espoused a non-doctrinaire

[9]Whittlesey was more well-known as a political geographer. This serves as a reminder that many geographers of the 1930s to 1950s combined systematic and regional interests and some appeared in quite different sections at successive congresses.

regionalism, influenced as much by Sauer's cultural geography as Hartshorne's rigorous areal differentiation. For Jobberns (2010: 109), the Washington congress was far "too large and too diverse for any one person to cope with". He believed regional geography should be accessible to any properly trained geographer. He was especially critical of James Speers (Cambridge) who "laid down the law" about the origins of a wave cut platform—"it must be very comfortable to be so sure" Jobberns (2010: 108) wrote. As an afterthought, he added that European geographers were somewhat disappointed with their American counterparts. The "profound scholarship in some special field" displayed by European geographers was at odds with the more "discursive and descriptive" geography of the Americans (Jobberns 2010: 109).

The 1956 Rio de Janeiro congress was the first to take place in the Southern Hemisphere. The presidential address by Britain's Dudley Stamp (1958) on "The measurement of land resources" was concerned with natural resources and economic development. There were now only nine presenters in the Regions section. Other regional papers were subsumed within the symposium on "Problems of grasslands in tropical regions". The paucity of presenters in the Regional section was glossed over by Charles Hitchcock in the *Geographical Review* who emphasised the vibrancy of the regional discussion: "a paper on the geographical regions of Canada by J. Lewis Robinson (Canada) evoked considerable discussion and a paper on Cuban landscapes by Salvador Massip (Cuba) also received favourable comment". Robinson emphasised the pragmatic value of the region as a distinctive disciplinary conceptual tool (Hitchcock 1957: 121), while Massip offered a five-fold classification of Cuban landscapes (Massip 1956).

Regional geography remained a major section heading at the 1960 (Norden), 1964 (London), 1968 (New Delhi), and 1972 (Montreal) congresses. But by 1960, as we will discuss in the next section, the new geography was gaining momentum and becoming on par if not exceeding regional geography. This was not immediately reflected in the 1960 congress programme although by the 1972 Montreal meeting only 35 or 7% of the papers presented were in the Regional Geography Section (Fairchild 1973).

The 1960 congress subsumed "regional geography" within Human and Economic Geography and accounted for 18% of those presentations (Fairchild 1961: 111). Significantly, it also provided the setting for a symbolic contest between regional geography and the new quantitative geography. The Soviet geographer Yulian Saushkin (1961: 40), later a founding member of the Commission on Mathematical Geography, attended two separate unscheduled talks, one by Hartshorne and the other by the arch quantifier cum regional scientist, Walter Isard. According to Saushkin (1961: 40), Hartshorne's "paper was declarative in character and gave the impression that the speaker [was] merely 'renovating the front to the building' without changing any of the concepts expounded in his well-known theoretical works, *The Nature of Geography* and *Perspective on the Nature of Geography*". Saushkin (1961: 41) also "heard negative comments about the 'schematic constructivism' and 'technicism' of Isard's work". In contrast, Saushkin (1961: 41) observed a "skeptical attitude among geographers of the middle and younger generation toward the views stated in *American Geography; Inventory and Prospect*, especially with regard to the concept of subjectivism in regionalization, which is so evident in that volume". Interestingly, while attendees of the Stockholm IGC received the regionally oriented *A Geography of Norden* (Somme 1961), those who showed up at the affiliated IGU Lund Symposium got something very different as we discuss in more detail below.

Clearly the tide was turning. At the 1968 New Delhi congress, only the paper by David Hooson (Berkeley), "Rejuvenating regional geography: ends and means", offered a conceptual discussion of the region. At the following 1972 IGC in Montreal, systematic geography dominated, as well as several quantitative papers (collected in Yeates 1974). Steele and Watson (1972: 141) observed that at the Montreal meeting there was a "move away from regional geography…. [C]ertainly, many geographers now seem more concerned with process than with place—and sometimes it would seem, more with statistical manipulations than with environmental reality". Nonetheless, the regional tradition was not yet dead. The local organisers at the Montreal IGC prepared six volumes of local regional description as well as several field trips for IGC attendees (Fairchild 1973: 94).

Regional geography never occupied a hegemonic position at congresses. Once IGCs were dominated by academics, there was always a mixture of regional and systematic geography. Early on IGCs served to foster especially a Vidalian conception of regionalism. But from the mid-to-late 1930s, IGCs increasingly became sites where regional geography and systematic geography in combination with quantitative methods sat albeit uneasily side-by-side. That continued through the IGCs of the 1950s and 1960s with regional geography more and more overshadowed by systematic geography. Regional geography continued to exist but as it circulated among IGC meetings it was amended, relabelled under other categories of geography. That said, because local organising committees were made up of established disciplinary figures who often specialised in regional geography, regional geography remained alive, and on occasion even gave a robust performance.

5.4 The New Geography, the IGU and the Commission on Quantitative Methods

Christaller's reception at the 1938 Amsterdam IGC might have been mixed, but some geographers already recognised the novelty of his work, welcoming its radical implications for the discipline and propelling its circulation to new international spaces (Fig. 5.2).

One of them was Estonia. The Estonian geographer Edgar Kant read Christaller's (1933) doctoral thesis within a year of its publication, recognising its enormous potential. By 1935, through Kant's influence at the University of Tartu, Christaller's central place theory was used to re-plan Estonia's administrative territories (Tammiksaar et al. 2018: 82). The Second World War changed everything, however. In 1940, Christaller left his university position at Freiburg to work for the Nazis, using his central place theory to plan the conquered eastern territories, especially the annexed region of Poland around Poznań (Barnes 2012; Olwig 2018).[10] Then in summer 1944, after a fraught period during which Estonia was occupied by the Soviets, followed by the Nazis, and finally by the Soviets again, Edgar Kant first by motorbike then by motorboat fled to Sweden as a political refugee, albeit taking his commitment to Christaller with him (Barnes and Abrahamsson 2017: 117). Finding work as an archivist, later as a researcher, at the Department of Geography, University of Lund, in 1947 Kant was assigned a first-year departmental graduate student, Torsten Hägerstrand, to help him with Swedish, one of the few European languages he could not speak (Barnes and Abrahamsson 2017: 117). Kant introduced Christaller's work to Hägerstrand, and Hägerstrand introduced Kant, and by proxy Christaller, to other Swedish geographers.

Christaller's work also circulated to another international space, almost as unlikely as Estonia: the US Pacific Northwest, specifically, the University of Washington, Seattle. Steeped in the regionalist tradition, in 1950 the Department of Geography appointed in retrospect a Trojan-

[10] Christaller lived a turbulent life. Born in 1893, he served in the German army during World War One. After a protracted undergraduate education, he wrote his famous 1933 Ph.D. thesis in only nine months at the University of Erlangen. A socialist, he fled briefly to France in 1934 as a political refugee from the Nazi regime that had come to power the previous year. But he was soon lured back to Germany to complete his Habilitation in 1937 at the University of Freiburg. After three years lecturing there, he joined the Nazi Party in 1940, recruited by Konrad Meyer, head of the Planning and Soil Office of Himmler's Reich Commission for the Strengthening of Germandom. Christaller's task was to use his central place theory to plan the occupied territories of the German East, one part of Himmler's *Generalplan Ost*. After the War, Christaller became a communist, accused at one point of spying for the Stasi. He never worked at another university, spending his post-war years as a travel agent. He died in 1969. On Christaller's life and work see Hottes et al. (1977), Rössler (1989), Preston (2009), Barnes (2012) and Olwig (2018).

Fig. 5.2 Edward Ullman (L) and Walter Christaller (R) at the IGU Symposium in Urban Geography, Lund, Sweden, 1960. Photo: Chauncy D. Harris. *Source* University of North Carolina, Department of Geography

horse candidate, the newly minted Assistant Professor, William Garrison,[11] and, a year later from Harvard, the more established, Edward Ullman. A decade earlier, Ullman (1941) as a graduate student at the University of Chicago had written a paper that introduced and summarised Christaller's central place theory for an English-speaking audience. Garrison's link to Christaller was more indirect, although by the end more influential, certainly in its impact on the IGU. It came through Garrison's role as muse, minder, teacher, paymaster and provocateur to a group of talented, ambitious and energetic male graduate students who pioneered the new geography. Nicknamed the "space cadets", the group's members who had all arrived in Seattle between 1954 and 1958 drew heavily on central place theory.[12] Side-by-side with their allegiance to Christaller went their derogation of the old descriptive conception of the region as unique, synthetic and holistic—"mere" description.

These two connected branches of an incipient new geography, Baltic and Pacific Northwest, met when Donald Hudson, chair of the Department of Geography at the University of Washington, wrote to Torsten Hägerstrand on 9th December 1957 inviting him to come to Seattle as a visiting professor.[13] Hägerstrand accepted and spent the spring quarter of 1959 in Seattle, later writing to Garrison that he had arrived as "a missionary" but departed as "a student".[14]

It was during the Seattle visit that Hägerstrand conceived the idea to organise with Edgar Kant the IGU Urban Symposium in Lund as an ancillary meeting to the main 1960 IGC held in Stockholm (Tammiksaar et al. 2018: 86).[15] The symposium would celebrate the new geography, and especially the alliance between Baltic and

[11]Garrison had given one of the only quantitative papers on the programme at the 1952 IGC in Washington, D.C. on "Quantitative changes in the structure of employment" (Freeman 1972: 207). His struggle to leave behind regional geography and become a new geographer at the University of Washington is described in his memoir (Garrison 2002: 102–110).

[12]An English translated version of Christaller's *Central Places in Southern Germany* existed on microfiche and was passed between cadets. It was Carl Baskin's 1957 Ph. D. thesis completed at the Department of Economics, University of Virginia.

[13]Donald Hudson to Torsten Hägerstrand, 7th December, 1957, Papers of Torsten Hägerstrand, volume 3, Lund University.

[14]Torsten Hägerstrand to William L. Garrison, 15th September, 1959, Papers of Torsten Hägerstrand, volume 5, Lund University.

[15]A brilliant as yet unpublished paper about the Lund IGU Urban Symposium has been written by Michiel van Meeteren, "Writing blue notes in the march of geographical history: revisiting the 1960 Lund Seminar in Urban Geography".

Pacific Northwest versions. What began at the IGC in Amsterdam returned to the IGU 22 years later. This time Christaller was given pride of place with many of the symposium's contributors chosen because of their connection with the circulation of his ideas: Kant, of course, as well as Hägerstrand along with other Swedish geographers and the Seattle group, both Garrison and Ullman and among the cadets, Brian Berry, Michael Dacey, Duane Marble, Bob Mayfield and Dick Morrill.

With 87 participants from 17 countries, held over five days, the Lund Symposium was tape recorded, its proceedings published in a 600-page volume (Norborg 1962). The symposium offered a stage for performances, although for the most part one needed to understand English to appreciate them (the dominant language for the event). There were polylingual participants like the American geographer Chauncy Harris and the British geographer Robert Dickinson who acted as linguistic bridges, but there was only so much they could do. Even the multi-lingual Kant was effectively shut out because of his limited English. That applied even more so to the star of the show but linguistically challenged, Christaller. It was the Americans and the space cadets who stole the limelight. Their version of the new geography became ascendant.

The Lund Symposium demonstrated that a new set of geographical concepts and ideas had arrived on the disciplinary scene turning on formal theorisation, quantification and statistical analysis. The IGU's response was to launch a new Commission, "Commission XV", on Quantitative Methods. In 1964, Garrison became its chair along with five "regular members": Dick Chorley, Torsten Hägerstrand, Akin Mabogunje, Vaddiparti Prakasa-Rao and Yulian Saushkin. The Commission's establishment was the IGU's formal recognition that the new geography had officially arrived.

It took some time before anything happened, though. It was not until the middle of 1966 that the Commission set out its objectives. Garrison thought the most important task was to facilitate "training" including the provision of "materials for use by students, special seminars and summer programs for dissemination of new ideas".[16] He believed this essential for consolidating the new geography because, given the prior impress of regional geography, few geographers were ever taught quantitative methods. Garrison's emphasis on training also likely arose from his 1961 participation in a National-Science-Foundation-funded summer institute in the United State. In effect, that was a "boot camp" for non-quantitative American geography professors and graduate students, offering intensive instruction in remedial math.

There is no indication that the Commission ever attempted to run such training sessions and, given the international institutional structure of the IGU, it is not clear that they could ever have been run anyway. The rationale of the IGU was as a forum of international exchange not of educational catch-up. Possibly Garrison realised this because in late 1966 he called for a discussion among the Commissioners to rethink aims.[17] Between 14th and 18th August 1967, a meeting was organised in the Department of Geography at Cambridge University, hosted by Chorley.[18] The minutes of the meeting suggest only a general discussion but there were some specific action items: the preparation of a handbook; the compilation of a bibliography highlighting non-published sources on quantitative methods; the promotion of sites of exchange between professionals and students; and planning for the 1968 Delhi IGC.[19] Only the last of those tasks was completed, but it became important for the new geography.

At the 21st IGC in Delhi, from 1st to 5th December 1968, there were two sessions put on by the Commission on Quantitative Methods,

[16]William L. Garrison to Hans Boesch, IGU Secretary, 27th June, 1966, p. 1. IGU Commission on Quantitative Methods, Volume 5.

[17]William L. Garrison to Members of the Commission on Quantitative Methods (IGU), 30th October, 1966. IGU Commission on Quantitative Methods, Volume 5.

[18]Only five of the six Commissioners attended. Saushkin was absent.

[19]Meeting of the Commission on Quantitative Geography, Cambridge, England, 14th–18th August, 1967. IGU Commission on Quantitative Methods, Volume 5.

one on the theory of those methods, the other on their application (Congress and ICA Programmes 1968, 49, 54–55). In addition, there were two joint sessions run with other commissions (ibid., 40–41). In total, 18 papers were given. Furthermore, between December 10th and 12th there was also a separate "Symposium in Quantitative Methods in Geography" run by R. P. Misra at Mysore University, Southern India. 16 papers were given, 12 by Indians and four by non-Indians (Proceedings of Symposium on Quantitative Methods in Geography 1971).

Just as importantly, at the same congress it was agreed that the Commission be "reconstituted for the four-year period 1968–72" with a former space cadet, Brian Berry, as the new chair, and Forrest "Woody" Pitts appointed in the new position of "Executive Secretary" (Berry and Pitts 1970: no page number). Berry and Pitts (1970: no page number) said that "the first step of the Commission was to try … to review the state of the art in terms readily comprehensible to a world-wide and non-technical audience". To meet that objective, a group of high-profile quantitative geographers, a third from the United States, were asked to be "corresponding members" of the Commission. In turn, it was they who in summer 1969 were asked to provide state-of-the-art reviews of the field in two parallel IGU sponsored symposia, one in London, the other in Ann Arbor, MI.[20] Pitts (2002: 286) went on to convene eight more international conferences, which according to him yielded "many influential publications … extend[ing] quantitative methods into many areas of traditional concern to geographers".

Perhaps the most important of those publications was the first, published in June 1970, as a special supplement to the flagship journal *Economic Geography*. It contained a suite of classic review papers gathered from the "Proceedings" of those two 1969 "IGU Commission on Quantitative Methods" symposia. The contributors were a who's who list of quantitative geographers including two former space cadets (three if Berry, the co-organiser, is counted). Striking, however, was the absence of any paper on central place theory. Indeed, there was barely mention of Christaller. Over the decade since the Lund symposium, the geographical revolution that Christaller partly sparked significantly widened and diversified. Also, that revolution had become technically recondite, producing a large chest full of new geographical tools and techniques and reviewed in the supplement: models of spatial diffusion, measures of spatial autocorrelation, Markov-Chain analysis, entropy maximisation, spatial filtering and discriminant analysis. More broadly, ideas from the IGU Lund symposium had continued to circulate, to provoke and to stimulate, to be augmented and extended, attracting new allies and resources. They then returned as new performances at future IGU events and publications such as in Delhi, or in Mysore, or in London, or in Ann Arbor, and became materially embodied in the June 1970 supplement to volume 46 of *Economic Geography,* which was sent around the world.

In sum, the IGU along with its congresses, symposia and publications were fundamentally entangled with the development and circulation of one of the most significant post-Second World War disciplinary approaches, the new geography. While the IGU did not cause the new geography, because of its existence and patterns of circulation it distinctively marked the new geography. The IGU made its applications broader, its justification more convincing, its influence wider, and its effectiveness in challenging regionalism likely stronger.

5.5 Conclusion

The purpose of this chapter is to illustrate the role of the IGU in the international circulation and dissemination of geographical ideas and concepts. We used, as case studies, regionalism and the new geography. In both cases, changing conceptions of the region and quantitative methods geographically circulated through

[20]The conferences were by invitation with roughly 30 geographers attending each venue and 32 papers given in total. The local organisers were Nigel Spence in London and Waldo Tobler in Ann Arbor (Berry and Pitts (1970: 57).

various kinds of performances at IGU congresses, symposia and publications, especially commission-sponsored proceedings. Those conceptions influenced the broader discipline which, through a process of circulation, later travelled to other IGU events and were further elaborated and amended.

The IGU's institutional structure shaped which geographers came to its events, the performances that were put on, and their subsequent circulations. In this sense, the IGU created its own geography rather than geography creating it. Until the 1956 Rio IGC, congresses were European events (14 of the first 17 IGCs were held in Europe). Even when the 1925 IGC was held in Cairo, 56.5% of delegates still originated from Europe (Kish 1972a: 49). Thus, participants likely knew one another, or at least knew of one another's work, shared basic assumptions about the discipline, and held senior positions (if you did not you would be unlikely to be invited or able to afford to go). Key German and French thinking about regions pre-dated the 1922 IGC or, in the case of Hartshorne, was codified in the US. There was no single towering figure performing regional geography at the IGC during the 1920s and 1930s. Rather, a second generation of internationally less renowned French geographers continued to articulate the regional case. They were joined by regional geographers in other countries, including leading national figures, some with French connections who continued to champion regional geography into the 1950s. The result was a relatively narrow, distinctively European, *ex cathedra* view of geography buttressing the status quo. That began to change from the mid-1950s, with especially Americans becoming more prominent—at the 1960 Stockholm IGC American geographers were the largest single national group (over 300 attended; Kish 1972a: 46). With cheapening international travel, the global expansion of geography and universities, and less patrician attitudes, that change has only accelerated.

Further shaping what was performed, and how, and what happened afterwards was a result of changing rules set internally by the IGU. Until 1952, presenters were required to address in their sections pre-circulated questions. The use of sections siloed debate, while the use of "Communications" effectively acted like a set of "shelves" that separated issues that might threaten the appearance of disciplinary cohesion. It was only because of the (fortuitous) assignment of the regionalist Charles Biermann as *président* at Christaller's session that produced the exchange between regionalism and the embryonic new geography at the 1938 Amsterdam's IGC. Then there were the problems inherent within all large, international conferences: the uneven quality of the research; poor presentation styles; insufficient discussion time; problems of communication because of either too many languages or too few (see Footnote 5); and national jingoism.

Despite these various constraints, since the beginning and until recently, the IGU has been one of the key institutional disciplinary forums allowing the geographical word to spread around the world. Ironically, however, as the world has become more international and the study of geography even more important, the IGU appears to be losing its significance as national organisations, especially the Association of American Geographers, becomes ever-more global and hegemonic. In stark contrast to its beginning, possibly the main role of the IGU should now be counter-hegemony, to use its congresses to celebrate the diversity of local geographies.

Acknowledgements We would like to thank the following for their help: Bruno Schelhaas, Federico Ferretti, Francisco Roque de Oliveira, Mariana Lamego and Michiel van Meeteren.

References

Anon (1928) The international geographical congress, Cambridge, 1928. Geogr J 72(3):259–267

Anon (1935) The fourteenth international geographical congress, Warsaw, 1934. Geogr Rev 25(1):142–148

Anon (1949) International geographical congress, Lisbon, 1949. Geogr J 114(1):76–82

Barnes TJ (2012) Notes from the underground—why the history of economic geography matters: the case of Central Place Theory. Econ Geogr 88(1):1–26

Barnes TJ, Abrahamsson C (2017) The imprecise wanderings of a precise idea: the travels of spatial analysis. In: Jöns H, Meusburger P, Heffernan M (eds) Mobilities of knowledge. Springer, Cham, pp 105–121

Berry BJL, Pitts FR (1970) Quantitative geography: international needs. In: Economic geography, vol 46. Supplement: proceedings. International geographical union. Commission on quantitative methods

Callon M (2008) What does it mean to say that economics is performative? In: MacKenzie D, Muniesa F, Siu L (eds) Do economists make markets? On the performativity of economics. Princeton University Press, Princeton, NJ, pp 311–356

Christaller W (1933) Die zentralen Orte in Süddeutschland: Eine ökonomisch-geographische Untersuchung über die Gesetzmäßigkeit der Verbreitung der Siedlungen mit städtischen Funktionen. Ph.D. thesis. University of Erlangen, Gustav Fischer, Jena

Christaller W (1938) Rapports fonctionnels entre les agglomérations urbaines et les campagnes. In: Comptes rendus du congrès internationale de géographie Amsterdam 1938. E. J. Brill, Leiden, pp 123–127

Clout H (2005a) Geographers in their ivory tower: academic geography and popular geography in Paris 1931. Geogr Ann B 87(1):15–29

Clout H (2005b) France, Poland and Europe: the experience of the XIVth international geographical congress, Warsaw 1934. Belgeo 6:435–444

Clout H (2012) French geographers under the international gaze: regional excursions for the XIIIth international geographical congress, 1931. Belgeo 1–2: 1–15

Congress and ICA Programme (1968). Survey of India Offices, New Delhi, 1–15 Dec 1968

Craggs R, Mahony M (2014) The geographies of conference knowledge, performance and protest. Geogr Compass 8(6):414–430

Djament G, Covindassamy M (2005) Traduire Christaller en français. Textes seuils, réception, récit de découverte. Cybergeo Rev Eur Géogr 298. https://doi.org/10.4000/cybergeo.3144

Driver F (2001) Geography militant: cultures of exploration and empire. Blackwell, Oxford

Edwards KC (KCE) (1953): Regional geography—international geographical congress, Washington, 1952. Geogr J 119(1):81–90

Fairchild W (1961) The nineteenth international geographical congress, Stockholm, 1960. Geogr Rev 51 (1):109–113

Fairchild W (1973) The twenty-second international geographical congress, Montreal, 1972. Geogr Rev 63(1):90–98

Freeman WT (1972) Geography from congress to congress. In: Geography through a century of international congresses. Imprimerie Ozanne, Caen, pp 197–213

Garrison WL (2002) Lessons from the design of life. In: Gould P, Pitts FR (eds) Geographical voices. Syracuse University Press, Syracuse, pp 99–123

Harris CD (2001) English as international language in geography: development and limitations. Geogr Rev 91(4):675–689

Hartshorne R (1939) The nature of geography. A critical survey of current thought in the light of the past. Association of American Geographers, Lancaster, PA

Hitchcock C (1957) The eighteenth international geographical congress, Rio de Janeiro, 1956. Geogr Rev 47(1):118–123

Hottes K, Hottes R, Schöller P (1977) Walter Christaller 1893–1969. In: Freeman TW, Oughton M, Pinchemel P (eds) Geographers: biobibliographical studies 7. Mansell Information, London, pp 11–18

Jackowski A, Bilska E, Sołjan I (2014) 80 years after the 14th congress of the international geographical union in Warsaw, 23–31 August 1934. Geogr Pol 87(2):295–308

Jobberns G (2010) The association of American geographers and international geographical union meetings 1952. In: Roche M (ed) A geographer by declaration: selected writings by George Jobberns. University of Canterbury Press, Christchurch, pp 103–110

Kish G (1972a) The participants. In: Geography through a century of international congresses. Imprimerie Ozanne, Caen, pp 35–49

Kish G (1972b) The communications. In: Geography through a century of international congresses. Imprimerie Ozanne, Caen, pp 83–86

Kish G (1979) Bibliography of international geographical congresses 1871–1976. Hall & Co, Boston

Latour B (1987) Science in action: how to follow scientists and engineers around in society. Harvard University Press, Cambridge, MA

Livingstone DN (1992) The geographical tradition: episodes in the history of a contested tradition. Blackwell, Oxford

Martin G (1992) One hundred and twenty-five years of geographical congresses and the formation of the international geographical union or, from Antwerp to the Hague. IGU Bull 47:5–26

Massip S (1956) Cuban landscapes. Rev Geogr 19(45):183

Norborg K (ed) (1962) Proceedings of the IGU symposium in urban geography lund 1960. In: Lund studies in geography series B. Human geography, vol 24. C. W. K. Gleerup, Lund

Ogilvie AG, Russell EJ (eds) (1930) Great Britain essays in regional geography. Cambridge University Press, Cambridge (1st edn, 1928)

Olwig KR (2018) Transcendent space, reactionary-modernism and the "diabolic" sublime. Walter Christaller, Edgar Kant, and geography's origins as a modern spatial science. GeoHumanities 4(1):1–25

Ophir A, Shapin S (1991) The place of knowledge a methodological survey. Sci Context 4(1):3–22

Pitts FR (2002) Sliding sideways into geography. In: Gould P, Pitts FR (eds) Geographical voices. Syracuse University Press, Syracuse, pp 269–292

Preston R (2009) Walter Christaller's research on regional and rural development planning during World War II. METAR—papers in metropolitan studies, vol 52, Freie Universität Berlin, Institut für Geographische Wissenschaften

Proceedings of Symposium on Quantitative Methods in Geography (1971) Held in Mysore on 10 to 12 December 1968. National Committee for Geography, Calcutta
Raj K (2013) Beyond postcolonialism … and postpositivism: circulation and the global history of science. Isis 104(2):337–347
Roche M (2019) From an "absent to a silent presence"—new Zealand at the international geographical congress 1875 to 1952. NZ Geogr 74(3):143–151
Rössler M (1989) Applied geography and area research in Nazi society: central place theory and planning, 1933–1945. Environ Plann D Soc Space 7(4):19–31
Saushkin YG (1961) Economic geography at the 1960 international geographical congress. Sov Geogr 2 (7):34–42
Secord J (2000) Knowledge in transit. Isis 95:654–672
Somme A (ed) (1961) A geography of Norden. Heinemann, London 3rd edn. [first published in Norway 1960]
Stamp LD (1958) The measurement of land resources. Geogr J 48(1):1–15
Stanković S (2006) Regional geography of the Professor Borivoje Ž. Milojevicć. Bull Serb Geograph Soc 86 (2):19–28
Star S, Griesemer J (1989) Institutional ecology, "translations" and boundary objects: amateurs and professionals in Berkeley's museum of vertebrate zoology, 1907–39. Soc Stud Sci 19(3):387–420
Steele RW, Watson JW (1972) Geography the United Kingdom 1968–72: report to the 22nd international geographical congress at Montreal, Canada, in August 1972. Geogr J 138(2):139–153
Tammiksaar E et al (2018) Edgar Kant, Estonian geography and the reception of Walter Christaller's central place theory, 1933–1960. J Hist Geogr 60(1):77–88
Ullman EL (1941) A theory of location for cities. Am J Sociol 46:853–864
Van Meeteren M (forthcoming) In the footsteps of the quantitative revolution? performing spatial science in the Netherlands. In: Gyuris F, Michel B, Paulus K (eds) Recalibrating the quantitative revolution in geography: travels, networks, translations. Routledge, London
Yeates M (ed) (1974) Proceedings of the 1972 IGU commission on quantitative geography. McGill-Queen's University Press, Montreal, Kingston

Trevor Barnes is Professor of Geography and Distinguished University Scholar at the Department of Geography, University of British Columbia, Vancouver, Canada. He is also a Fellow of the Royal Society of Canada and a Fellow of the British Academy. Most of his work is on economic geography and the history of especially American geography from the Second World War.

Michael Roche is Professor of Geography in the School of People environment and Planning at Massey University, Palmerston North, New Zealand. He completed his Ph.D. at the University of Canterbury and has written mainly on historic and contemporary aspects of agriculture and forestry in New Zealand. In recent years he has published on the history of university geography in New Zealand.

The 'International' in Geography: Concepts, Actors, Challenges

6

Heike Jöns

Abstract

This chapter critically interrogates the notion of the 'international' in the discipline of geography. Drawing on interdisciplinary conceptual debates about internationalisation strategies in higher education, the analysis compares the geographical reach of the four International Geographical Congresses (IGCs) in Paris 1984, Sydney 1988, Cologne 2012, and Beijing 2016. This chapter shows that between the last decade of the Cold War and the greater geopolitical multiverse in the first decade of the twenty-first century, geographical knowledge production and exchange not only diversified and decentralised on a global scale but also experienced a profound shift from a distinct Anglo-American internationalism towards a more complex multicultural internationalism. Consequently, I argue that the international nature of geographical knowledge production and exchange is relational because international conference experiences vary by the geopolitical, socio-economic, cultural, linguistic, and academic positionality of the event and its participants. Policy-relevant conclusions highlight the great value of the IGCs for facilitating international experiences for growing numbers of attendees from the events' host countries; they stress the important politics of choosing host cities in different world regions and offering flexible conference delivery formats and registration options; and they call for a greater emphasis on the development of intercultural skills to achieve more equality, diversity, and inclusion (EDI) in geography.

Keywords

Internationalism · Geography · Conference · Geography of science · Geography of higher education · International geographical congress

6.1 Introduction

This chapter discusses the nature and role of international encounters in the discipline of geography since the 1980s as a contribution to a wider research agenda of examining the histories and geographies of internationalism (Hodder et al. 2015). The focus is on a comparison of the geographical reach of four International Geographical Congresses (IGCs) organised within the International Geographical Union (IGU): Paris 1984, Sydney 1988, Cologne 2012, and Beijing 2016. This comparative and geographical perspective on the global reach of the IGU's international conference series at four moments

H. Jöns (✉)
School of Social Sciences and Humanities, Loughborough University, Loughborough, UK
e-mail: h.jons@lboro.ac.uk

V. Kolosov et al. (eds.), *A Geographical Century*,
https://doi.org/10.1007/978-3-031-05419-8_6

in time enables a critical examination of how changing geopolitics and varying conference locations in Europe and Asia–Pacific impacted on the nature of international and intercultural encounters in the discipline of geography at a time of a prevalent but eroding Anglo-American academic hegemony. More specifically, this chapter compares IGC attendance during the very different political contexts of the Cold War in the 1980s and the greater global geopolitical multiverse of the 2010s, a decade shaped by considerable economic and scientific growth in China (Gui et al. 2020). Previous publications have commented upon the role of the IGU for international academic exchange and collaboration, including the politics of conference locations and language choices (e.g. Robic et al. 1996; Lamego 2020; Meadows 2020), and analysed the geographies of conference attendance at the first 21 IGCs (Kish 1972). Yet, the global geographies of IGC attendance have neither been analysed for the final Cold War and the most recent decade nor previously been visualised on world maps.

Since the foundation of the IGU in 1922, IGCs have been held about every four years (Harris 2001). The comprehensive analysis by George Kish (1972) of the geographical origin of conference participants at the first 21 IGCs—from Antwerp 1871 to New Delhi 1968—has reconstructed the growth and geographical decentralisation of geographical knowledge exchange away from Europe, with European geographers having accounted for 98% in Antwerp 1871 and 66% in London 1964. Over the same period, the share of US participation increased from 1 to 22% and that of participants from other continents from 4 to 13%. According to Kish (1972, 49), the largest shares of conference attendees came from outside Europe and the United States at the IGCs of Cairo 1925 (40%), Rio de Janeiro 1956 (69%), and New Delhi 1968 (50%). This insight underlines the pivotal importance of moving the conference location to different world regions for a greater integration of a more diverse set of national geographical communities on an international scale and suggests a comparative perspective for identifying wider global shifts in more recent IGC participation.

The choice of four case study IGCs held three decades apart in different parts of the world is necessarily partial but allows for a comparison of international geographical knowledge exchange from a long-term perspective. Based on this analysis, I develop a threefold argument: first, geographical knowledge production and exchange diversified and decentralised globally —along with the development of more complex global power-relations, wider economic globalisation processes, and proliferating internationalisation strategies in higher education; second, the IGCs experienced a considerable shift from an Anglo-American internationalism towards a multicultural internationalism; and third, the international nature of academic conferences in geography is relational because international conference experiences vary depending on the geopolitical, socioeconomic, cultural, linguistic, and academic positionality of the event and its participants. In the first of four parts of this chapter, I discuss different conceptual understandings of the 'international' in higher education and research. In the second part, I critically examine the geographical diversity of the four case study IGCs by comparing the participants' countries of work. In the third part, I outline the main challenges to international dialogues and therefore also to more international equality, diversity, and inclusion (EDI) in the discipline of geography. In the fourth and final part, I offer some conclusions on how geographical knowledge production can benefit from more decentralised and diversified international perspectives.

6.2 Concepts

A widely used definition of internationalisation processes in higher education, proposed and developed by higher education scholar Jane Knight since the early 1990s, refers to the "integration of international, global, intercultural, and comparative perspectives into the teaching/learning process and program content" provided in higher education institutions, either abroad or at home (Knight 2012, 27). Due to an increasing role of online communication in international higher education since the 1990s, a team of researchers involving geographers Ashley Gunter and Parvati Raghuram has differentiated three main types of internationalisation—abroad, at home, and at a distance (Mittelmeier et al. 2021). These three types of internationalisation in higher education are linked to different generations of cross-border mobilities of people, programmes, and providers (Knight 2014; see also Waters 2012; Rye 2014; Bauder 2015; Waters and Leung 2017; Gunter and Raghuram 2018). The latter have proliferated since the General Agreement on Trade in Services (GATS) was implemented by the World Trade Organisation (WTO) in 1995, with the aim of making the international provision of services such as higher education economically more attractive by creating a global education market through trade regulations (Robertson 2017). Table 6.1 suggests that such a multidimensional conceptualisation of internationalisation in higher education could usefully include international events, such as the IGCs as a quadrennial event series with changing conference locations.

This conceptual understanding of internationalisation processes in higher education takes into consideration not only the international movements of people but much more diverse material, dynamically hybrid, and immaterial mobilities that constitute academic institutions, practices, and events (Jöns et al. 2017, 5). The acquisition of international experiences in higher education does thus not necessarily involve one's own geographical mobility but can also be obtained by the majority of internationally immobile students, researchers, and academics through the international mobility of peers, programmes, and providers and by accessing online education provided at a distance from their home country (Knight 2014). In this chapter, the notion of home country/countries of internationally mobile individuals—as opposed to their host country/countries—is flexibly understood as the current country/countries of education or work, rather than the country of birth, citizenship, and/or residence, because institutional affiliations are documented in the lists of conference participants. In regard to internationally mobile conference series, the location(s) of the individual or institutional organiser(s) usually determine(s) the home country/countries, whereas the place(s) of delivery serve(s) as host country/countries. Yet, in the case of the IGU Secretariat, which resides where the Secretary-General (or President) is located, all event locations are addressed as host countries.

6.2.1 International and Intercultural Experiences

In her well-known definition of internationalisation in higher education, Knight clarified that the term "international" would be used for "the sense of relationships between and among nations, cultures, or countries", whereas "internationalisation relating to the diversity of cultures that exists within countries, communities, and institutions," would be addressed by the term "intercultural" (Knight 2004, 11). This chapter emphasises that a geographical analysis of internationalisation processes through the lens of the IGCs held in different countries and cultural contexts requires the differentiation of international and intercultural experiences for *both* participants from the event's host country and visitors from abroad. Comparing international and intercultural experiences in this way, as discussed by Tebbett et al. (2021, 5) for internationalisation processes at home, is important because encounters with culturally diverse conference participants may provide intercultural but not necessarily international experiences for delegates from the same home country, given that in some countries a larger presence of racialised and other ethnic minorities provide more culturally diverse environments than in others. Internationally mobile conference participants may also gather different international and intercultural experiences depending on the degree to which their interactions focus on fellow nationals and delegates who speak the same first language (Derudder and Liu 2016). Finally, intercultural challenges can impact in very

Table 6.1 Conceptualising internationalisation in higher education (HE) by research focus

Internationalisation via:	I. People	II. Provisions	III. Events
1. Abroad [*through outgoing mobilities to host country*]	• International students (module/credit/degree) • International academics (visiting/temporarily employed/permanent) [*from abroad*]	• International degree programme • International branch campus • Other international HE provision • International placement employer • International publications [*from abroad*/in situ]	• International conference travel[a] • I.1 and I.3 in conference • Travelling international conference series[a] [*from abroad*/in situ]
2. At a distance [*through the involvement of at least some international distance learning*]	• Students/academics in online/other distant HE provision [*from abroad*]	• Academic communication • Project collaboration • Blended learning course • Online/other distant degree programme • Virtual university [*from abroad*/in situ]	• Online/phone project meetings • Online/phone seminar/lecture • Online conference [*from abroad*/in situ]
3. At home [*through incoming mobilities to home country*]	• International students (module/credit/degree) • International academics (visiting/temporarily employed/permanent) [*from abroad*]	• International degree programme • International branch campus • Other international HE provision • International placement employer • International publications [*from abroad*/in situ]	• International conference travel[a] • I.1 and I.3 in conference • Travelling international conference series[a] [*from abroad*/in situ]

[a]Research focus in this study
Source Own design building on Knight (2014) and Mittelmeier et al. (2021)

different ways on monolingual geographers' international experiences, knowledge exchanges, and collaborative networks when compared to linguistically better positioned or more versatile peers (Aalbers and Rossi 2007). Attending an international conference, with at least some delegates working in different countries, thus not only constitutes an important aspect of what Clare Madge, Parvati Raghuram, and Patricia Noxolo called "international study" (Madge et al. 2015, 684) for both internationally mobile and immobile conference participants but may also provide—to varying degrees—an intercultural experience. Hence, I suggest that comparing international *and* intercultural experiences helps to understand the *relational* nature of the 'international'.

6.2.2 International Events as Centres of Circulation

International conferences have been conceptualised as temporary clusters of diverse actors and resources in different economic sectors and political contexts, facilitated through heterogenous mobilities and aiming at knowledge exchange and networking, sometimes in highly politicised and controversial ways (e.g. Maskell et al. 2006; Craggs 2014). Annual business conferences have been characterised as cyclical clusters because of their constitutive role in the yearly operational rhythm of firms (Power and Jansson 2008), which can be compared to the organisational rhythm of the academic year when conceptualising academic conference series as cyclical clusters. For example, the Annual Meeting of the American Association of Geographers (AAG) usually takes place in the spring, the Annual International Conference of the Royal Geographical Society (with the Institute of British Geographers) in the late summer, and the Biannual German Congress for Geography in the autumn. From a geographical perspective, academic conferences have also been conceptualised as "centres of circulation" with varying material, social, and intellectual geographies (Momm and Jöns 2020, 106–107). Viewed from this triadic thought perspective, conferences constitute nodes of incoming and outgoing flows of people, dynamically hybrid and (im)material resource ensembles whose wider geographical reach can be analysed from multidimensional thematical and multiscalar geographical perspectives.[1] Further studies have analysed the politics of local movements, performances, and exchanges within the conference settings (Weisser and Müller-Mahn 2017) and uncovered strikingly gendered behaviour in conference sessions at the German Congress for Geography (Aufenvenne et al. 2021). Based on a conceptual understanding of the IGCs as temporary and cyclical centres of circulation, this chapter examines the international reach and intercultural diversity of four IGCs in two different decades through a comparison of the conference locations and participants' countries of work.

6.2.3 The Importance of a Contextual Approach

Historical geographical analyses have underlined the particular significance ascribed to personal gatherings in the discipline of geography because exploring the event locations and wider regions in organised and private excursions has contributed to at least three important epistemic, social, and political outcomes of international

[1] Analysing the national reach of two academic conference series in urban and regional planning in Brazil from 2004 to 2013 has shown that the geographies of these biannually organised cyclical centres of circulation decentralised on the national level over time, but that this process varied in regard to the conferences' material, social, and intellectual dimensions (Momm and Jöns 2020). Materially, the conference locations of the mobile event series focused on state capitals in coastal states and that of the immobile conference series on a medium-sized city in the most southern state. Socially, the workplaces of conference paper authors acquired national reach but clustered in the scientifically most resourceful and productive southeastern and southern regions. Intellectually, geographical knowledges and imaginations communicated in conference presentations showed the widest geographical reach within the country but were shaped by similar east-west and south-north disparities as the authors' workplaces.

conference participation beyond the immediate conference setting: first, the production of place-specific geographical knowledges and imaginations; second, the (re)production of academic networks; and third, the encouragement of friendly international relations (Collignon 1996; Withers 2010; Roche 2019). Prior to the global coronavirus pandemic of 2020–21, there was a widespread expectation that participation in international geography conferences required physical co-presence in specific conference locations, even if remote participation would have been technically possible, occurred occasionally, and was increasingly discussed in calls for less in-person and more online participation as a strategy for mitigating climate change through reduced conference travel (Hopkins et al. 2016). This expectation changed during the coronavirus pandemic lockdowns when restrictions on travel and in-person interactions meant that conferences were organised either entirely online, or as hybrid, dual-delivery events with limited in-person interaction alongside online events. It is likely that the hybrid conference, successfully introduced in recent months, will become the dominant format in the future, including for future IGCs.[2] In this new context of profoundly changed modes of international academic communication, the second decade of the twenty-first century, analysed in this chapter, will most likely be the last decade in which in-person attendance at international conferences was the norm rather than an alternative to online participation (see section on challenges below). Yet, in conceptual terms, it is worth noting that the notion of centres of circulation accommodates the geographies of in-person, dual-mode delivery, and online conferences because this concept considers various ontologies of mobilities, ranging from material via human to virtual (Momm and Jöns 2020, 106).

As large-scale international and cyclical events with a long history and varying conference locations, the IGCs are particularly well-suited for interrogating how the international nature of geography as a discipline and practice has changed over time. The following analysis adopts a contextual approach that critically examines geographical knowledge production and exchange in multiscalar individual, institutional, national, and international contexts from multidimensional political, economic, environmental, social, cultural, and intellectual perspectives. Such a contextual approach has been prominently developed by historians of geography (e.g. Livingstone 1992; McEwan 2000; Withers 2010; Ferretti 2019) to complement existing approaches to the history of geography, such as Anne Buttimer's (1983) widely praised collection of autobiographical essays in the international dialogues project, but it has been less often used for interpreting geography's more recent global histories. Informed by critical geopolitics, the following contextual analysis thus provides an account of changing yet contingent international practices in geography (Agnew 2001).

6.3 Actors

The IGU's eleven statutory articles focus on the term 'international' and include international standardisation as an explicit aim, when characterising the purposes of this institution and its events. More varied understandings of 'intercultural', 'global', 'comparative', and 'transnational' relationships that feature in broader definitions of internationalisation processes in higher education (Knight 2012, 27) are not included in the IGU statutes. At a time of post-structural and postcolonial diversity in Anglophone geographical thought and praxis (Cresswell 2013), as well as international efforts of decolonising the geography curriculum (Esson 2020; Schelhaas et al. 2020), it can therefore be argued that these new sensibilities in geographical research and praxis could be considered more explicitly by referring to intercultural experiences and exchanges in the IGU statutes in order to create a greater awareness about (1) cultural

[2] The 34th IGC in Istanbul, scheduled as an in-person meeting in August 2020, was first postponed due to the COVID-19 pandemic by one year and then held virtually from 16 to 18 August 2021. For details, see Turkey Geographical Society (2021).

diversity in geographical theory and praxis; (2) the advantages of using different languages in the production and communication of geographical knowledges; and (3) the emergence of epistemological pluralism as an alternative to epistemological orthodoxy.

Yet, some linguistic diversity is assured in the IGU statutes through the identification of English and French as the working languages of the IGU and at the IGCs, with the possibility of including other conference languages if translation facilities are provided (International Geographical Union 2021, Statute VIII.C and Statute IX.E). Cultural respect and understanding are also explicitly requested: "There shall be no discrimination as regards race, ethnic group, citizenship, religion, sex or political opinion within the Union or in the meetings organized by it or held on its behalf" (International Geographical Union 2021, Statute VIII.D). The IGU statutes further contain key ideas of a wider human rights agenda that are represented by a commitment to EDI principles and practice, such as "the free circulation of scientists" (International Geographical Union 2021, Statute VIII.E) as a principle conveying the ideal of equal access to the IGCs across the world.

The following analysis focuses on conference participants because comparable data were available for all four IGCs.[3] Yet, one limitation of the analysis has to be considered: the statistics for Beijing 2016 include only the number of participants for countries with at least ten conference participants. This chapter's first world map thus provides to some extent a partial representation of the geographical reach of the four IGCs, but mapping the number of conference participants from countries with at least ten delegates enables the visualisation of changing clusters of geographers involved in international geographical knowledge exchange and thus helps to identify broader and shifting national interests in IGC participation at a global scale.[4] The second world map provides a geographically more inclusive perspective by comparing countries of work for all participants at the IGCs in Paris 1984 and Cologne 2012.

6.3.1 Paris 1984 and Sydney 1988

The 1980s were shaped by resonant Americanisation that followed upon American economic high hegemony, or capacious Americanisation (1945–1970), and resulted in the profound political, economic, and cultural influences of American ideas, practices, and products across the westernised world (Taylor 1999). This situation found its expression in an Anglo-American academic internationalism that shaped the global geographies of IGC conference attendance and the languages of conference papers. US geographer Harris (2001) analysed the use of different languages for conference presentations at the IGCs from 1871 to 2000. Based on his analysis, it is possible to argue that a shift from German to English as the lingua franca in geography is already discernible in a comparison of the share of English-language papers presented at the IGCs in London 1899 (55%), Berlin 1909 (19%), and Washington 1904 (82%). Yet, six languages shaped the first three post-World War II IGCs in Lisbon 1949 (28% English-language papers), Washington 1952 (70%), and Rio de Janeiro 1956 (36%), namely English, French, German, Italian, Portuguese, and Spanish. According to Harris

[3] The data sources used for the analysis list conference participants: Comité International d'Organisation (1985, 204–208), International Geographical Union (1988, 61–94), Butsch (2021) on checked-in participants at the IGC Cologne 2012 (see also Butsch 2015, 136–137 on registered participants before the conference on 15 July 2012), and Geographical Society of China (2016, 4–5).

[4] This methodology of a comparison of countries with at least ten participants was applied in reports on the IGC Paris 1984 (Dalmasso 1986, 157) and the IGC Beijing 2016 (Geographical Society of China 2016, 4–5). Different numbers of participants at one and the same IGC may be linked to varying pre- and post-conference registration numbers (Kish 1972, 35). At the IGC 2012 in Cologne, 310 of 2,864 registered conference participants (11%) did not visit the congress (Butsch 2015, 136), but this chapter analyses checked-in candidates. The terminology suggests that the published lists of participants at the IGCs in Paris 1984 and Sydney 1988 and the statistics on participants at the IGC 2016 in Beijing also refer to actual conference attendees (see note 3).

(2001, 675–677), English and French became the official IGC languages only in Stockholm 1960 (76%), resulting in 85–97% of English-language papers at almost all IGCs between 1964 and 2000 except Paris 1984 (68%).

The IGCs in Paris 1984 and Sydney 1988 allow for a comparison of conference attendance in the IGU's two main language areas of French and English and of meeting locations in the northern and southern hemispheres during the Cold War. These different linguistic and geographical positionalities led to varying participation numbers—1820 delegates in Paris and 1117 in Sydney, from 88 and 65 countries of work, respectively (Comité International d'Organisation 1985, 204–208; International Geographical Union 1988, 61–94). These events were large and medium-sized conferences when compared to the earlier IGCs in the period 1871–1992 (Collignon 1996, 92). Despite a flourishing Anglo-American academic hegemony in science and higher education during the second half of the twentieth century (Paasi 2005), the greater accessibility and attractiveness of Paris as a host city, and the strong appeal of this event to both Anglophone and Francophone geographers, made Paris 1984 a larger conference than Sydney 1988. The IGC 1984 was also culturally much more diverse than the IGC 1988 because in Paris, only 24% of all participants arrived from the six English-language countries of the United States, Canada, the United Kingdom, Ireland, Australia, and New Zealand—the countries that have driven an Anglo-American academic hegemony—but this share reached 51% at the IGC 1988 in Sydney.

These geographies of international conference attendance were considerably shaped by the Cold War, with many more conference participants arriving from the United States of America and its allied member countries of the North Atlantic Treaty Organization (NATO) than from the Soviet Union (USSR) and its supporting member states of the Council for Mutual Economic Assistance (COMECON) (Fig. 6.1). This global rivalry between a western and eastern geopolitical bloc in the early 1980s had prominently impacted on the cultural sphere by resulting in a wide-ranging political boycott of the Olympic Games 1980 in Moscow, led by the United States in response to the Soviet invasion of Afghanistan. In turn, the subsequent 1984 Summer Games in Los Angeles were boycotted by the USSR and other eastern bloc countries, citing security concerns (Gold and Gold 2007, 38). Although the IGCs coincided with the quadrennial rhythm of the Summer Games from 1952 to 2016 (before both events were postponed due to the coronavirus pandemic from 2020 to 2021) and the IGC 1984, held 27–31 August 1984, took place only four weeks after the start of the boycotted Summer Games (28 July–12 August 1984), the Olympic boycott of 1984 did not find its expression in international conference attendance. On the contrary, the ten represented COMECON countries showed a greater presence in Paris at the IGC 1984 (7.9% of overseas participants) than in Sydney at the IGC 1988 (3.6%).[5]

Highly uneven global geographies of international geography conferences also resulted from an underrepresentation of delegates from global south countries, especially in Central and South America and Africa (Table 6.2). The IGC 1984 in Paris was very attractive to attend for geographers from Africa (7%) when compared to the IGC Sydney 1988 (4%), most likely because of a combination of lower conference attendance costs and the Francophone higher education context. A similar trend can be identified in regard to the overall very low representation of geographers from Central and South America in those two international conferences (4% vs. 1%). According to geographer Roche (2019), IGC participation by geographers from New Zealand, as an economically well positioned but remote country in Oceania, was shaped so much by cost considerations that Professor Kenneth Cumberland (Auckland) regarded his upcoming election as an IGU Vice

[5] The COMECON participants of 1984 worked in Bulgaria, Cuba, Czechoslovakia, the German Democratic Republic, Hungary, Poland, Romania, the USSR, Vietnam, and Yugoslavia, whereas in 1988, out of these ten COMECON countries, delegates did not participate from Cuba, Romania, and Vietnam.

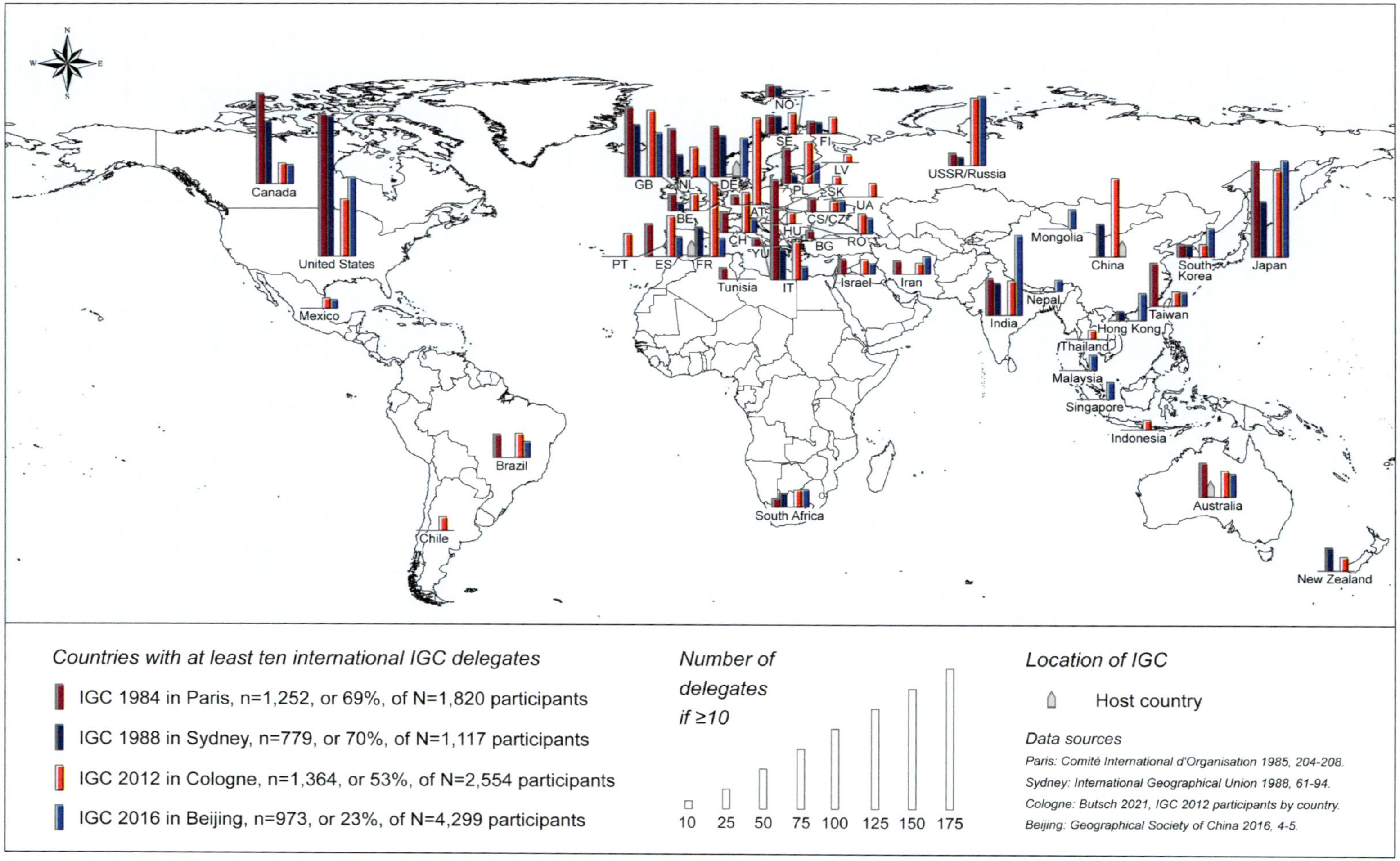

Fig. 6.1 Changing geographical clusters of at least ten international conference participants at four International Geographical Congresses (IGCs) by country of work. *Source* Comité International d'Organisation (1985, 204–208), International Geographical Union (1988, 61–94), Butsch (2021), and Geographical Society of China (2016, 4–5)

Table 6.2 Geographical origins of IGC participants by world region

Geographical region[a]	Paris 1984		Sydney 1988		Cologne 2012	
	N	%	N	%	N	%
(1) Overseas	1443	*79*	897	*80*	1502	*59*
a. North America	293	*20*	255	*28*	98	*7*
b. Central and South America	61	*4*	10	*1*	75	*5*
c. Europe	614	*43*	331	*37*	768	*51*
d. Africa	106	*7*	35	*4*	45	*3*
e. Asia	321	*22*	230	*26*	466	*31*
f. Oceania	48	*3*	36	*4*	50	*3*
(2) IGC host country	375	*21*	220	*20*	1052	*41*
(3) Unknown	2	*0.1*	0	*0*	0	*0*
(4) Total	1820	*100*	1117	*100*	2554	*100*
(5) Number of countries	88		65		84	

[a](1)–(3) Add up to 100% of all participants and a.–f. to 100% of overseas participants (subject to rounding)

Source Comité International d'Organisation (1985, 204–208), International Geographical Union (1988, 61–94), Butsch (2021)

President at the IGC 1960 in Stockholm as a chance of securing travelling expenses to attend future IGCs as an academic representative of New Zealand geography (Roche 2019, 54). The highly uneven global geographies of IGC attendance in the 1980s, especially between global north and global south countries, can therefore be explained to a large extent by different language areas, political restrictions, and the unequal availability and access to economic resources for undertaking research and travelling to international conferences.

Overall, the varying global geographies of international conference attendance confirm that the IGC 1984 in Paris was culturally, linguistically, and politically more diverse than the IGC 1988 in Sydney. Since geographical knowledge production is often place-specific and context-dependent due to the empirical nature and fieldwork components of most research in both physical and human geography, it can be argued that more diverse geographies of conference attendance, as empirically shown for the two conference series on urban and regional development and planning in Brazil (Momm and Jöns 2020), also indicate a greater diversity of study sites, and thus more diverse and heterogeneous epistemological spaces of geographical knowledge, offering more nuanced understandings for context-specific policy recommendations. Yet, this wider argument about ontologically varying geographies of conferences remains to be empirically verified for the IGCs. What this analysis has demonstrated so far is that the international nature of geographical knowledge production and exchange, resulting from paper presentations, session attendance, and networking practices of conference participants, has continued to vary considerably by the conference's host country during the 1980s, thus offering very different, event-specific international and intercultural perspectives on geography that were shaped by an overarching Anglo-American academic internationalism in the final Cold War decade.

6.3.2 Cologne 2012 and Beijing 2016

In the 24 years between the IGC Sydney 1988 and the IGC Cologne 2012, the world's rigid geopolitical order of the Cold War was transformed without a global war into what could be regarded as a post-Cold War political multiverse.

About one year after the IGC in Sydney, the literal fall of the Iron Curtain started with a first opening of the border between Hungary and Austria in August 1989 and led to the first free passing of east Germans through a gate of the Berlin Wall from the German Democratic Republic into the Federal Republic of Germany on 9 November 1989. German reunification followed suit on 3 October 1990 and by 1991, the former eastern European socialist states with centrally planned economies were transforming into democratic states with capitalist market economies (e.g. Bradshaw and Stenning 2004; Lang et al. 2015; Drummond and Young 2020). The new freedom of international travel between eastern and western European states, as well as from the successor states of the Soviet Union, impacted significantly on global science and scholarship, including geography. In the subsequent three decades, international academic mobilities and co-publication patterns decentralised and globalised considerably. At the same time, economic growth in China went hand-in-hand with more scientific productivity in the natural and technical sciences and a global shift of international academic mobilities and scientific collaboration towards Asia–Pacific (Jöns 2015; Maisonobe et al. 2017; Gui et al. 2019).

The comparison of the IGCs 1984, 1988, 2012, and 2016 shows that this international community of geographers has grown since the 1980s and that many more geographers of the host country have benefitted through their attendance from international and intercultural perspectives provided by conference presentations and informal exchanges (Table 6.2). In the official conference reports, the IGCs in Cologne 2012 and Beijing 2016 were both presented as the largest IGCs of all times, with 3199 and 4299 attendees, respectively (Butsch 2015, 136; Geographical Society of China 2016, 4). Overseas participants were more numerous, too, but showed much lower shares than in Paris 1984 (79%) and Sydney 1988 (80%), with 1502 delegates in Cologne 2012 (59% of 2554 checked-in participants; Butsch 2021) and 1950 in Beijing 2016 (45%; Geographical Society of China 2016, 4). The difference between Paris 1984 and Cologne 2012 amounted to only 59 more overseas geographers in Cologne, which underlines growing attendance from the national host community. In this context, it is important to remember that IGCs do not only cater for university academics but also for schoolteachers in search of new geographical insights, teaching materials, and fieldtrip experiences (Roche 2019, 60).

When comparing international attendance at the IGCs during the last decade of the Cold War with the second decade of the twenty-first century, the geographical perspective reveals a profound global shift of delegates' countries of work away from established centres of academic knowledge production in the United States and Canada towards China, India, and Russia (Fig. 6.1). A wider shift of academic mobility and collaboration from North America to East Asia has also been observed in other contexts of international academic collaboration and exchange across different disciplines, such as academic mobility to Germany from 1954 to 2010 (Jöns 2015) and international co-authorship patterns (Maisonobe et al. 2017; Gui et al. 2019). Yet, the crucial difference in geography is a much sharper decline of North American participation in the two recent European and East Asian IGCs. As previously assumed by IGU President Meadows (2020, 495), this particular situation in geography contrasts with the development of increasing transpacific exchanges and knowledge networks in other disciplines, which means that the considerable retreat of North American geographers from the IGCs has prevented a larger growth of overseas attendance.

The Anglo-American academic internationalism that shaped the IGCs in Paris 1984 and Sydney 1988 has thus given rise to a stronger multicultural academic internationalism at the IGCs in Cologne 2012 and Beijing 2016, with important geographical variations. This transformation of the IGCs' academic internationalisms is first and foremost expressed in considerably lower shares of Anglo-American geographers from the six countries of the United States, Canada, the United Kingdom, Ireland, Australia, and New Zealand among all overseas participants in Cologne 2012 (16%) than in Paris 1984 (30%), the two European and non-Anglophone conference locations analysed in this chapter. Many western European countries were prominently represented at the IGC 2012 in Cologne, notably Austria, France, Poland, Spain, Switzerland, Italy, and the Netherlands, but less so at the IGC 2016 in Beijing, except the United Kingdom and Germany. This situation confirms that geographical distance still matters for international conference attendance, even if Japan continued to show a strong presence at both IGCs. New clusters of geographers on the global stage were represented at the IGC 2012 in Cologne by ten or more delegates from Portugal, Romania, Chile, Ukraine, Hungary, Mexico, Indonesia, Thailand, and Latvia; at the IGC 2016 in Beijing, this applied to ten or more geographers from Mongolia, Singapore, Malaysia, and Nepal. In very similar ways to other disciplines (Gui et al. 2019), geography has thus also experienced a notable decentralisation of global clusters of international knowledge production and exchange (Fig. 6.1).

The comparison of shares of overseas attendance from different parts of the world focuses in this section on overseas participants only because the nature of the IGCs profoundly changed through a greater participation from the events' host countries in Cologne and Beijing.[6] This shows that the decline of North American geographers' presence at the IGCs was more than compensated for by an almost inverse increase of overseas participation from Asia (Paris: 22%; Cologne: 31%). IGC attendance from Central and South America, notably from Brazil, Chile, and Mexico, rose slightly between Paris 1984 (4%) and Cologne 2012 (5%), and thus also contributed to the globalisation of international geographical knowledge production and exchange despite strong regional and transnational networks based on well-developed conference circuits in Spanish and Portuguese. The number of South African geographers attending IGCs grew steadily but much more could be done, as discussed in the final section below, to increase the share of other geographers working across African countries, given that African participation showed a considerable absolute and relative decrease between Paris 1984 (7%) and Cologne 2012 (3%). Finally, a geographical perspective reveals that the IGC community could have expanded more over the past three decades, if US and Canadian geographers had attended the IGCs in larger numbers, as they did in the 1980s. This insight is particular striking because the AAG 2013 in Los Angeles was attended by 6125 US geographers (Derudder and Liu 2016, 320), yet only 72 went to the IGC 2012 in Cologne and 99 to the IGC 2016 in Beijing.

6.4 Challenges

Three main challenges for IGC participation can be identified based on the preceding analysis: first, the costs of international academic exchanges; second, the demand for more intercultural skills; and finally, the rise of online conference participation as an alternative to in-person attendance. The first main challenge is of an economic nature because globally highly uneven access to material resources in science and higher education, such as funds for advanced studies, research, and travelling, has encouraged many students and academics either to visit or migrate to established and emerging centres of academic research and teaching for their education, training, and work (Mbah 2017). Yet, vibrant geographical communities exist across the world, as reflected in the more than 60 member states of the IGU as of 2020 (Meadows 2020, 497). Given that a country's full membership is linked to membership fees, uneven socioeconomic circumstances are already reflected in the global geography of IGU membership, with the potential for more South American, African, Western Asian, and South-East Asian countries to join the IGU (Meadows 2020, 496). This situation highlights the importance of the IGU's travel grant scheme for encouraging IGC attendance in underrepresented countries, even if visa restrictions might

[6] The IGC 2012 in Cologne replaced the Biannual Meeting of German Geographers scheduled for 2011 and thus encouraged a large contingent of domestic geographers to attend.

occasionally prevent IGC attendance despite a travel grant (Meadows 2020, 502). Offering more travel grants at institutional, national, and international levels, especially for doctoral researchers and early career academics, seems to be of prime importance for assuring a wide geographical reach and sustainable future of the IGC event series. Figure 6.2 compares all IGC participants' countries of work for Paris 1984 and Cologne 2012, and thus may help to identify underrepresented geographical communities that are more in need of travel grant support than others because of a profound lack of economic resources. This economic challenge also requires the addressing of major inequalities between geographers of different genders and across other axes of social difference because historically, female geographers travelled less often to conferences, and received less financial support for their academic travels than men, and thereby missed out on important positive feedback effects for their academic careers (Jöns 2017).

After a career in geography that spanned more than 60 years, Chauncy D. Harris concisely addressed a second main challenge of internationalisation processes in geography beyond an Anglo-Americanisation of knowledge exchange by drawing attention to the benefits of knowing more than one language: "Cultures can be fully understood only if one knows the language of the culture. Furthermore, in my own view, one cannot fully appreciate one's own culture unless one can also see it through the lens of another culture" (Harris 2001, 686). The tension between linguistic diversity and homogeneity in international knowledge exchange remains a major challenge for the interconnection of more than 60 national geographical communities through the IGU and its IGCs because the widespread use of English as lingua franca facilitates international knowledge exchange, but at the same time, it privileges students, researchers, and academics who are either native speakers or were socialised in countries that use English as an important language of instruction in secondary and tertiary education. In the context of Mitchell's (2018, 46) argument that a promotion of multiculturalism in US education is linked to a neoliberal agenda of state actors striving for competitiveness in the global knowledge economy as part of a broader "transnational capitalist logic", it seems as if university graduates educated in the centres of Anglo-American scholarship may often feel prepared through their English-language school and university education for a globalised English-language labour market but might not always have acquired the intercultural knowledge, skills, and understanding that makes engaging with people from different countries, language areas, and cultural backgrounds a rewarding experience (Tebbett et al. 2021, 536). Hence the widely discussed demand for the development of more language skills in Anglo-American universities, especially in highly place-specific academic fields such as human geography (Garcia-Ramon 2003; Minca 2018), should be extended to include other intercultural skills. More diverse language skills and cultural knowledges could not only increase intercultural understanding but, as this analysis suggests, prevent the proliferation of different Anglo-American and multicultural international conference circuits in geography, as exemplified by the considerable varying geographies of the AAGs and IGCs.

New challenges have been created by the COVID-19 pandemic that has required online conferences and dual mode delivery of lectures and conference papers in hybrid in-person and online spaces of interaction. Online participation speaks to current EDI agendas in science and higher education and should therefore be considered as a permanent alternative mode of international conference attendance, especially because online conference attendance can also be regarded as a strategy for reducing one's carbon footprint and thus for climate change mitigation (Hopkins et al. 2016). Yet, the professional significance of in-person networking practices for academics, especially for the visibility of research presented by early career researchers at conferences (Storme et al. 2017) and the attractiveness as well as pedagogic benefits of conference fieldtrips as a learning experience (Roche 2019), provide strong arguments for the great value of in-person conference attendance within and beyond geography.

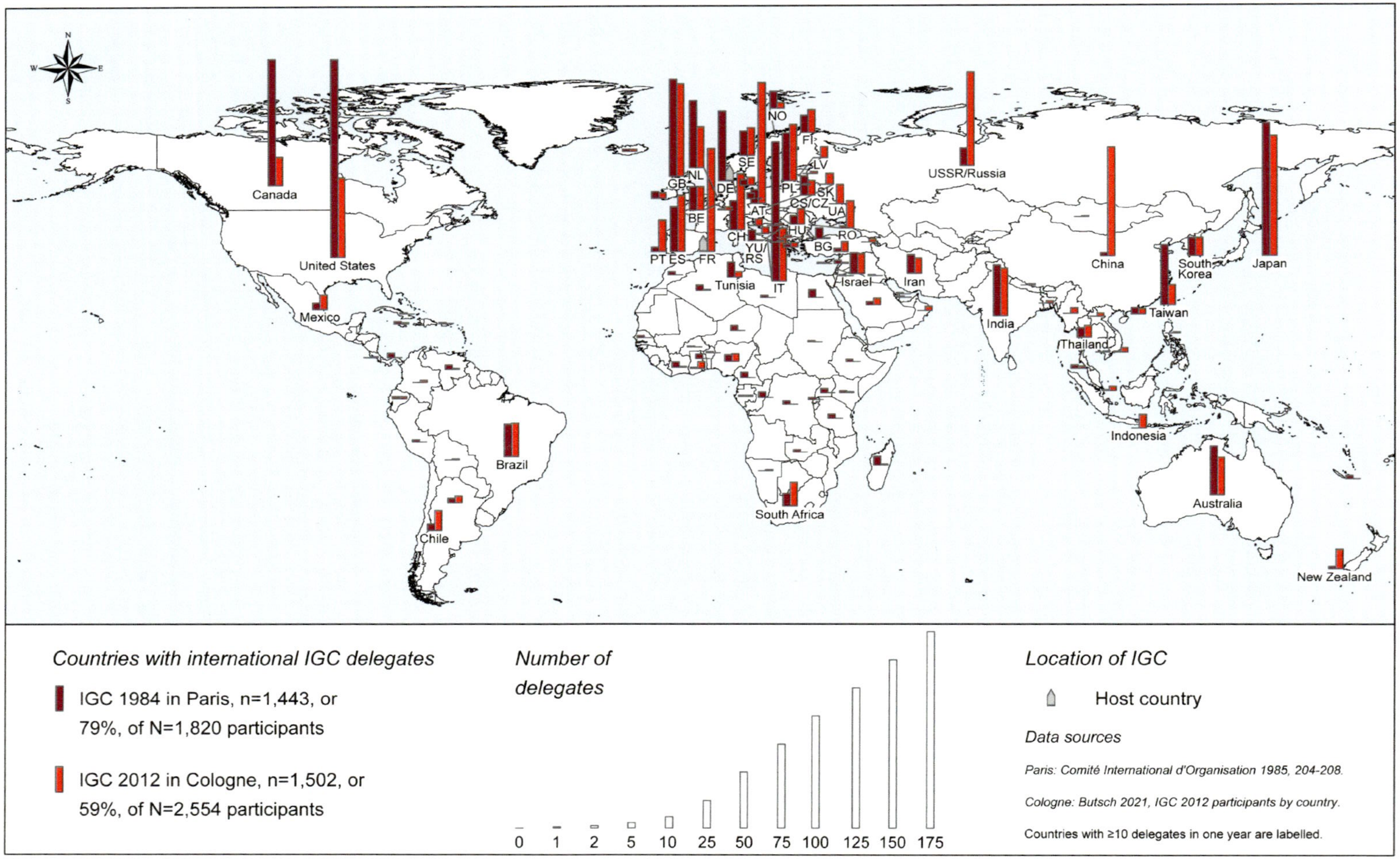

Fig. 6.2 Geographical origins of the international conference participants at the International Geographical Congresses (IGCs) in Paris 1984 and Cologne 2012 by country of work. *Source* Comité International d'Organisation (1985, 204–208) and Butsch (2021)

With the aim of addressing pertinent debates about environmental sustainability, global economic inequality, and EDI issues across different axes of social difference, I would argue that this third challenge requires the creation of diverse and flexible options of conference attendance that consider various positionalities of potential participants, such as in-person/hybrid/online and low/medium/high-cost. In other words, ensuring a sustainable international and intercultural diversity at future IGCs requires a greater variety of conference registration options and the active advertising of the IGCs.

Reflecting on these three challenges of internationalisation in geography suggests that geographers working in different countries are facing very different challenges shaped by uneven socioeconomic circumstances, varying geopolitical positionalities, diverse cultural and linguistic contexts, and differing geographical traditions (see also Kitchin 2007). These multiple intersectional positionalities of geographers interacting internationally have often led to complex challenges that may require context-specific responses rather than more standardised solutions. Further research needs to be done in order to understand the positive/neutral/negative, the planned/planned-but-not-implemented/unplanned, as well as the material/dynamically hybrid/immaterial effects/impacts/legacies of the IGCs as large-scale international events at micro/meso/macro scales, in similar ways as a burgeoning interest in the multidimensional and multiscalar outcomes of sport mega events has recently developed wider research agendas on event leveraging and legacies (e.g. Gold and Gold 2007; Dawson and Jöns 2018).

6.5 Conclusions

Throughout this chapter, I have argued that the nature of the 'international' in geography is relational. This point has emerged from a comparison of international conference attendance at four IGCs in the international conference series of the IGU, namely Paris 1984, Sydney 1988, Cologne 2012, and Beijing 2016, because the international and intercultural experiences offered at these events vary profoundly by the geopolitical, socio-economic, cultural, linguistic, and academic positionalities of the conference locations and the conference participants. This chapter has further shown that the global geographies of conference attendance have been considerably shaped by spatial proximity to the host region, mainly due to less costs of conference attendance, and by different geopolitical contexts as important enablers of and obstacles to international conference travel. Geographical, sociocultural, and geopolitical proximities have, variously, helped to mobilise conference participants with relevant cultural and social capital, such as language skills and personal networks. Both aspects can be seen to have provided an important basis, as Growe (2019) has argued, for the (re)production of cognitive and affective trust in professional meetings as a foundation of future interactions and collaborations. IGC attendance has also been influenced by geographers' regional expertise and their professional interest in exploring unfamiliar locations and countries, which explains the great importance of organised conference fieldtrips for attracting participants and creating positive event memories (Collignon 1996; Roche 2019).

The geographical perspective on international conference attendance has identified a profound shift in the international nature of the IGCs from an Anglo-American academic internationalism in the 1980s to a greater multicultural academic internationalism in the 2010s. This shift has been shaped by wider geopolitical and socioeconomic transformations since the end of the Cold War and through considerable economic growth in China, resulting in globally more decentralised conference attendance and many more participants from China, India, and Russia. A sharp decline in the number of North American geographers at recent IGCs seems to be linked to a growing differentiation of international geographical discourses at the quadrennial IGCs and the annual AAGs. Over twice as many overseas delegates attended the AAG 2013 in Los Angeles (Derudder and Liu 2016, 320) than the IGC 2012 in Cologne (Butsch 2021). On the one hand, this situation highlights the ongoing appeal of

American geography and an Anglo-American academic internationalism in the early twenty-first century, but on the other hand, a renewed call is appropriate, as geographers Harris (2001), Garcia-Ramon (2003), Minca (2018) and Lamego (2020) have previously argued, for more Anglo-American geographers to embrace the required intercultural knowledge and skills and to invest the time and effort needed for making the most of international exchanges in geography, and thus of the multicultural academic internationalism that has characterised the IGC experience—for themselves and for the discipline of geography.

In the interplay of internationalisation at home, at a distance, and abroad (Mittelmeier et al. 2021), it is crucial to stress that geographers in the host country of large-scale international conferences such as the IGCs benefit significantly from gaining international and intercultural experiences at home without being transnationally mobile. This considerable positive impact of the IGCs' multicultural internationalism on the internationalisation of geographers at home results in an enormous social and cultural inclusivity of geographical knowledge production and exchange by enabling diverse members of the different national geographical host communities—teachers, students, doctoral researchers, postdocs, academics, and professionals in other sectors of the economy—to engage with various international and intercultural perspectives. The IGCs thus possess considerable value for facilitating international geographical knowledge exchange that bridges both physical and human geographical discourses in Anglo-American and many other language areas. Yet, this chapter has also emphasised the great significance of the politics underpinning the choice of host cities, the flexibility of delivery formats and registration options, and the institutional and national encouragement and support of attendance at past and future IGCs for a sustainable diversity and inclusiveness of international geographical knowledge production and exchange.

Acknowledgements I gratefully acknowledge the support of Bruno Schelhaas who kindly made available materials from the Archive for Geography at the Leibniz-Institute of Regional Geography in Leipzig when it was inaccessible during the COVID-19 pandemic. Sincere thanks are also due to Carsten Butsch, for his exemplary generosity of sharing statistics on the IGC 2012 in Cologne, and to Federico Ferretti, Vladimir Kolosov, and Bruno Schelhaas for their constructive comments on earlier versions of this text.

References

Aalbers MB, Rossi U (2007) A coming community: young geographers coping with multi-tier spaces of academic publishing across Europe. Soc Cult Geogr 8:283–302

Agnew JA (2001) Disputing the nature of the international in political geography. The Hettner-Lecture in human geography. Geogr Z 89:1–16

Aufenvenne P et al (2021) Participation and communication behaviour at academic conferences: an empirical gender study at the German Congress of Geography 2019. Geoforum 126:192–204

Bauder H (2015) The international mobility of academics: a labour market perspective. Int Migr 53:83–96

Bradshaw M, Stenning A (2004) East Central Europe and the Former Soviet Union: the post-socialist states. Pearson Education, Harlow

Butsch C (2015) IGC 2012: facts and figures. In: Kraas F et al (eds) IGC Cologne 2012: down to earth: documenting the 32nd International Geographical Congress in Cologne 26–30 August 2012 (=Kölner Geographische Arbeiten 95). Geographisches Institut der Universität zu Köln, Köln, pp 136–159

Butsch C (2021) Number of checked-in participants at the IGC Cologne 2012 by country of registration. Email communication by PD Dr Carsten Butsch, Department of Geography, University of Bonn. Received 28 Sept 2021

Buttimer A (1983) The practice of geography. Longman, London

Collignon B (1996) Congrès et conférences regionales. In: Robic M-C, Briend A-M, Rössler M (eds) Géographes face au monde: l'union géographique internationale et les congrès internationaux de géographie. L'Harmattan, Paris, pp 87–116

Comité International d'Organisation (1985) Congress proceedings: 25th International Geographical Congress Paris 1984. Comité International d'Organisation, Caen

Craggs R (2014) Postcolonial geographies, decolonization, and the performance of geopolitics at Commonwealth conferences. Singap J Trop Geogr 35:39–55

Cresswell T (2013) Geographic thought: a critical introduction. Wiley-Blackwell, Malden

Dalmasso E (1986) Le XXVe congrés international de géographie: Paris—Les Alpes (27–31 août 1984). Ann Géo 528:148–167
Dawson J, Jöns H (2018) Unravelling legacy: a triadic actor-network theory approach to understanding the outcomes of mega events. J Sport Tour 22:43–65
Derudder B, Liu X (2016) How international is the annual meeting of the Association of American Geographers? A social network analysis perspective. Environ Plan A Econ Socd 48:309–329
Drummond LBW, Young D (2020) Socialist and post-socialist urbanisms: critical reflections from a global perspective. University of Toronto Press, Toronto
Esson J (2020) 'The why and the white': racism and curriculum reform in British geography. Area 52:708–715
Ferretti F (2019) Anarchy and geography: Reclus and Kropotkin in the UK. Routledge, New York
Garcia-Ramon M-D (2003) Globalization and international geography: the questions of languages and scholarly traditions. Prog Hum Geogr 27:1–5
Geographical Society of China (2016) The 33rd International Geographical Congress, Beijing, China, 21–25 Aug 2016: "Shaping our harmonious worlds" provided by the local organising committee of the IGU 2016. Geographical Society of China. http://igu-online.org/wp-content/uploads/2016/11/Report-of-The-33rd-International-Geographical-Congress.pdf. Accessed 28 Sept 2021
Gold JR, Gold MM (2007) Athens to Athens: the summer Olympics, 1896–2004. In: Gold JR, Gold MM (eds) Olympic cities: city agendas, planning, and the world's games, 1896–2012. Routledge, London, pp 15–47
Growe A (2019) Developing trust in face-to-face interaction of knowledge-intensive business services (KIBS). Reg Stud 53:720–730
Gui Q, Liu C, Du D (2019) Globalization of science and international scientific collaboration: a network perspective. Geoforum 105:1–12
Gunter A, Raghuram P (2018) International study in the global south: linking institutional, staff, student, and knowledge mobilities. Glob Soc Educ 16:192–207
Harris CD (2001) English as international language in geography: development and limitations. Geogr Rev 91:675–689
Hodder J, Legg S, Heffernan M (2015) Introduction: historical geographies of internationalism, 1900–1950. Polit Geogr 49:1–6
Hopkins D, Higham J, Tapp S, Duncan T (2016) Academic mobility in the Anthropocene era: a comparative study of university policy at three New Zealand institutions. J Sustain Tour 24:376–397
International Geographical Union (1988) Final report. In: 26th International Geographical Congress Sydney 1988. International Geographical Union, Sydney
International Geographical Union (2021) Statutes. https://igu-online.org/organization/statutes/. Accessed 28 Sept 2021
Jöns H (2015) Talent mobility and the shifting geographies of Latourian knowledge hubs. Popul Space Place 21:372–389
Jöns H (2017) Feminizing the university: the mobilities, careers, and contributions of early female academics in the University of Cambridge, 1926–1955. Prof Geogr 69:670–682
Jöns H, Heffernan M, Meusburger P (2017) Mobilities of knowledge: an introduction. In: Jöns H, Meusburger P, Heffernan M (eds) Mobilities of knowledge. Springer, Cham, pp 1–19
Kish G (1972) The participants. In: IGU Commission on History of Geographical Thought (eds) Geography through a century of international congresses. UNESCO, Paris, pp 35–49
Kitchin R (ed) (2007) Mapping worlds: international perspectives on social and cultural geographies. Routledge, New York
Knight J (2004) Internationalization remodeled: definition, approaches, and rationales. J Stud Int Educ 8:5–31
Knight J (2012) Concepts, rationales, and interpretive frameworks in the internationalization of higher education. In: Deardorff DK, de Wit H, Heyl JD, Adams T (eds) The SAGE handbook of international higher education. Sage, Thousand Oaks, pp 27–42
Knight J (2014) Three generations of crossborder higher education: new developments, issues and challenges. In: Streitwieser B (ed) Internationalisation of higher education and global mobility. Symposium Books, Oxford, pp 43–58
Lamego M (2020) How international was the International Geographical Congress in Rio de Janeiro 1956? On location and language politics. In: Schelhaas B et al (eds) Decolonising and internationalising geography: essays in the history of contested science. Springer, Cham, pp 113–126
Lang T et al (2015) Understanding geographies of polarization and peripheralization: perspectives from central and eastern Europe and beyond. Palgrave Macmillan, New York
Livingstone DN (1992) The geographical tradition: episodes in the history of a contested enterprise. Blackwell, London
Madge C, Raghuram P, Noxolo P (2015) Conceptualizing international education: from international student to international study. Prog Hum Geogr 39:681–701
Maisonobe M et al (2017) The global geography of scientific visibility: a deconcentration process (1999–2011). Scientometrics 113:479–493
Maskell P, Bathelt H, Malmberg A (2006) Building global knowledge pipelines: the role of temporary clusters. Eur Plan Stud 14:997–1013
Mbah M (2017) Formal education as a facilitator of migration and integration: a case study of Nigerian university graduates. In: Jöns H, Meusburger P, Heffernan M (eds) Mobilities of knowledge. Springer, Cham, pp 247–268

McEwan C (2000) Gender, geography and empire: Victorian women travellers in West Africa. Ashgate, Aldershot
Meadows ME (2020) Perspectives on internationalism in geography. Geogr Pol 93:493–503
Minca C (2018) The cosmopolitan geographer's dilemma: or, will national geographies survive neo-liberalism? Geogr Z 106:4–15
Mitchell K (2018) Making workers: radical geographies of education. Pluto Press, London
Mittelmeier J et al (2021) Conceptualizing internationalization at a distance: a "third category" of university internationalization. J Stud Int Educ 25:266–282
Momm CF, Jöns H (2020) Decentralized concentration through cyclical events: the geographies of academic conferences in urban and regional development and planning in Brazil, 2004–2013. Geoforum 112:104–117
Paasi A (2005) Globalisation, academic capitalism, and the uneven geographies of international journal publishing spaces. Environ Plann Econ Soc 37:769–789
Power D, Jansson J (2008) Cyclical clusters in global circuits: overlapping spaces in furniture trade fairs. Econ Geogr 84:423–448
Robertson SL (2017) Making education markets through global trade agreements. Glob Soc Educ 15:296–308
Robic M-C, Briend A-M, Rössler M (eds) (1996) Géographes face au monde: l'union géographique internationale et les congrès internationaux de géographie. L'Harmattan, Paris
Roche M (2019) New Zealand geography and the International Geographical Union part II: audience to stage, 1956–1974. NZ Geogr 75:51–62
Rye SA (2014) The educational space of global online higher education. Geoforum 51:6–14
Schelhaas B et al (eds) (2020) Decolonising and internationalising geography: essays in the history of contested science. Springer, Cham
Storme T et al (2017) Mobility and professional networks in academia: an exploration of the obligations of presence. Mobilities 12:405–424
Taylor PJ (1999) Modernities: a geohistorical interpretation. Polity Press, Cambridge
Tebbett N, Jöns H, Hoyler M (2021) Openness towards diversity? Cultural homophily in student perceptions of teaching and learning provided by international and home academics. Glob Soc Educ 19:522–544
Turkey Geographical Society (2021) 34th International Geographical Congress. https://www.igc2020.org/. Accessed 28 Sept 2021
Waters JL (2012) Geographies of international education: mobilities and the reproduction of social (dis)advantage. Geogr Compass 6:123–136
Waters JL, Leung M (2017) Trans-knowledge? Geography, mobility, and knowledge in transnational education. In: Jöns H, Meusburger P, Heffernan M (eds) Mobilities of knowledge. Springer, Cham, pp 269–285
Weisser F, Müller-Mahn D (2017) No place for the political: micro-geographies of the Paris climate conference 2015. Antipode 49:802–820
Withers CWJ (2010) Geography and science in Britain, 1831–1939: a study of the British Association for the Advancement of Science. Manchester University Press, Manchester

Heike Jöns is Professor of Geography at Loughborough University. She was awarded her Ph.D. in Geography at the University of Heidelberg (2002) and has subsequently worked at the universities of Heidelberg, Nottingham, and Loughborough. Heike has published widely on the development, geographies, and politics of academic mobilities and knowledge production, transatlantic knowledge networks, and internationalisation and globalisation processes in the university, as well as on the benefits of triadic thought for geographical research. Heike's books include the research monograph *Grenzüberschreitende Mobilität und Kooperation in den Wissenschaften* (2003) and the edited volumes *Geographies of Science* (2010) and *Mobilities of Knowledge* (2017).

7 Internationalisation in the International Geographical Union: Landmarks, Periods and Personalities

Joos Droogleever Fortuijn

Abstract

The central question in this chapter is: How international is the International Geographical Union (IGU)? The chapter reports an analysis of the membership of the IGU Executive Committee (1922–2022), the steering committees of the IGU Commissions and Task Forces (2004–2022) and the IGU Commission and Task Force members (2014–2019). The literature on internationalisation in geography focuses on the Anglo-American hegemony and the dominance of the English language. The analysis in this chapter demonstrates that IGU, in terms of membership, has never experienced a dominance of Anglophone countries. IGU has been established 100 years ago as a predominantly European organisation and developed gradually into a truly global organisation in which a shift took place from Europe into other continents, Asia in particular. Women members are still more often from European countries than men. The English language, however, has become the dominant language. While IGU was initially multilingual, IGU, officially bilingual, developed into a monolingual, Anglophone organisation. The chapter concludes with some recommendations to increase diversity in terms of geography, language, gender and age.

Keywords

Internationalisation · History of IGU membership · Anglophone hegemony · Women in geography

7.1 Introduction

The International Geographical Union (IGU) was founded one hundred years ago as a community of geographers worldwide. After the end of the First World War, representatives of the Academies of Sciences from allied countries established the International Research Council (IRC) and to encourage the (re)creation of disciplinary unions (Hinks 1922; Robic 1996). The second General Assembly of the IRC met in Brussels in July 1921 and representatives of the national geographical societies of Belgium, France, Italy, Japan, Portugal, Spain and the UK, a mixture of military and civil servants and a minority of academic geographers, decided to establish the IGU. One of the most important tasks of the proposed IGU was to organise the International Geographical Congresses (IGCs). In the words of the first Secretary-General of the IGU, Sir Charles Close, who was also President of the Royal Geographical Society in Great

J. D. Fortuijn (✉)
Department of Geography, Planning and International Development Studies, University of Amsterdam, Amsterdam, Netherlands
e-mail: j.c.droogleeverfortuijn@uva.nl

V. Kolosov et al. (eds.), *A Geographical Century*,
https://doi.org/10.1007/978-3-031-05419-8_7

Britain: "Science is essentially international, and every worker finds, from time to time, the need of freeing himself from the intellectual preoccupations of his fellow-countrymen. This is especially the case with geography, which, of all branches of knowledge, requires most to be studied from the standpoint of a citizen of the world" (Close 1928, 100; see also Chap. 2, by Heffernan).

An international community of geographers existed already fifty years before the foundation of the IGU. In 1871, the municipality of Antwerp invited members of geographical societies from several countries to the first IGC and another nine congresses took place before the foundation of the IGU in 1922 (Table 7.1). From the IGC in Venice (1891) onwards, the foundation of an international geographical organisation has been discussed (Robic 1996). The IGU was eventually established in 1922 under the auspices of the IRC. Members of the IGU are national committees representing geographers in their country. The chairs of the national committees form the General Assembly that elects President, Secretary-General and Vice-Presidents and decides on the establishment, continuation and dissolution of sub disciplinary Commissions and Task Forces (CTFs).

The founding countries were six European countries (Belgium, France, Italy, Portugal, Spain and the UK) and Japan. The IRC was initiated by the scientists from allied countries, and members from the Central Powers were excluded from the IRC and from the disciplinary unions. Although German geographers participated actively in the IGCs before the First World War, they were not allowed to join the IGU until 1934 (Robic 1996; though see Chap. 2, by Heffernan).

In the period before World War Two, IGU membership rose from 19 members in 1928 to 30 in 1938 (Volle 1996). After 1945, the geographical diversity of the IGU's membership expanded further. By 1956, the IGU had 74 member countries of which a minority (30%) were European. In 1984, 29% of the 89 members was European (Kish 1992). The IGU currently

Table 7.1 Location of the International Geographical Congresses, 1871–2021 and Planned Congresses 2022–2028

I	1871	Antwerp	XX	1964	London
II	1875	Paris	XXI	1968	New Delhi
III	1881	Venice	XXII	1972	Montreal
IV	1889	Paris	XXIII	1976	Moscow
V	1891	Bern	XXIV	1980	Tokyo
VI	1895	London	XXV	1984	Paris
VII	1899	Berlin	XXVI	1988	Sydney
VIII	1904	Washington	XXVII	1992	Washington
IX	1908	Geneva	XXVIII	1996	The Hague
X	1913	Rome	XXIX	2000	Seoul
XI	1925	Cairo	XXX	2004	Glasgow
XII	1928	Cambridge	XXXI	2008	Tunis
XIII	1931	Paris	XXXII	2012	Cologne
XIV	1934	Warsaw	XXXIII	2016	Beijing
XV	1938	Amsterdam	XXXIV	2021	Istanbul
XVI	1949	Lisbon	Centennial	2022	Paris
XVII	1952	Washington	XXXV	2024	Dublin
XVIII	1956	Rio de Janeiro	XXXVI	2028	Melbourne
XIX	1960	Stockholm			

Source Kish (1992), www.igu-online.org

has 110 country members from all parts of the world (www.igu-online.org), 32% of them are European. As Volle (1996) argues, European domination has increased since 1990, partly as a result of the geopolitical fragmentation of Europe after the fall of the Berlin wall and partly because of the strict financial membership criteria, at the expense of African, Asian and Latin American countries. Nowadays, 32% of all IGU members and 48% of the full (voting) members are European (www.igu-online.org). The central question in this chapter is: how international is the International Geographical Union?

7.2 Internationalisation in Geography

In the past decades, several articles have been published about the internationalisation in the discipline of geography. Authors have analysed "the geopolitics of knowledge production" (Paasi 2015) and "skewed transnationalisation" (Van Meeteren 2019). According to Paasi "there is an uneven geopolitics of knowledge embedded in communication. Academic fields largely exist through publication forums which are structural asymmetrically in global space. Scholars operating in 'peripheral' writing spaces often face substantial hindrances in publishing" (Paasi 2015, p. 511). Some authors refer to the increased dominance of the English language in geography, while others focus on the hegemony of the Anglo-American geographical traditions. However, as Garcia-Ramon (2003), Gregson et al. (2003) and Van Meeteren (2019) argue, languages and academic traditions are strongly interrelated. Using the English language in academic work implies adoption of Anglo-American academic traditions. Minca (2000) discusses the position of geographers from "peripheral" and "silent" European communities who perform a balancing act: "all are forced continually and inescapably to dialogue/work on two parallel levels—within the context of their own national geographies, with their rules, logics and languages, but also within the broader international (read Anglo-American) context, with its own logics and its own particular lingua franca" (Minca 2000, p. 287). A few authors, however, argue that the position of English as lingua franca in geography is a prerequisite for communication across language communities and promotes diversity instead of uniformity (Rodriguez-Pose 2006). Others emphasise that the idea of an Anglo-American hegemony denies the heterogeneity within Anglo-American geography (Samers and Sidaway 2000).

The majority of the publications about the Anglo-American hegemony focus on publications and the dominance of Anglo-American journals. The predominantly Anglo-American editorial boards of the high impact "international" journals (Banski and Ferenc 2013), published by a few large commercial publishers, define what is relevant for publication. Authors from non-Anglophone countries are disadvantaged and their work is merely seen as "just case studies" instead of core contributions to theory development (Aalbers 2004; Gregson et al. 2003; Meadows et al. 2016; Minca 2000; Simonsen 2003; Timar 2004). University management, funding organisations and assessment systems in non-Anglophone countries perform a critical role in the Anglo-American dominance: they require scholars to publish in high ranked international journals, in practice Anglo-American journals, and they qualify publications in the national language and non-Anglo-American journals as less relevant.

Journal publications are an important dimension of internationalisation, but there are other dimensions as well. De Rudder and Liu (2016) discuss the internationalisation in the participation in academic conferences in an analysis of the Annual Meeting of the American Association of Geographers (AAG). The Annual Meetings of the AAG are—with more than 10,000 participants—the largest conferences in the field of geography, and the participation has become more and more international. In 2005, 77% of the participants were based at an American university, while more than one-third of the participants in 2013 came from a university outside the USA. The number of countries represented at the Annual Meeting increased from 44 in 2005 to 90 in 2013. In 2005, 93% of the participants came from one of the Anglophone countries, the USA, Canada, UK, Ireland, Australia and New Zealand, while 21% came from outside these Anglophone countries in 2013 (De Rudder and Liu 2016). The analysis of De Rudder and Liu focuses not so much on participants in the Annual Meetings, but on the diversity within paper and panel sessions. They conclude that the increased internationalisation of the conference participants is not reflected in the sessions. International interaction is limited (for example, a session on Chinese cities in which only Chinese geographers presented papers) and international interaction in sessions did not increase between 2005 and 2013.

Participation in international conferences is an important dimension of internationalisation. However, geographers from low-income countries are usually under-represented. I have been responsible for the travel grant program of the IGU that enables (young) geographers from low-income countries to participate at the International Geographical Congresses and IGU Regional Conferences. For each congress and conference, 25–30 geographers were awarded a travel grant, but several of the awardees were not able to participate because of visa problems and problems with additional funding.

This chapter discusses the internationalisation of the IGU drawing on data on the individuals who are responsible for the governance of the IGU and its Commissions and Task Forces and on the Commission and Task Force members. The core questions are: how international is the International Geographical Union? Is it still a European dominated organisation? Is there an Anglo-American hegemony? How international was the IGU in the different periods of its existence?

7.3 Data

The data include, first, the complete list of Presidents, Secretaries-Generals and Vice-Presidents since the foundation of the IGU (1922–2022), as published in IGU Bulletins, the IGU website and other publications on the history of the IGU (Appendix). Second, the data includes the (almost) complete lists of chairs and steering committee members of the more than 40 Commissions and Task Forces (CTFs) from 2004 to 2022, as published in the annual reports and websites of the CTFs and lists provided by Mike Meadows, President and former Secretary-General of the IGU. The list is not fully complete, because some commissions did not provide an annual report or list every year. Third, I collected data on CTF membership in 2014 and 2019. The majority of the CTFs publish the number of their members in each country and a few CTFs publish a list of names or email addresses in their annual reports. I contacted the chairs of those CTFs that did not publish a list of members in the annual report. Some chairs responded positively and sent the full list of names or email addresses, others sent the lists of steering committee members only.

Lists of email addresses were provided by seven CTFs in 2014 and five in 2019. The countries in which members work were identified through the ISO country code in the email address, though many members have email addresses without a country code (.com or .org). I was able to identify the university or country of about 90% of IGU members using a simple Google search. Some members have more than one email address. Duplications were deleted as far as possible. However, the final data still contain an unknown number of duplications. Another source of duplications is the multiple membership: people can be a member of more than one CTF. Finally, some people use an email address of a country other than where they are based (for example, several members of the Commission on Mediterranean Basin from North-African countries have a French email address). As a result, the total CTF membership of the IGU of 10,700 in 2014 and 14,100 in 2019 is an overestimate of the actual membership insofar as this includes duplications from email addresses and multiple membership, but is probably also an underestimate because the data is also incomplete for four CTFs in 2014 and seven CTFs in 2019.

7.4 The Geography of IGU Membership: The Executive Committee, 1922–2022

The IGU Executive Committee (EC) consisted of five European geographers and one non-European at its foundation (Appendix). The first President of the IGU was Prince Roland Bonaparte, from France, and the first Secretary-General was Charles Close, from Britain. The first vice-presidents were the Italian general Nicola Vacchelli, the Belgian commandant Adrien De Gerlache, the Spanish general Severo Gómez Nuñez, and a Japanese professor, Naomasa Yamasaki. Five of these men had military backgrounds or were currently serving in the armed forces. Only one was a professional academic (Hinks 1922). The (West) European dominance in the EC was in line with the 50-year history of the IGCs before the foundation of the IGU. Between 1871 and 1922, the IGCs were located in ten European and one non-European cities (Table 7.1).

In the 100 years of the existence of the IGU, a total of 33 countries from all continents were represented in the EC, half of them European, mainly West European (Fig. 7.1; see also Volle 1996). The USA has been represented in almost all ECs since 1928 (over a continuous period of 88 years). France (69 years), Italy (59 years), Japan (53 years), UK (51 years) and USSR/Russia (50 years) were also represented on the IGU EC for at least half of its existence.

Four periods can be identified in the internationalisation of the IGU:

- From 1922 to 1949, when Europe dominated;
- From 1949 to 1976, when all continents were represented but with strong European domination;
- From 1976 to 2000, when a global organisation emerged in which Europe was less dominant but other Anglophone countries in Europe, America and Oceania featured more prominently;
- From 2000 to 2022, when the dominance of Europe and Anglophone countries decreased and Asian countries became more prominent.

7.4.1 1922–1949

In this period, the EC was an almost exclusively (West) European committee. Presidents and Secretaries-General came from a limited number of countries (Table 7.2) and no more than nine European and three non-European countries were represented in the EC. In this period 63–89% of the EC members were European and 13–38% came from Anglophone countries. The IGCs in this period took place exclusively in European cities, with the exception of the IGC in Cairo in 1925. Six European languages—French, English, German, Italian, Spanish and Portuguese—were used as languages of communication at these congresses, but French remained the dominant language until the Second World War (Harris 2001, p. 676). French was also the language of communication in the EC.

7.4.2 1949–1976

In the post-Second World War period, the IGU became more and more a global organisation, although still European dominated. Presidents came mainly from European countries, but also from non-European countries such as the USA and India. Secretaries-General were mainly from the USA and Canada. A total of 21 countries from all continents were represented in the EC, but still 50–67% of the EC members were still European; about a quarter of the EC members came from Anglophone countries. The IGCs were more widely spread around the globe. Until 1956, six languages were used during the congresses; after 1960 the IGCs became bilingual French–English. English was already the dominant language: 76% of the paper presentations in 1960 were English spoken (Harris 2001, p. 676). The EC has communicated in English since 1949.

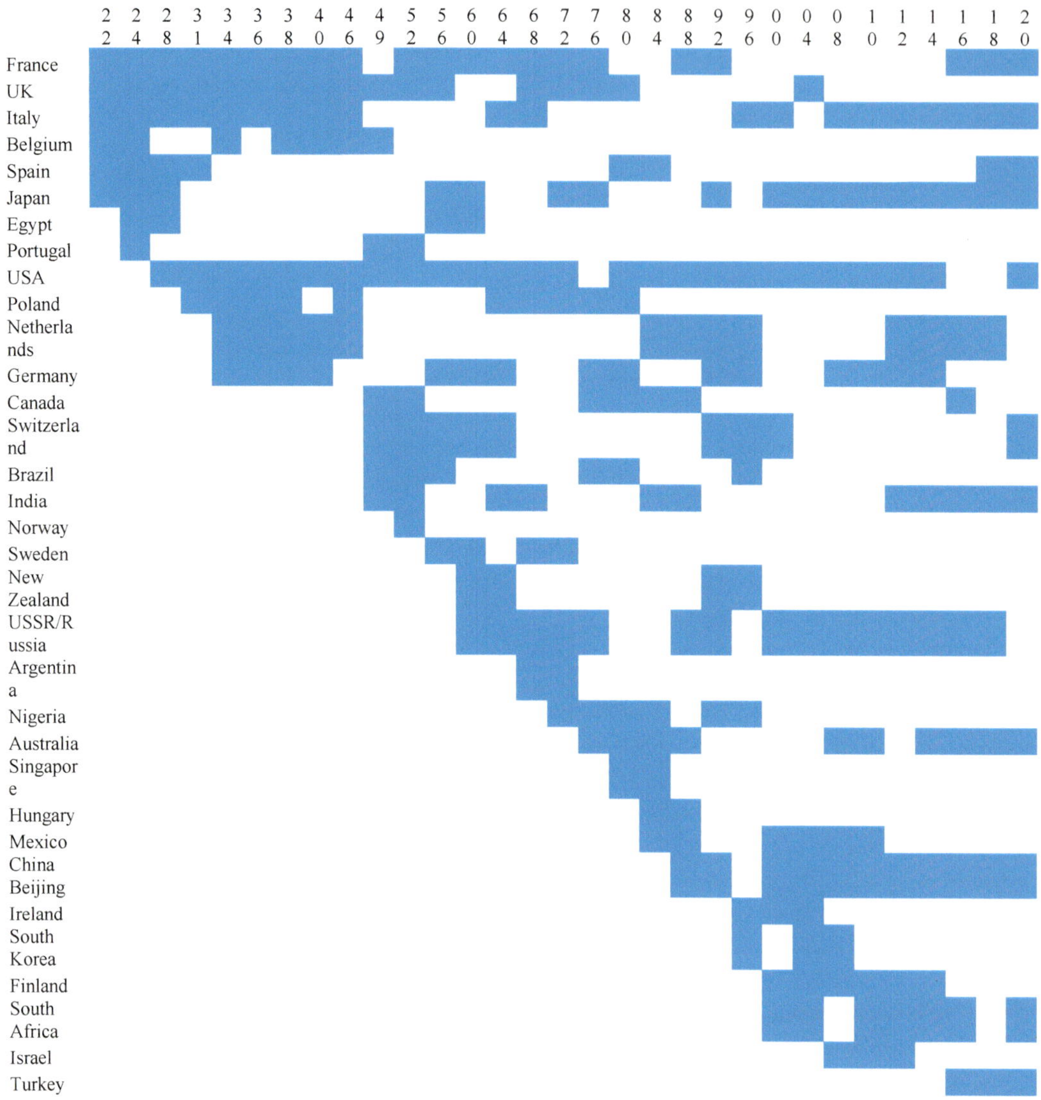

Fig. 7.1 Countries represented in the IGU Executive Committee, 1922–2022. *Source* Reports of the International Geographical Union, 1922–1938, Reports of the International Geographical Congresses 1928 and 1949, Birdseye (1939), Do Amaral (1968), Newsletters of the International Geographical Union 1950–1956, Bulletins of the International Geographical Union 1956–2016, www.igu-online.org

7.4.3 1976–2000

After 1976, the IGU became less dominated by representatives from European countries. Half of the Presidents and Secretaries-General came from European countries; the share of European EC members was 30–50%. Anglophone countries became more prominent. Half of the Presidents and 20–40% of the EC members came from Anglophone countries. The IGCs were located in European as well as non-European cities. Although officially bilingual, English remained the dominant language of communication in the EC and during the congresses. Since

Table 7.2 Presidents and Secretaries-General of the IGU 1922–2022

President	Secretary-General
1922–1924 Roland Bonaparte, France	1922–1928 Charles Close, UK
1924–1928 Nicola Vacchelli, Italy	1928–1931 Filippo de Fillippi, Italy
1928–1931 Robert Bourgeois, France	1931–1938 Emmanuel de Martonne, France
1931–1934 Isaiah Bowman, USA	1938–1949 Paul Michotte/Marguerite Lefèvre, Belgium
1934–1938 Charles Close, UK	1949–1956 George Kimble, Canada
1938–1949 Emmanuel de Martonne, France	1956–1968 Hans Boesch, Switzerland
1949–1952 George Babcock Cressey, USA	1968–1976 Chauncy Harris, USA
1952–1956 Laurence Dudley Stamp, UK	1976–1984 Walther Manshard, Germany
1956–1960 Hans Ahlman, Sweden	1984–1992 Leszek Kosínski, Canada
1960–1964 Carl Troll, Germany	1992–2000 Eckart Ehlers, Germany
1964–1968 Shiba Chatterjee, India	2000–2008 Ronald Abler, USA
1968–1972 Stanislaw Leszcycki, Poland	2008–2010 Woo-ik Yu, South Korea
1972–1976 Jean Dresch, France	2010–2018 Michael Meadows, South Africa
1976–1980 Michael John Wise, UK	2018–2021 RB Singh, India
1980–1984 Akin Mabogunje, Nigeria	2021– Barbaros Gönençgil, Turkey
1984–1988 Peter Scott, Australia	
1988–1992 Roland Fuchs, USA	
1992–1996 Herman Verstappen, Netherlands	
1996–2000 Bruno Messerli, Switzerland	
2000–2004 Anne Buttimer, Ireland	
2004–2006 Adalberto Vallega, Italy	
2006–2008 José Luis Palacio-Prieto, Mexico	
2008–2012 Ronald Abler, USA	
2012–2016 Vladimir Kolosov, Russia	
2016–2020 Yukio Himiyama, Japan	
2020– Michael Meadows, South Africa	

Source www.IGU-online.org

1980, more than 90% of the paper presentations were in English, with the exception of the Paris congress in 1984 (Harris 2001, p. 676).

7.4.4 2000–2022

In this period, the EC was no longer dominated by representatives from Europe and the share of members from Anglophone countries was substantially lower than in the period 1976–2000. Presidents were more often from outside Europe than from European countries, and the four Secretaries-General were all from outside Europe. 33–50% of the EC members were European and 9–22% came from Anglophone countries. Asian countries in particular were more represented in the EC and formed 36 percent of the EC membership between 2016 and 2020. The majority of the IGCs were located in non-European countries.

7.5 The Geography of IGU Membership: Commission and Task Force Steering Committees, 2004–2024

In the steering committees of the Commissions and Task Forces (CTFs) between 2004 and 2022, European countries are more dominant than in the Executive Committee. More than 40% of the CTF steering committee are European, mainly from West European countries, although the share of members from Eastern Europe, from Poland and Russia in particular, increased slightly (Table 7.3). The presence of Asian, Latin American and African countries in the steering committees increased, from China Beijing, India, Brazil and South Africa in particular. More prominent, however, is the diminishing share of steering committee members from Anglophone countries.

Table 7.3 Steering Committee members of the IGU Commissions and Task Forces by continent, 2004–2022

	2004–2008	2008–2012	2012–2016	2016–2020	2020–2022
Europe	162 (41.3%)	179 (45.3%)	191 (48.0%)	194 (43.2%)	196 (41.6%)
Asia	86 (21.9%)	83 (21.0%)	84 (21.1%)	109 (24.3%)	111 (23.6%)
Latin America	21 (5.4%)	23 (5.8%)	26 (6.5%)	40 (8.9%)	44 (9.3%)
North America	61 (15.6%)	53 (13.4%)	46 (11.6%)	45 (10.0%)	48 (10.2%)
Africa	28 (7.1%)	28 (7.1%)	20 (5.0%)	35 (7.8%)	44 (9.3%)
Oceania	34 (8.7%)	29 (7.3%)	31 (7.8%)	26 (5.8%)	28 (5.9%)
Total[a]	392 (100.0%)	395 (100.0%)	398 (100.0%)	449 (100.0%)	471 (100.0%)
Anglophone[b]	123 (31.4%)	114 (28.9%)	106 (26.6%)	94 (20.9%)	97 (20.6%)

[a] Data are missing of 3 Commissions and Task Forces in 2004–2008, 2 in 2008–2012 and 3 in 2012–2016
[b] Australia, Canada, Ireland, New Zealand, UK and USA
Sources Lists provided by the President of the IGU; Annual Reports and websites of the IGU Commissions and Task Forces

Several new CTFs have been formed since 2004, among them the regional focused Commission on African Studies and on Latin American and Caribbean Studies. The new CTFs have a larger share of steering committee members from Latin America and Africa and a lower percentage of steering committee members from Europe, Asia and Anglophone countries than the older CTFs.

Physical geography Commissions have a larger share of steering committee members from Asian countries and less from Europe and Anglophone countries, while human geography Commissions have less steering committee members from Asia and—due to the regional focused Commissions Mediterranean Basin, African Studies and Latin American and Caribbean Studies—more from Africa and Latin America (Table 7.4).

7.6 The Geography of IGU Membership: Commission and Task Force Members, 2014 and 2019

The total number of CTF members increased substantially between 2014 and 2019: from 10,700 to 14,100, an increase of 32%. The growth is the result of the formation of new CTFs as well as an increased membership in the older CFTs. The growth in membership goes hand in hand with a rising number of countries: from 139 countries in 2014 to 161 in 2019, in Asia, the Caribbean and Africa in particular. However, "white spots" can still be identified (Fig. 7.2). Parts of Africa, Latin America, the Caribbean,

Table 7.4 Steering Committee members of IGU Commissions and Task Forces by continent and subdiscipline, 2020–2022

	Physical geography	Human geography	Mixed
Europe	48 (37.2%)	98 (42.2%)	50 (45.5%)
Asia	42 (32.6%)	43 (18.5%)	26 (23.6%)
Latin America	13 (10.1%)	25 (10.8%)	6 (5.5%)
North America	13 (10.1%)	23 (9.9%)	12 (10.9%)
Africa	7 (5.4%)	28 (12.1%)	9 (8.2%)
Oceania	6 (4.7%)	15 (6.5%)	7 (6.4%)
Total	129 (100.0%)	232 (100.0%)	110 (100.0%)
Anglophone[a]	21 (16.3%)	50 (21.6%)	26 (23.6%)

[a] Australia, Canada, Ireland, New Zealand, UK and USA
Source see Table 7.3

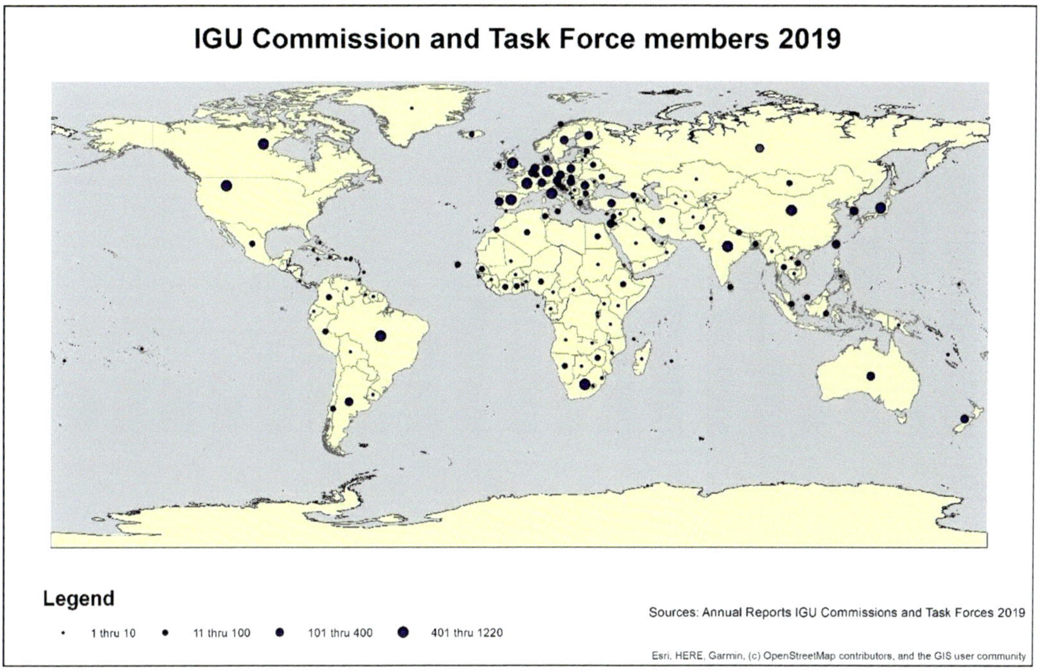

Fig. 7.2 IGU Commission and Task Force members 2019

Central Asia and the Middle East are regions with few or no members.

The USA, the UK, Germany, China Beijing and India are the "big five" in terms of CTF membership in 2019 (Fig. 7.3). USA, UK, Germany, Japan and Canada were the top-five in 2014, but the membership in Japan and Canada decreased, while membership in China Beijing and India (and South Africa and Portugal) grew by more than 100%.

The number of CTF members increased in all continents except Oceania (Table 7.5). Europe and North America had a moderate increase and Asia, Latin America and Africa a substantial growth in membership. The membership in the Anglophone countries declined.

The IGU is still European dominated in terms of CTF membership in 2019. Almost half of the CTF members are European. Around a quarter of the members come from Anglophone countries. European and Anglophone domination decreased between 2014 and 2019. This might be a result of the growing internationalisation of the Annual Meeting of the American Association of Geographers that has increasingly become the most important international meeting for many geographers from Anglophone countries and Europe (De Rudder and Liu 2016). Many geographers from these countries with a budget for one international meeting prefer the AAG meetings. At the same time, the international orientation of geographers from Asian and African countries has increased in the past few years.

Although CTF membership can be characterised as global, not all CTFs demonstrate a diverse geography. Membership of some Commissions is dominated by one or two countries only, for example the Commission on Environmental Evolution with 59% members from Russia, and the Commission on Biogeography and Biodiversity with 49% members from Germany and 16% from India. Other Commissions are more diverse, for example the Commission on Marginalisation, Globalisation and Regional and Local Responses and the Commission on Geography of Governance with no more than 10% members from one country. European domination is relatively low in the new regional

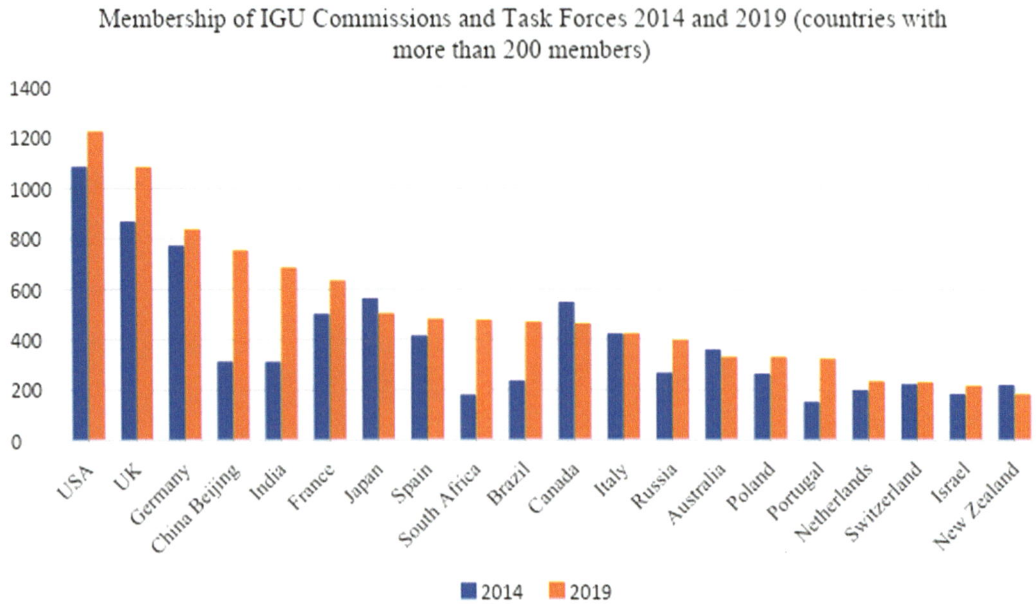

Fig. 7.3 Membership IGU Commissions and Task Forces 2014 and 2019 (countries with more than 200 members). *Source* Annual Reports of the IGU Commissions and Task Forces 2014 and 2019

Table 7.5 Members and membership increase of IGU Commissions and Task Forces by continent, 2014 and 2019

	Members 2014	Members 2019	Increase 2014–2019 (%)
Europe	5288 (51.2%)	6463 (48.6%)	22.2
Asia	2044 (19.8%)	3022 (22.7%)	47.8
Latin America	545 (5.3%)	818 (6.1%)	50.1
North America	1544 (14.9%)	1679 (12.6%)	8.7
Africa	337 (3.3%)	809 (6.1%)	140.1
Oceania	577 (5.6%)	513 (3.9%)	−11.1
Total[a]	10,335 (100.0%)	13,304 (100.0%)	28.7
Anglophone[b]	3159 (30.6%)	3108 (23.4%)	−1.6

[a] Country unknown: 395 in 2014, 873 in 2019
[b] Australia, Canada, Ireland, New Zealand, UK and USA
Source See Fig. 7.3

focused Commissions (African Studies and Latin American and Caribbean Studies), in some physical geography Commissions and the Young and Early Career Task Force (Fig. 7.4). Dominance of Anglophone countries can be found in the Commissions on Applied Geography and on Dynamics of Economic Spaces, while physical geography Commissions usually have relatively few members from Anglophone countries (Fig. 7.5).

7.7 Women in the IGU

In the first decades of its existence, the IGU was a male-dominated organisation. As Rössler (1996) reports, only a few women participated in the IGCs of 1891 and 1904 (see also Close 1928). The list of participants of the Rome congress (1913) includes six women among several hundreds of male participants. "Women

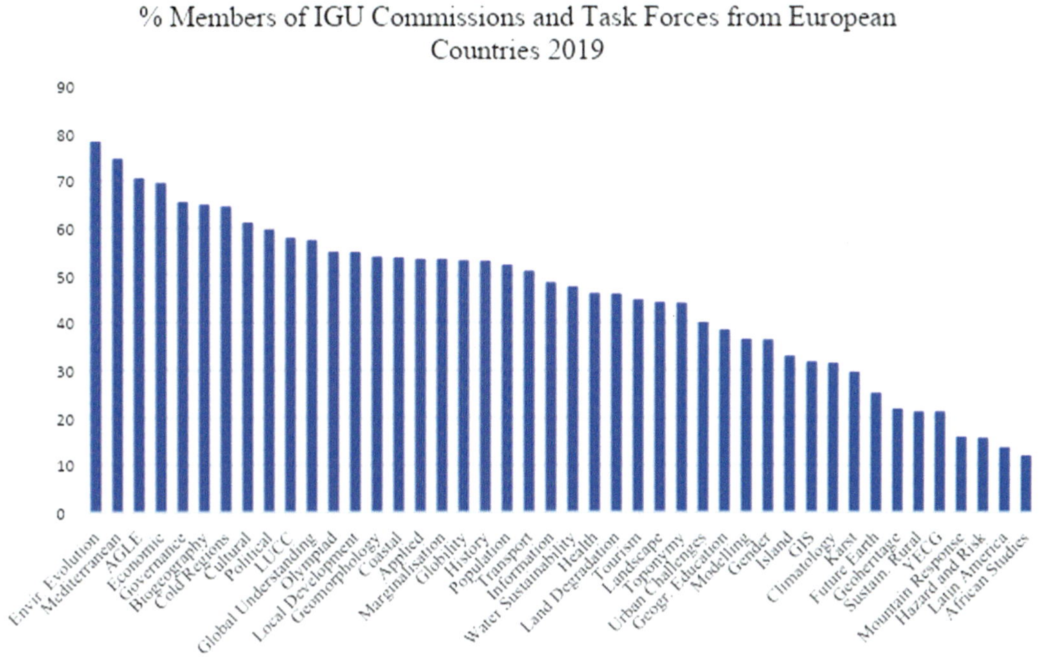

Fig. 7.4 Members of IGU Commissions and Task Forces from European countries 2019 (percentages). *Source* See Fig. 7.2

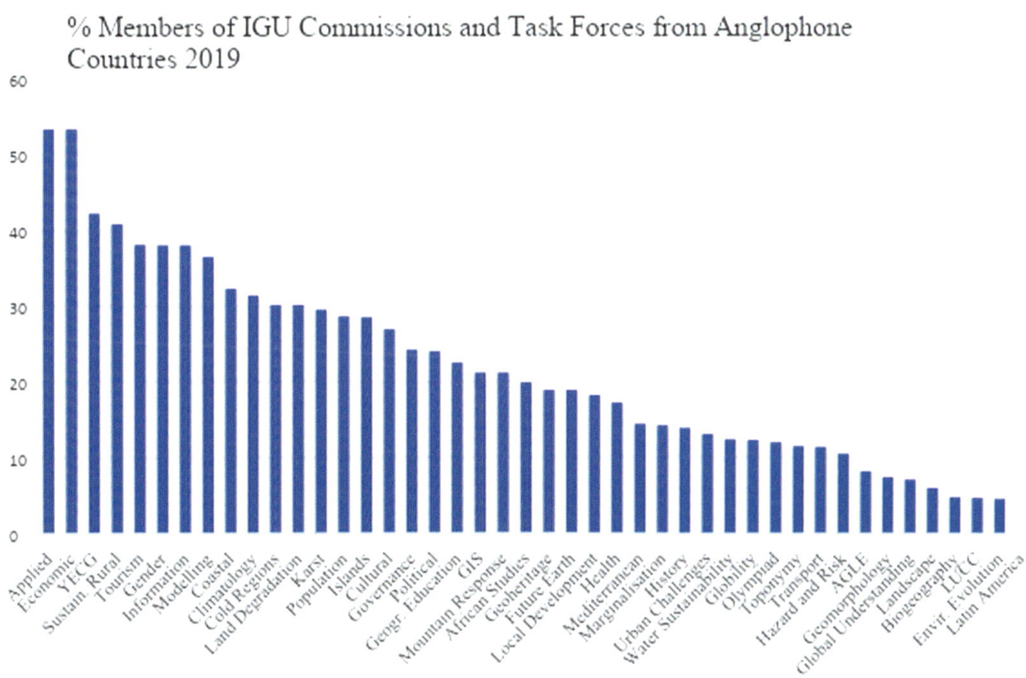

Fig. 7.5 Members of IGU Commissions and Task Forces from Anglophone countries 2019 (percentages). *Source* See Fig. 7.2

in sciences existed in two forms: outside the scientific institutions and universities or as wives, daughters or secretaries accompanying male scientists" (Rössler 1996, p. 259; see also Monk 2003) participating in the "ladies programme" of the congress. The people who took the initiative to establish the IGU were all men, several of them high-ranking members of the military.

Prior to 1984, the only woman member of the EC was Marguerite Lefèvre from Belgium who served as Secretary-General from 1938 to 1949 (BESTOR 2021). Lefèvre had a remarkable academic career, different from male geographers but rather typical for women geographers in early twentieth century (Droogleever Fortuijn 2019; Monk 2004). Before entering an academic career, she trained as a primary school teacher and was employed in that capacity when she became secretary to Professor Paul Michotte at the Catholic University of Louvain. She studied geography initially in Louvain and later in Liège where she was allowed, for the first time, to take courses in physical geography. After graduation, she attended geography courses at the Sorbonne in Paris and later completed her Ph.D. at the same institution. She returned to Louvain and was elected as IGU Secretary-General jointly with Michotte. In 1940, Paul Michotte, at that time Director of the Geographical Institute at the Catholic University of Louvain, died but Marguerite Lefèvre retained her position as IGU Secretary-General and later succeeded Michotte as Director of the Geographical Institute. In 1960, she became the first female full professor at Louvain. In 1984, Maria Gutiérrez de Macgregor from Mexico became the first woman Vice-President of the IGU, and from then onwards all ECs had one or two female members, with three women serving continuously in the 1996–2000. The current EC is the first to include four women.

The IGU had a total of 25 Presidents since the foundation in 1922, and only one of them was a woman: Anne Buttimer from Ireland, who was President in 2000–2004. Her academic career was exceptional: after graduation from University College in Cork, she became a nun in the Dominican order in Seattle and completed her PhD at Washington University in that city. She had a cosmopolitan career in Belgium, France, Canada, UK and Sweden before she returned to Ireland where she became full professor at University College Dublin in 1991 (Ferretti and Jones 2018).

The representation of women in the current steering committees of the Commissions and Task Forces is similar to the EC: one-third of the members are women. The percentage of women has increased since 2004 (Table 7.6). Steering committee members from Europe and Latin American are more often women than steering committee members from Asia, Africa, North America and Oceania (Table 7.7). Nine CTFs have a majority of women in the steering committee. The share of women is lower in physical than in human geography Commissions (Fig. 7.6).

Table 7.6 Steering Committee members of IGU Commissions and Task Forces by gender, 2004–2022

	Women	Men	All[a]
2004–2008	98 (25.0%)	294 (75.0%)	392 (100.0%)
2008–2012	102 (25.9%)	292 (74.1%)	394 (100.0%)
2012–2016	108 (27.1%)	290 (72.9%)	399 (100.0%)
2016–2020	141 (31.4%)	308 (68.6%)	449 (100.0%)
2020–2022	163 (34.6%)	308 (65.4%)	471 (100.0%)

[a] Missing data: see Table 7.4; gender unknown of 1 Steering Committee member in 2008–2012

Source See Table 7.3

Table 7.7 Steering Committee of IGU Commissions and Task Forces by gender and continent, 2020–2022

	Women	Men	All
Europe	83 (42.3%)	113 (57.7%)	196 (100.0%)
Asia	28 (25.2%)	83 (74.8%)	111 (100.0%)
Latin America	17 (38.6%)	27 (61.4%)	44 (100.0%)
North America	15 (31.3%)	33 (68.8%)	48 (100.0%)
Africa	13 (29.5%)	31 (70.5%)	44 (100.0%)
Oceania	7 (25.0%)	21 (75.0%)	28 (100.0%)
Total	163 (34.6%)	308 (65.4%)	471 (100.0%)
Anglophone[a]	35 (36.1%)	62 (63.9%)	97 (100.0%)

[a] Australia, Canada, Ireland, New Zealand, UK and USA
Source See Table 7.3

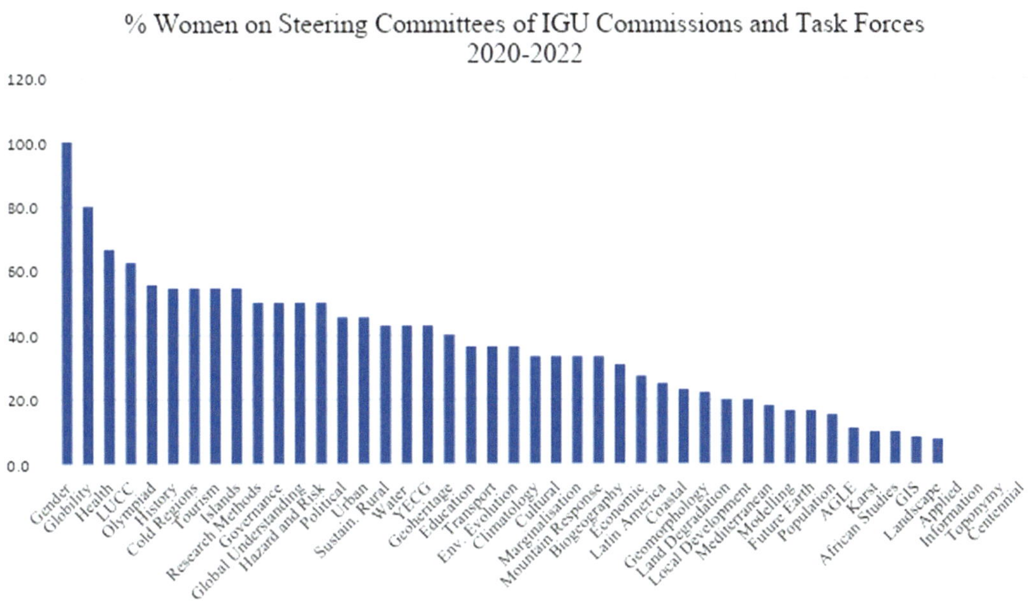

Fig. 7.6 Percentage of women in Steering Committees of IGU Commissions and Task Forces 2020–2022. *Source* See Table 7.3

7.8 Conclusions

The IGU was established a century ago as a predominantly European project. Six of the seven founding countries and five of the six members of the first EC were European. The central question addressed in this analysis of IGU membership is: how international was and is the IGU?

The IGU remained very much European dominated until the end of the Second World War. After 1949, the IGU became more and more a global community in which all continents were represented, although Europeans were still dominant until 1976. From 1976 onwards, Europeans formed a minority in the Executive Committee and in the 1976–2000 period the Anglophone countries from Europe, North America and Oceania became more prominent.

Since 2000, a shift has taken place from European and Anglophone countries to other continents, Asia in particular. These shifts can be observed in EC membership as well as in the steering committees of the CTFs between 2004 and 2022 and in the CTF membership between 2014 and 2019. At the same time, diversity grew with the increase in the share of women geographers on the EC and CTF steering committees. Women members, however, are more often from European countries than men who are more prominent among members from Asia and Africa.

At its centenary, the IGU has become, in the words of former President Vladimir Kolosov, "the only truly global organisation reuniting geographers from about 100 countries" (Kolosov 2014, p. 73). However, Europeans still form half of the CTF members and around 40% of the EC and CTF steering committees.

In terms of membership, the IGU has never experienced Anglophone dominance. The IGU can be characterised more as a European than an Anglophone organisation. However, according to Hodder et al. (2015, p. 3) "international … is almost invariably used as an adjective rather than a political concept worthy of interrogation in its own right". The IGU and the early IGCs started as a multilingual project in which geographers communicated in French, English, German, Italian, Spanish and Portuguese (Harris 2001). Nowadays, the IGU is formally bilingual—French and English—but English has become the main or only language of communication. As argued by Garcia-Ramon (2003), Gregson et al. (2003) and Van Meeteren (2019), language and academic traditions are interrelated. In line with Minca (2000), the IGU could consider initiating multilingual experiments—not only in English and French—and special sessions to discuss how diverse national geographies contribute to geographical knowledge and theories.

The IGU aims to strengthen its role as a global organisation with a diverse community of geographers in terms of geographical background, gender and age (Kolosov 2014). Women are still under-represented in IGU membership, especially from low-income countries (except Latin America). The challenge for the IGU in the coming years is to increase diversity in terms of geography as well as gender and age. The election of a woman as President (after 25 years) or as Secretary-General (after 86 years) would make progress in terms of gender diversity. The creation of instruments such as travel grants, paper awards and training workshops that are accessible and attractive for women, young generations and people from low-income countries can be important in making the IGU a truly global and inclusive community of geographers.

Acknowledgements I would like to thank Bruno Schelhaas for his support in compiling the list of members of the IGU Executive Committee 1922–2022 (Appendix) and Sjoerd de Vos for designing Fig. 7.2.

Appendix

List of IGU Executive Committee members 1922–2022

	Name	Country	Gender	
1922–1924				
President	Roland BONAPARTE	France	M	
Secretary-General	Charles CLOSE	UK	M	
First Vice-President	Nicola VACCHELLI	Italy	M	
Vice-President	Adrien DE GERLACHE	Belgium	M	
Vice-President	Severo GÓMEZ NUŇEZ	Spain	M	
Vice-President	Naomasa YAMASAKI	Japan	M	
1924–1928				
President	Nicola VACCHELLI	Italy	M	
Secretary-General	Charles CLOSE	UK	M	
First Vice-President	Robert BOURGEOIS	France	M	
Vice-President	Adrien DE GERLACHE	Belgium	M	
Vice-President	Severo GÓMEZ NUŇEZ	Spain	M	
Vice-President	Naomasa YAMASAKI	Japan	M	
Vice-President	Ahmed HASSANEIN BEY	Egypt	M	
Vice-President	Ernesto DE VASCONCELLOS	Portugal	M	
1928–1931				
President	Robert BOURGEOIS	France	M	
Secretary-General	Filippo DE FILLIPPI	Italy	M	
First Vice-President	Nicola VACCHELLI	Italy	M	
Vice-President	Charles CLOSE	UK	M	
Vice-President	Naomasa YAMASAKI	Japan	M	
Vice-President	Ahmed HASSANEIN BEY	Egypt	M	
Vice-President	Isaiah BOWMAN	USA	M	
Vice-President	Severo GÓMEZ NUŇEZ	Spain	M	
1931–1934				
President	Isaiah BOWMAN	USA	M	
Secretary-General	Emmanuel DE MARTONNE	France	M	
First Vice-President	Robert BOURGEOIS	France	M	
Vice-President	Severo GÓMEZ NUŇEZ	Spain	M	
Vice-President	Eugeniusz ROMER	Poland	M	
Vice-President	Nicola VACCHELLI	Italy	M	
Vice-President	Harold St. John Lloyd WINTERBOTHAM	UK	M	
1934–1936				
President	Charles CLOSE	UK	M	
Secretary-General	Emmanuel DE MARTONNE	France	M	
First Vice-President	Nicola VACCHELLI	Italy	M	

(continued)

	Name	Country	Gender	
Vice-President	Isaiah BOWMAN	USA	M	
Vice-President	Willem Everhard BOERMAN	Netherlands	M	
Vice-President	Giotto DAINELLI	Italy	M	
Vice-President	Ludwig MECKING	Germany	M	
Vice-President	Eugeniusz ROMER	Poland	M	
Vice-President	Harold St. John Lloyd WINTERBOTHAM	UK	M	
1936–1938				
President	Charles CLOSE	UK	M	
Secretary-General	Emmanuel DE MARTONNE	France	M	
First Vice-President	Isaiah BOWMAN	USA	M	
Vice-President	Willem Everhard BOERMAN	Netherlands	M	
Vice-President	Giotto DAINELLI	Italy	M	
Vice-President	Ludwig MECKING	Germany	M	
Vice-President	Eugeniusz ROMER	Poland	M	
Vice-President	Harold St. John Lloyd WINTERBOTHAM	UK	M	
1938–1940				
President	Emmanuel DE MARTONNE	France	M	
Secretary-General	Paul MICHOTTE/Marguerite Alice LEFÈVRE	Belgium	M	Michotte ł 1940
First Vice-President	Charles CLOSE	UK	M	
Vice-President	Willem Everhard BOERMAN	Netherlands	M	
Vice-President	Claude Hayle BIRDSEYE	USA	M	
Vice-President	Stanislaw PAWLOWSKY	Poland	M	ł 1940
Vice-President	Antonio Renato TONIOLO	Italy	M	
Vice-President	Ludwig MECKING	Germany	M	
1940–1946				
President	Emmanuel DE MARTONNE	France	M	
Secretary-General	Marguerite Alice LEFÈVRE	Belgium	F	
First Vice-President	Charles CLOSE	UK	M	
Vice-President	Claude Hayle BIRDSEYE	USA	M	ł 1941
Vice-President	Willem Everhard BOERMAN	Netherlands	M	
Vice-President	Ludwig MECKING	Germany	M	
Vice-President	Antonio Renato TONIOLO	Italy	M	ł in Second World War
1946–1949				
President	Emmanuel DE MARTONNE	France	M	
Secretary-General	Marguerite Alice LEFÈVRE	Belgium	F	
First Vice-President	Herbert John FLEURE	UK	M	

(continued)

	Name	Country	Gender	
Vice-President	Roberto ALMAGIÀ	Italy	M	
Vice-President	Willem Everhard BOERMAN	Netherlands	M	
Vice-President	George Babcock CRESSEY	USA	M	
Vice-President	Eugeniusz ROMER	Poland	M	
Vice-President	John K. WRIGHT	USA	M	
1949–1952				
President	George Babcock CRESSEY	USA	M	
Secretary-General	George Herbert Tinley KIMBLE	Canada, USA	M	
First Vice-President	Marguerite Alice LEFÈVRE	Belgium	F	
Vice-President	Hans BOESCH	Switzerland	M	
Vice-President	Christovam LEITE DE CASTRO	Brazil	M	
Vice-President	Orlando RIBEIRO	Portugal	M	
Vice-President	George KURIYAN	India	M	
Vice-President	Laurence Dudley STAMP	UK	M	
1952–1956				
President	Laurence Dudley STAMP	UK	M	
Secretary-General	George Herbert Tinley KIMBLE	Canada/USA	M	
First Vice-President	Orlando RIBEIRO	Portugal	M	
Vice-President	Hans AHLMAN	Norway	M	
Vice-President	Hans BOESCH	Switzerland	M	
Vice-President	George CRESSEY	USA	M	
Vice-President	George KURIYAN	India	M	
Vice-President	Maximilien SORRE	France	M	
Vice-President	Hilgard O'REILLY STERNBERG	Brazil	M	
1956–1960				
President	Hans AHLMAN	Sweden	M	
Secretary-General	Hans BOESCH	Switzerland	M	
First Vice-President	Hilgard O'REILLY STERNBERG	Brazil	M	
Vice-President	Hassan AWAD	Egypt	M	
Vice-President	Chauncy HARRIS	USA	M	
Vice-President	Maximilien SORRE	France	M	
Vice-President	Dudley STAMP	UK	M	
Vice-President	Fumio TADA	Japan	M	
Vice-President	Carl TROLL	Germany	M	
1960–1964				
President	Carl TROLL	Germany	M	
Secretary-General	Hans BOESCH	Switzerland	M	
First Vice-President	Hans AHLMAN	Sweden	M	
Vice-President	Hassan AWAD	Egypt	M	
Vice-President	Kenneth CUMBERLAND	New Zealand	M	

(continued)

	Name	Country	Gender	
Vice-President	Innokenti P. GERASSIMOV	USSR	M	
Vice-President	Chauncy HARRIS	USA	M	
Vice-President	Pierre MONBEIG	France	M	
Vice-President	Fumio TADA	Japan	M	
1964–1968				
President	Shiba CHATTERJEE	India	M	
Secretary-General	Hans BOESCH	Switzerland	M	
First Vice-President	Innokenti P. GERASSIMOV	USSR	M	
Vice-President	Kenneth CUMBERLAND	New Zealand	M	
Vice-President	Arch GERLACH	USA	M	
Vice-President	Ferdinando GRIBAUDI	Italy	M	
Vice-President	Stanislaw LESZCZYCKI	Poland	M	
Vice-President	Pierre MONBEIG	France	M	
Vice-President	Carl TROLL	Germany	M	
1968–1972				
President	Stanislaw LESZCYCKI	Poland	M	
Secretary-General	Chauncy HARRIS	USA	M	
First Vice-President	Ferdinando GRIBAUDI	Italy	M	
Vice-President	Shiba CHATTERJEE	India	M	
Vice-President	Jean DRESCH	France	M	
Vice-President	Torsten HÄGERSTRAND	Sweden	M	
Vice-President	Stanislav KALESNIK	USSR	M	
Vice-President	Michael John WISE	UK	M	
Vice-President	Mariano ZAMORANO	Argentina	M	
1972–1976				
President	Jean DRESCH	France	M	
Secretary-General	Chauncy HARRIS	USA	M	
First Vice-President	Michael John WISE	UK	M	
Vice-President	Feofan Farneevich DAVITAYA	USSR (Georgia)	M	
Vice-President	Torsten HÄGERSTRAND	Sweden	M	
Vice-President	Shinzo KIUCHI	Japan	M	
Vice-President	Stanislaw LESZCZYCKI	Poland	M	
Vice-President	Akin MABOGUNJE	Nigeria	M	
Vice-President	Mariano ZAMORANO	Argentina	M	
1976–1980				
President	Michael John WISE	UK	M	
Secretary-General	Walther MANSHARD	Germany	M	
First Vice-President	Akin MABOGUNJE	Nigeria	M	
Vice-President	Feofan Farneevich DAVITAYA	USSR (Georgia)	M	

(continued)

	Name	Country	Gender	
Vice-President	Jean DRESCH	France	M	
Vice-President	Speridião FAISSOL	Brazil	M	
Vice-President	Shinzo KIUCHI	Japan	M	
Vice-President	Jerzy KOSTROWICKI	Poland	M	
Vice-President	John ROSS MACKAY	Canada	M	
Vice-President	Peter SCOTT	Australia	M	
1980–1984				
President	Akin MABOGUNJE	Nigeria	M	
Secretary-General	Walther MANSHARD	Germany	M	
First Vice-President	Peter SCOTT	Australia	M	
Vice-President	Speridião FAISSOL	Brazil	M	
Vice-President	Roland FUCHS	USA	M	
Vice-President	Jerzy KOSTROWICKI	Poland	M	
Vice-President	John ROSS MACKAY	Canada	M	
Vice-President	Ooi Jin BEE	Singapore	M	
Vice-President	Juan VILÁ-VALENTÌ	Spain	M	
Vice-President	Michael John WISE	UK	M	
1984–1988				
President	Peter SCOTT	Australia	M	
Secretary-General	Leszek KOSIŃSKI	Canada	M	
Past President	Akin MABOGUNJE	Nigeria	M	
First Vice-President	Roland FUCHS	USA	M	
Vice-President	György ENYEDI	Hungary	M	
Vice-President	María GUTIÉRREZ DE MACGREGOR	Mexico	F	
Vice-President	Ooi Jin BEE	Singapore	M	
Vice-President	Mohammad SHAFI	India	M	
Vice-President	Herman VERSTAPPEN	Netherlands	M	
Vice-President	Juan VILÁ-VALENTÌ	Spain	M	
1988–1992				
President	Roland FUCHS	Japan	M	
Secretary-General	Leszek KOSIŃSKI	Canada	M	
Past President	Peter SCOTT	Australia	M	
First Vice-President	Herman VERSTAPPEN	Netherlands	M	
Vice-President	Etienne DALMASSO	France	M	
Vice-President	György ENYEDI	Hungary	M	
Vice-President	María GUTIÉRREZ DE MACGREGOR	Mexico	F	
Vice-President	Vladimir KOTLYAKOV	USSR	M	
Vice-President	Mohammad SHAFI	India	M	

(continued)

	Name	Country	Gender	
Vice-President	Chuan-jun WU	China Beijing	M	
1992–1996				
President	Herman VERSTAPPEN	Netherlands	M	
Secretary-General	Eckart EHLERS	Germany	M	
Past President	Roland FUCHS	USA	M	
First Vice-President	Vladimir KOTLYAKOV	Russia	M	
Vice-President	Folasade IYUN	Nigeria	F	
Vice-President	Bruno MESSERLI	Switzerland	M	
Vice-President	Alain METTON	France	M	
Vice-President	Warren MORAN	New Zealand	M	
Vice-President	Chuan-jun WU	China Beijing	M	
Vice-President	Masatoshi YOSHINO	Japan	M	
1996–2000				
President	Bruno MESSERLI	Switzerland	M	
Secretary-General	Eckart EHLERS	Germany	M	
Past President	Herman VERSTAPPEN	Netherlands	M	
Vice-President	Warren MORAN	New Zealand	M	
Vice-President	Ronald ABLER	USA	M	
Vice-President	Bertha BECKER	Brazil	F	
Vice-President	Anne BUTTIMER	Ireland	F	
Vice-President	Folasade IYUN	Nigeria	F	
Vice-President	Chan LEE	South Korea	M	
Vice-President	Adalberto VALLEGA	Italy	M	
2000–2004				
President	Anne BUTTIMER	Ireland	F	
Secretary-General	Ronald ABLER	USA	M	
Past President	Bruno MESSERLI	Switzerland	M	
First Vice-President	Adalberto VALLEGA	Italy	M	
Vice-President	Nikita GLAZOVSKY	Russia	M	
Vice-President	Changming LIU	China Beijing	M	
Vice-President	Markku LÖYTÖNEN	Finland	M	
Vice-President	Lindisizwe MAGI	South Africa	M	
Vice-President	José Luis PALACIO-PRIETO	Mexico	M	
Vice-President	Hiroshi TANABE	Japan	M	
2004–2006				
President	Adalberto VALLEGA	Italy	M	Ɨ 2006
Secretary-General	Ronald ABLER	USA	M	
Past President	Anne BUTTIMER	Ireland	F	

(continued)

	Name	Country	Gender	
First Vice-President	José Luis PALACIO-PRIETO	Mexico	M	
Vice-President	Nikita GLAZOVSKY	Russia	M	
Vice-President	Changming LIU	China Beijing	M	
Vice-President	Markku LÖYTÖNEN	Finland	M	
Vice-President	Lindisizwe MAGI	South Africa	M	
Vice-President	Hiroshi TANABE	Japan	M	
Vice-President	Woo-ik YU	South Korea	M	
2006–2008				
President	José Luis PALACIO-PRIETO	Mexico	M	
Secretary-General	Ronald ABLER	USA	M	
Past President	Anne BUTTIMER	Ireland	F	
First Vice-President	Vladimir KOLOSOV	Russia	M	
Vice-President	Changming LIU	China Beijing	M	
Vice-President	Markku LÖYTÖNEN	Finland	M	
Vice-President	Lindisizwe MAGI	South Africa	M	
Vice-President	Hiroshi TANABE	Japan	M	
Vice-President	Woo-ik YU	South Korea	M	
2008–2010				
President	Ronald ABLER	USA	M	
Secretary-General	Woo-ik YU	South Korea	M	
First Vice-President	Vladimir KOLOSOV	Russia	M	
Vice-President	Irasema ALCÁNTARA AYALA	Mexico	F	
Vice-President	Giuliano BELLEZZA	Italy	M	
Vice-President	Ruth FINCHER	Australia	F	
Vice-President	Yukio HIMIYAMA	Japan	M	
Vice-President	Aharon KELLERMAN	Israel	M	
Vice-President	Markku LÖYTÖNEN	Finland	M	
Vice-President	Dahe QIN	China Beijing	M	
Vice-President	Dietrich SOYEZ	Germany	M	
2010–2012				
President	Ronald ABLER	USA	M	
Secretary-General	Michael MEADOWS	South Africa	M	
First Vice-President	Vladimir KOLOSOV	Russia	M	
Vice-President	Giuliano BELLEZZA	Italy	M	
Vice-President	Aharon KELLERMAN	Israel	M	
Vice-President	Ruth FINCHER	Australia	F	
Vice-President	Irasema ALCÁNTARA AYALA	Mexico	F	
Vice-President	Dietrich SOYEZ	Germany	M	

(continued)

	Name	Country	Gender	
Vice-President	Yukio HIMIYAMA	Japan	M	
Vice-President	Dahe QIN	China Beijing	M	
Vice-President	Markku LÖYTÖNEN	Finland	M	
2012–2014				
President	Vladimir KOLOSOV	Russia	M	
Secretary-General	Michael MEADOWS	South Africa	M	
Past President	Ronald ABLER	USA	M	
First Vice-President	Dietrich SOYEZ	Germany	M	
Vice-President	Giuliano BELLEZZA	Italy	M	
Vice-President	Yukio HIMIYAMA	Japan	M	
Vice-President	Joos DROOGLEEVER FORTUIJN	Netherlands	F	
Vice-President	Aharon KELLERMAN	Israel	M	
Vice-President	Qin DAHE	China Beijing	M	
Vice-President	RB SINGH	India	M	
Vice-President	Jarkko SAARINEN	Finland	M	
2014–2016				
President	Vladimir KOLOSOV	Russia	M	
Secretary-General	Michael MEADOWS	South Africa	M	
Past President	Ronald ABLER	USA	M	
First Vice-President	Dietrich SOYEZ	Germany	M	
Vice-President	Joos DROOGLEEVER FORTUIJN	Netherlands	F	
Vice-President	Jarkko SAARINEN	Finland	M	
Vice-President	Elena DELL’AGNESE	Italy	F	
Vice-President	Chenggu ZHOU	China Beijing	M	
Vice-President	Iain HAY	Australia	M	
Vice-President	RB SINGH	India	M	
Vice-President	Yukio HIMIYAMA	Japan	M	
2016–2018				
President	Yukio HIMIYAMA	Japan	M	
Secretary-General	Michael MEADOWS	South Africa	M	
Past President	Vladimir KOLOSOV	Russia	M	
First Vice-President	Joos DROOGLEEVER FORTUIJN	Netherlands	F	
Vice-President	Chenggu ZHOU	China Beijing	M	
Vice-President	Elena DELL’AGNESE	Italy	F	
Vice-President	Nathalie LEMARCHAND	France	F	
Vice-President	Iain HAY	Australia	M	
Vice-President	RB SINGH	India	M	

(continued)

	Name	Country	Gender	
Vice-President	Barbaros GÖNENÇGIL	Turkey	M	
Vice-President	Rémy TREMBLAY	Canada	M	
2018–2020				
President	Yukio HIMIYAMA	Japan	M	
Secretary-General	RB SINGH	India	M	
Past President	Vladimir KOLOSOV	Russia	M	
First Vice-President	Joos DROOGLEEVER FORTUIJN	Netherlands	F	
Vice-President	Elena DELL'AGNESE	Italy	F	
Vice-President	Nathalie LEMARCHAND	France	F	
Vice-President	Iain HAY	Australia	M	
Vice-President	Barbaros GÖNENÇGIL	Turkey	M	
Vice-President	Bojie FU	China Beijing	M	
Vice-President	Rubén Camilo LOIS GONZÁLEZ	Spain	M	
2020–2021				
President	Michael MEADOWS	South Africa	M	
Secretary-General	RB SINGH	India	M	ƚ 2021
Past President	Yukio HIMIYAMA	Japan	M	
First Vice-President	Iain HAY	Australia	M	
Vice-President	Nathalie LEMARCHAND	France	F	
Vice-President	Elena DELL'AGNESE	Italy	F	
Vice-President	Barbaros GÖNENÇGIL	Turkey	M	
Vice-President	Bojie FU	China Beijing	M	
Vice-President	Rubén Camilo LOIS GONZÁLEZ	Spain	M	
Vice-President	Holly BARCOS	USA	F	
Vice-President	Céline ROZENBLAT	Switzerland	F	
2021–2022				
President	Michael MEADOWS	South Africa	M	
Interim Secretary-General	Barbaros GÖNENÇGIL	Turkey	M	
Past President	Yukio HIMIYAMA	Japan	M	
First Vice-President	Iain HAY	Australia	M	
Vice-President	Nathalie LEMARCHAND	France	F	
Vice-President	Elena DELL'AGNESE	Italy	F	
Vice-President	Bojie FU	China Beijing	M	
Vice-President	Rubén Camilo LOIS GONZÁLEZ	Spain	M	
Vice-President	Holly BARCOS	USA	F	
Vice-President	Céline ROZENBLAT	Switzerland	F	

References

Aalbers MB (2004) Creative destruction through the Anglo-American hegemony: a non-Anglo-American view on publications, referees and language. Area 36 (3):319–322

Banski J, Ferenc M (2013) "International" or "Anglo-American" journals of geography? Geoforum 45:285–295

BESTOR (2021) Lefèvre, Marguerite (1894–1967). https://www.bestor.be/wiki_nl/index.php/Lef%C3%A8vre,_Marguerite_(1894-1967). Accessed on 6 March 2021

Birdseye CH (1939) The fifteenth international Geographical Congress at Amsterdam, July 1938. Ann Assoc Am Geogr 29(1):62–63

Close C (1928) Address at the anniversary meeting of the Society held on 18 June 1928. Geogr J 72(2):97–116

De Rudder B, Liu X (2016) How international is the annual meeting of the Association of American Geographers? A social network analysis perspective. Environ Plann A Econ Space 48(2):309–329

Do Amaral I (1968) A Geografia através dos seus congressos internacionals. Finisterra 5:84–101

Droogleever Fortuijn J (2019) Women geographers and the International Geographical Union. Boletín De La Asociación De Geógrafos Españoles 81(2819):1–16

Ferretti F, Jones A (2018) Anne Buttimer. 1938–2017. Geogr Biobibliographical Stud 37(37):13–40

Garcia-Ramon MD (2003) Globalization and international geography: the questions of languages and scholarly traditions. Prog Hum Geogr 27(1):1–5

Gregson N, Simonsen K, Vaiou D (2003) Writing (across) Europe: on writing spaces and writing practices. Eur Urban Reg Stud 10(1):5–22

Harris C (2001) English as international language in geography: development and limitations. Geogr Rev 91(4):675–689

Hinks AR [ARH] (1922) The International Geographical Union. Geogr J 60(4):291–294

Hodder J, Legg S, Heffernan (2015) Introduction: historical geographies of internationalism,1900–1950. Polit Geogr 49:1–6

Kish G (1992) International Geographical Union: a brief history. GeoJournal 26(2):224–228

Kolosov V (2014) The International Geographical Union before its centennial: new challenges and developments. J-Reading J Res Didactics Geogr 1(3):73–79

Meadows M, Dietz T, Vandermotten C (2016) A perspective on problems and prospects for academic publishing in geography. Geo Geogr Environ 3(1): e00016

Minca C (2000) Venetian geographical praxis. Environ Plann D Soc Space 18(3):285–289

Monk J (2003) Women's worlds at the American Geographical Society. Geogr Rev 93(2):237–257

Monk J (2004) Women, gender and the histories of American geography. Ann Assoc Am Geogr 94(1):1–22

Paasi A (2015) Academic capitalism and the geopolitics of knowledge. In: Agnew JA et al (eds) Companion to political geography. Wiley Blackwell, Hoboken, NJ, pp 509–523

Robic MC (1996) La naissance de l'Union Géographique Internationale. In: Robic MC, Briend AM, Rössler M (eds) Géographes face au monde. L'Union Géographique Internationale et les Congrès Internationaux de Géographie. Histoire de Sciences Humaines, Paris, pp 23–39

Rodriguez-Pose A (2006) Is there an 'Anglo-American' domination in human geography? And, is it bad? Environ Plann A Econ Space 38(4):603–610

Rössler M (1996) From the ladies' program, to the feminist session. In: Robic MC, Briend AM, Rössler M (eds) Géographes face au monde. L'Union Géographique Internationale et les Congrès Internationaux de Géographie. Histoire de Sciences Humaines, Paris, pp 259–267

Samers M, Sidaway J (2000) Exclusions, inclusions, and occlusions in 'Anglo-American Geography': reflections on Minca's "Venetian geographical praxis." Environ Plann D Soc Space 18:663–666

Simonsen K (2003) Global-local ambivalence. Trans Inst Br Geogr 27(4):391–394

Timar J (2004) More than 'Anglo-American', it is 'Western': hegemony from a Hungarian perspective. Geoforum 35(5):533–538

Van Meeteren M (2019) On geography's skewed transnationalization, anglophone hegemony, and qualified optimism toward an engaged pluralist future: a reply to Hassink, Gong and Marques. Int J Urban Sci 23 (2):181–190

Volle D (1996) La carte des états: vers le couverture du monde? In: Robic MC, Briend AM, Rössler M (eds) Géographes face au monde. L'Union Géographique Internationale et les Congrès Internationaux de Géographie. Histoire de Sciences Humaines, Paris, pp 41–62

Joos Droogleever Fortuijn was associate professor and head of the Department of Geography, Planning and International Development Studies of the University of Amsterdam. She published on gender, ageing, urban geography, rural geography and geography in higher education. She was steering committee member and chair of the IGU Commission on Gender and Geography and Vice-President and First Vice-President of the IGU.

Part II

The Challenges of International Geography

The Role of Internet and Open Access Journals in the Internationalisation of Geography

8

Denise Pumain and Christine Kosmopoulos

Abstract

The IGU took advantage of the Internet for collecting and promoting geographical information all over the world and developing its international networks. The use of remote sensing and the development of computerised models in spatial analysis preceded in dedicated IGU Commissions the explosion of georeferenced information generated through GPS and Internet. This evolution of the last three decades deeply modified the conceptions and practices among geographers and greatly amplified the internationalisation of geography to which IGU is dedicated over one century of existence. In order to reduce existing inequalities in the access to geographical scholarship and especially digital divide, open scientific journals may bring new opportunities for sharing geographical resources. Combined with initiatives of the many scholars involved in a diversity of IGU commissions, the development of an open geographical science is a promising avenue that will increase synergies and ethical practices for addressing the ambitious challenges in the future of IGU.

Keywords

Internet · Geographical information · Geocomputation · Global thinking · Open access journal · Open geographical science

8.1 Introduction

The introduction of the Internet in the 1990s is equivalent in terms of disruptive innovation to the spread of movable type and printing press, as early as the eleventh century in China, in the fifteenth century in Europe. The Internet would be the fourth "cognitive revolution" in the history of humankind (after the emergence of language, the advent of writing and the so-called Gutenberg galaxy) (Harnad 1991). Even if thirty years later less than 60% of the world's population has access to the Internet, the digital communication network has brought about considerable qualitative changes, directly and indirectly, in many human activities and in all parts of the world. In science, the Internet has not only facilitated access to publications, while contributing to their

D. Pumain (✉)
Emeritus Professor, University Paris I Panthéon-Sorbonne, Paris, France
e-mail: pumain@parisgeo.cnrs.fr

D. Pumain
CNRS Laboratory, Géographie-cités, Aubervilliers, France

C. Kosmopoulos
Chief Editor, Cybergeo, European Journal of Geography, CNRS Laboratory, Géographie-cités, Aubervilliers, France
e-mail: ck@parisgeo.cnrs.fr

V. Kolosov et al. (eds.), *A Geographical Century*,
https://doi.org/10.1007/978-3-031-05419-8_8

exponential multiplication, but introduced subtle epistemological changes in the interactions between theory and practices. In less than one generation, the Internet launched new sources of data, new methods and new concepts, for displaying, collecting and handling information, in repeated surges of innovation. Geography is concerned in a very particular way by these upheavals (see Chap. 16). The influx of georeferenced data and the speed of their dissemination with the smartphone technology and algorithmic applications have placed the discipline in its broadest acceptance at the heart of global information sharing activities. More significantly, geography became commonplace by widely disseminating its basic mapped productions in all continents, and at the same time required more and more high-level skills to extract meaningful knowledge from the massive quantity of newly collected data. Without being able to deal with this immense subject in this brief chapter, we give a few examples of the changes brought about by the Internet in the practices of geography and in its internationalisation, by detailing a little more the question of open access journals.

8.2 Internet and the Triumph of Geography: How Geolocation of Information Accelerates the Trends Towards Internationalisation

The growth of the Internet as a global network is still in its exponential stage. Some 4.7 billion people are connected to the web in 2021, while there were less than one billion in 2005. The rapid diffusion of smartphones since 2000 accelerated the connections, even in less wealthy regions of the world such as Africa. However, Africa remains well behind the global level with only 12% of its population connected to the Internet. The number of websites was close to one billion in 2014 and is above 1.8 billion in 2021.[1] The proliferation and multiple uses of the so-called social networks online added to the overwhelming diffusion of the innovation.

Three major trends in digital economy, sometimes interpreted as "empirical laws" characterise this extraordinary development: "Moore's law", enunciated in 1965, describes the exponential growth of power and decrease of costs for computer components; "Metcalfe's law", from 1980, measures the value of a telecommunication network as the square of the number of connected terminals; and "Reeds' law", in 2000, reveals the added value growth through networks of collaborative and cooperative work (Reed 2004). The nonlinear amplification of the value of these infrastructures and the diversity of their uses led towards an oligopolistic configuration of very large groups at their head. Combined with the previously engaged trends in globalisation of trade and expansion of multinational firms (Thrift and Leyshon 1994; Agnew 2009), this gave rise to the exceptional concentration of the financial value and power to innovate among a few groups such as GAFAM (Google, Amazon, Facebook, Apple and Microsoft) in USA and BHATX (Baidu, Huawei, Alibaba, Tencent and Xiaomi) in China. Geographers participate in this evolution because geolocation is among the main technical devices implicit in networking digital activities and provides an enormous quantity and diversity of new data for them. They develop new concepts and methods for handling digitised web traces and also provide reflexive views about the emergence of a Cyberspace or Geospatial web that challenge the geopolitical economy as well as generating more diverse networks at all scales.

8.2.1 Internet and the Digital Revolution: Three Steps for Geography

Although computerisation per se is not directly responsible for the internationalisation of geography and the related epistemological changes, it certainly opened the way to a broader diffusion of new ideas and technical skills between

[1] https://www.internetlivestats.com/total-number-of-websites/.

continents. During the period 1960–1990, the use of computers enabled geography to draw full benefits from concepts and methods that had already been invented sometimes decades earlier and promoted under the expression of theoretical and quantitative geography (Tobler 1959; Burton 1963; Haggett 1990; Unwin 1999; Sheppard 2001). While in the nineteenth century the IGU's precursor International Geographical Congresses developed commissions on "mathematical geography" with reference to topographical surveys and map projections, from the 1970s several IGU commissions have focused on quantitative geography and mathematical modelling across physical and human geography (Yeates 1974).[2] Besides GIS (Longley et al. 2005), other IGU Commissions using remote sensing data, such as the Commission on Land Cover and Land Use, also had an uninterrupted longevity.

The digitisation of geographic information preceded by at least three decades the arrival of the Internet. However, its widespread use was relatively slow, limited by a lack of equipment or skills among users. Shifting from printed maps towards digital databases, with new tools for processing and visualising information, took about a single generation. Digital data were at first extensively used for research and regional planning through remote sensing. After the launch of the first Sputnik satellite in 1957 by the Soviet Union, and the production of images for military or meteorological purposes, the National Aeronautics and Space Administration (NASA) launched satellites for earth observation in the 1970s. The Landsat program was initially intended for mapping, for geology and agronomy. Subsequently, Soviet (Resurs), French (SPOT) and Indian (IRS) satellites developed their uses in the 1980s and China followed after 2000 (Yaogan, Gaofen). For many years, remote sensing work has been focused on identifying observations between satellite signal and field and negotiating the interpretation of land use and land cover data for designing harmonised international nomenclatures. However, it was the Internet that triggered the development of geographic data infrastructures in the 1990s. In 1992, the Federal Geographic Data Committee in the USA published its standards, and in 1994, those of the ISO/TC 211 technical committee "Geographic Information/Geomatics". The Open Geospatial Consortium (OGC) created in 1994 to ensure the interoperability of geographic information systems now brings together 500 government agencies, companies and universities to negotiate standards for geomatics. Today, the Global Geospatial Information Management (UN-GGIM, established in 2011) and the United Nations Geospatial Information Section provide cartographic products for the United Nations Secretariat. The group devoted to earth observation (GEO) includes a structured partnership of more than 100 member states and 100 participating international and non-governmental organisations that coordinate the development of a Global Earth Observation System of Systems (GEOSS) (Giuliani et al. 2017). Between 2007 and 2017, the European Union established a comprehensive project to standardise national and regional map repositories with the implementation of the INSPIRE Directive. This first step summarises two essential characteristics that deeply transformed the geographer's work during these decades: maps become digital databases, potentially dynamic, and these data can more easily expand at transnational scales.

A second step in this evolution started in the 2000s and is clearly linked to the web 2.0 whose technology facilitates the contribution and exchanges of information. The civil use of the Global Positioning System (GPS) initiated in 1978 has considerably developed the production and use of georeferenced information including by laymen.[3] The applications Google Maps launched in 2004 and Google Earth in 2005 were described by Goodchild (2007, 213) as "democratization of GIS". The Beidou system used

[2]The longevity of this quantitative geography IGU commission is remarkable, though it has operated under slightly variable names, i.e. Mathematical Modelling Group and nowadays Modelling Geographical Systems.

[3]Since then, other geolocation systems have been launched by Russia (Glonass), Europe (Galileo) and China (Beidou).

in China is expanding similar applications in neighbouring countries. The IGU took its part in that history by developing in the 2000s a Commission on Geographical Information Science that is now co-chaired from China and USA. Integrating data from different sources or adding information on web maps (according to a cartographic mashup) enabled to considerably diversify the visualisation tools used by geographers and enlarge their abilities to create new types of representation (Cöltekin et al 2018; Cura 2019). Very quickly the diffusion of smartphones equipped with GPS enabled people to contribute to topographic mapping, giving birth to a "volunteered geography" that is "a special case of the more general Web phenomenon of user-generated content" (Goodchild 2007, p. 212). The success of an international operation such as OpenStreetMap (Bennett 2010; Mericksay and Roche 2011; Burini 2016; Duféal and Noucher 2017) is remarkable in a field previously confined to specialised national institutions. The difficulty of producing harmonised maps is well known at IGU where the project aiming at the 1:1,000,000 mapping of the planet decided at the end of nineteenth century was hardly half achieved by the Second World War (Pinchemel 1972; Robic et al. 1996; Pearson and Heffernan 2015). Such a project can now be done at all scales with a high precision and updated any time. The enlarged participation of the public to geographic work also is another consequence of the multiple uses of the Internet (Lopez and Olvera-Lobo 2018).

The third step fully developing the potentialities of computerised geography was anticipated by a visionary pioneer such as Tobler (1959). He early detailed the potential of computers to process geographic information in its spatial and temporal dimensions, to produce animated and interactive maps and explore dynamic models through simulations. Thanks to the Internet, it is now possible to monitor crisis situations in real time in almost all parts of the world, to anticipate the effects of disasters through simulations (as well illustrated from the IGU Commission on Hazards and Risks), to anticipate the possible consequences of planning decisions or to produce interactive multiscale atlases. There are numerous examples of these, either dedicated to mapping forests in Cameroon (Mertens et al. 2012), explaining results of votes, displaying the internal social structure of cities (https://atlas.brussels/), their daily mobility (https://mobiliscope.cnrs.fr/) or migrations in India linked with the Covid19 pandemic (Denis et al. 2020), to quote only a few… It is likely that people's spatial cognition (Couclelis et al. 1987) is undergoing upheavals induced by these new instruments. A new "geography of software" is emerging (Thrift and French 2002). Georeferencing has generated and organised a new global information space, the Geospatialweb or Geoweb (Joliveau 2011). The Geoweb is built on a multiform material infrastructure, and completed by semantics it produces a new kind of geographical information, consciously by many persons on websites, blogs, social networks or platforms or passively collected from the traces they leave on fixed or mobile sensors (Goodchild 2007). Not only physical movements of individuals can be precisely mapped and located but a wide range of individual and social behaviours, activities or opinions that before required tedious survey's work now become accessible for building more geographical knowledge. New geographies are emerging from the exploratory analysis of deep learning tools of social networks messages, such as Weibo (Liu and Wang 2015) or Twitter (Takhteyev et al. 2012). The technologies of web scraping extract huge volumes of georeferenced information that require non classical processing techniques for handling these *big data*. Indeed, methods for using the Internet in surveys already raised bestseller books (for instance Dillman (2011) for designing surveys or Mitchell (2018) for web scraping). Related ethical problems are emerging: "The right of everyone to access geospatial data, but also the right of everyone to preserve the confidentiality of the traces they leave on the Geoweb, are becoming major themes in geospatial data management" (Joliveau 2011, p. 162).

8.2.2 Managing Big Data Towards Multipolar Global Thinking in Geography

The "avalanche" or "deluge" of georeferenced Big Data has certainly encouraged not only geographers but also many other scientific disciplines, including the hard sciences, to take an interest in classical geographical questions. The contributions of these new participations to the geographical knowledge and to a greater international networking of geography are very unequal. A new light has been thrown on spatial interactions at the micro-geographic (individual) level that was little documented before. A good example was provided by Ratti et al. (2010) mapping the network of telephone exchanges in Great Britain and finding a configuration that is very close to the delimitation of administrative regions. A country such as Estonia, previously little visible in international geography, was able to come to the forefront, because its cell phone operators allowed access to personal data related to individual travel that is usually difficult to obtain (Calabrese et al. 2010). However, the first uses of GPS-tracked mobility have often only "rediscovered" basic knowledge about spatial interaction patterns that were already well established in geography (González et al. 2008).

As Goodchild (2016) points out, the novelties that remain a challenge for geographers are not so much the volume of these data, nor their variety, but their velocity and undoubtedly also the veracity or validity that it is possible to attribute to them. As far as velocity is concerned, "real-time" applications have been designed, for example, to locate crowds gathered for major events using cell phone traces (Lucchini et al. 2011), to analyse the pulses of urban life (Fen-Chong 2012; Pucci et al. 2015) or to quickly restore maps after catastrophic accidents—the 2010 earthquake in Haiti was one of the first examples. Combining "more than 24 million satellite photos, two petabytes of data, and 2 million hours of CPU processing time to create a 4.4-terapixel interactive view", the recent "Timelapse" application of Google Earth showing changes in landscapes over a 30 years period until 2020 is advertised as "a powerful and interactive way [for] a closer understanding of the kind of impact humans and natural phenomena have on our planet".[4] Observations of the 2020 pandemic, however, have shown how difficult it is yet to establish data quality and comparability in order to predict or even analyse and understand its paths (Boulos and Geraghty 2020).

The United Nations agencies have taken on the task of producing all types of data that must be used to solve major global issues, including development inequalities and environmental problems. Since 2015, the Sustainable Development Goals (SDG) are the global reference framework for most international, national and local policies. The SDGs are a set of 17 goals (broken down into 169 targets). The SDG is currently referred to by most of the IGU Commissions in their yearly reports. The integration of localised databases is conceived to produce and continuously update global indicators for sustainable development (Rohat et al 2018). After the pioneer attempts to escape the national boundaries in designing a global grid map of population (Tobler et al. 1997), a new challenge is to disaggregate global representations at infra-national scale and even to integrate sporadic information to deal with socio-spatial transformation associated with climatic change or environmental problems. A recent IGU book explores new possible contributions to sustainable development goals (Sikka et al. 2021). The use of big data will be a great help for entering this new way of displaying and processing geographical information.

A very promising exploitation of big data in geography is that of synthetic data generated by simulation models, which allow the validation of agent-based models used in the social sciences as a substitute for experimentation. Big data for observation and synthesis thus become complementary instruments for the advancement of geographic science. In the field of urban studies, a new "science of cities" has emerged (Batty

[4]https://gizmodo.com/google-earths-new-3d-time-lapse-feature-shows-how-human-1846685339.

2013). This seminal book referred to advances in the fields of geocomputation and geodemography for the morphological and societal urban analysis, including fractals and networks on the quantitative side and concentration of stakeholders among methods for planning design. Less than one decade later, very significant advances are reported in urban modelling methods, including debates about scaling laws (Cottineau et al. 2017; Arcaute and Hatna 2020; Finance and Swerts 2020) and the validation of simulations through genetic algorithms and distributed computing (Pumain and Reuillon 2017; Pumain 2020). Huge progress has been made in measuring network effects and co-evolution regimes in the dynamics of urban systems (Raimbault 2020).

Such advances in the theories and models of geography are now conceptualised in fully international and interdisciplinary research environments. Compared to previous national histories of geography, the geographers' ambition to "cover the world" (Robic et al. 2006) is now achieved in international teams. Among recent books about trends in world's urbanisation one may quote Rozenblat et al. (2018) that was published in a series directed by R. B. Singh from IGU in collaboration with the UN Habitat Division. The content relies on global data bases and the contributions bring together geographers and urban scientists from all regions of the world in chapters designed for international comparison. Such advances support a multiplicity of theories that are tested through a careful checking of evidence rather than imposing a unified vision. Debates initiated from a "Western" perspective have questioned the universality of urbanisation processes, such as Brenner and Schmid (2012). Scott and Storper (2015) rightly called for a greater multiplicity of viewpoints and the necessity of more contextualised knowledge (Sheppard et al. 2013). The Internet has facilitated the propagation of these debates and has given rise to responses illustrating the inanity of a concept such as the "Global South" and underlining the diversity of forms of subaltern urbanisation and "ordinary towns" under similar global pressures (Denis and Zérah 2017). The global thinking included in urban theories today enables to question and measure the relative originality of genuine local processes and demonstrate to what extent the cultural idiosyncrasies, historical path dependent trajectories or political incentives may or not influence the territorial development (Wu 2020).

Although the uses of artificial intelligence and deep learning methods are still in infancy and until now restricted to limited domains of geography, for tackling climate, environmental or agronomy problems (Kanevski et al. 2008; Grekousis 2019; Amato et al. 2020), the Internet-related computerised technologies may introduce new visions of the nature and production of geographical space. Relational spaces defined by all possible kinds of relationships or semantic similarities are making the word "map" becoming a common metaphor for multiplied ontologies of our sensible environment, possibly leading towards a "non-representational" theory of geography (Thrift 2008) and suggesting to rethink cartography as "processual rather than representational science" (Dodge et al. 2009).

The new stage of massive production of georeferenced data has given rise to the parallel development of new questions. The hypothesis of a "data-driven" geography received critical observations (Kitchin 2014; Miller and Goodchild 2015). Some authors see big data as transforming the epistemological bases of scientific work, by providing a substitute for the classic approach of surveys and experiments: the mass of information would compensate for its imperfections, and its fine granularity (in general, at a micro-geographic or "individual" level) would make it possible to revise the foundations of theories constructed for other levels of observation. However, most of the big data collected from web scraping strategies are not produced for research purposes and many of their applications to social sciences are not well adapted to sound scientific inquiries and experimental protocols. The many problems remaining to solve before making a fully meaningful use of the increasing flows of big geodata, fortunately, are not favouring the prospect of a data-driven geography in the near future (Graham and Shelton 2013).

8.3 Towards Open Access and Open Science

The more than exponential multiplication of data through the Internet also is manifested by the proliferation of scientific publications. This strong and uncontrollable growth encourages the development of open science to make publications accessible to all scholars having Internet access in the world. We detail an example of the viability of open access publishing for geography.

8.3.1 Explosion of Scientific Publications

The increase in visibility of scientific works through the Internet accompanied the growing number of scholars, the spirit of competition exacerbated by the race for research funding (the "publish or perish" rule) and the attractive strategies of editorial platforms and scientific social networks. In the sixties, it was still possible for a university professor in France to declare having read all doctorate thesis in geography, whatever physical or human, three or four major geographical journals published each less than ten articles on paper per trimester and the language barriers reduced the interest for other publications to the production of merely three other countries, even if she attended the IGU congresses that already managed to organise really global scholarly meetings in its sections and commissions. Nowadays, maintaining an updated review of the literature in one's own specialised subfield of geography has become a superhuman effort. We face a plethora of books, and not a handful but thousands of scientific journals. For all fields of social sciences and humanities (SSH) the pioneering tool JournalBase enumerates about 15,000 journals, among which 1300 are in open access (Dassa et al. 2010).[5] There are some 683 geography journals in JournalBase that are accessible with the Internet.

[5]https://journalbase.cnrs.fr/.

The community of geographers is still far from fully exploiting the opportunities offered by the digitisation of these online publications. Constructing any state of the art has become a hazardous exercise because of the proliferation of scientific papers and journals, the blurring of disciplines boundaries and the explosion of distinctive vocabulary choices in contemporary social sciences. The rapid review of the contributions of the Internet to the internationalisation of geography that we attempted in the first part of this chapter would probably have been more representative of the most frequent practices and influential works if we had used an advanced technique for that bibliographic exploration, either through citations networks or searching for phylomemetic patterns (Chavalarias and Cointet 2013). Such scientometric methods have been used for exploring the literature in specialised domains such as natural hazards (Emmer 2018) or transportation geography (Liu and Gui 2016). Peris et al. (2018) reconstituted the genealogy of studies on systems of cities and Raimbault (2019) explored the scientific landscape of urban sciences. Indeed, spatial scientometrics has emerged as a new research programme in geography and already provided significant results (Maisonobe et al. 2017). Web scraping and semantic mapping for detecting paradigmatic similarities, convergences or intellectual niche creation could become a compulsory step not only when starting a new research but all along the research process (Zhuang et al. 2020). Moreover, and unfortunately, new "intelligent" publication strategies also could become driven from the observation of these scientometric networks such as possibly revealed by analysing successful scholar trajectories (Sebastian and Chen 2021).

8.3.2 The Open Access Movement

The development of the Internet has been a "breakthrough" innovation to facilitate and accelerate exchanges, which has renewed models and practices for transmitting scientific information. What determines the value of scientific knowledge, as opposed to opinion or ideology, is the possibility of its reproduction, its repetition, its validation by "peers". Peers are competent people, certainly in the context of the paradigm of the moment, of a state of scientific knowledge that by definition is provisional and revisable. When scientific work is well done, its statements gradually become organised in a coherent manner, scientific knowledge becomes cumulative, not in the form of sedimentary layers that would cover each other without eroding, but in the manner of a spiral that rises while revisiting, reformulating in a more precise and/or more encompassing manner existing knowledge.

In a time of knowledge-based economies (David and Foray 2003) the processes of circulating and sharing scientific information become of crucial importance for explaining and possibly monitoring social change (Lane et al. 2009). Scientific knowledge has a value that can eventually be monetised, but it is not an economic good like any other. It does not lose its value when it is shared. On the contrary, the exchange of knowledge is often what allows new knowledge to be created (Leeuw 2020). A very large part of scientific research is carried out mainly through public funding, through public programs. Its evaluation is almost entirely carried out by researchers, most of whom practice it freely, transparently and in an impartial way, without remuneration (Garfield 1963; Harnad 1991). The ArXiv platform, created by Paul H. Ginsparg in 1991, offered scientists the possibility of depositing documents of various kinds and formats and making them freely accessible to all Internet users.[6] In 2009, the European Commission launched the OpenAire platform (Open Access Infrastructure for Research in Europe).[7] To date, OpenAire2020 has harvested over 5700 directories worldwide and the number of open access publications collected is close to 18 million. On 17 July 2012, the European Commission put forward a recommendation on scientific information, based on the assumption that wide free access to publicly funded research is "speeding up scientific progress". Four years later, on 17 May 2016, the Council of the European Union recommended rolling out this model to all Member States in order to make open access a "default option by 2020" (Langlais 2016).

For fighting the formalisation of the open access movement, the major scientific publishing houses (Elsevier, Taylor and Francis, Springer, etc.) have diverted that label from its initial objective. "In fact, there are now in what is defined as "Gold Open Access" hundreds of journals in which authors (and institutions) are required to pay to be published in free access, these so-called 'Article Processing Charges' (APC) supposed to cover all the costs of posting online articles to which access is open […]. Moreover, there are also the so-called 'hybrid journals', i.e. partially in open access (for articles whose authors have paid) and whose subscriptions are still charged to institutions! The financial impact is enormous […] imposing huge costs on taxpayers and citizens, knowing that these benefits do not go back to research, but mainly to financial investors. […] Quantitative evaluation scoring, measures based on reputation (altmetrics), or "professional" social networks, etc. have been set up to complete the system around these private distribution monopolies" (Kosmopoulos and Pumain 2018 16–17).

8.3.3 Diamond Open Access as a True Alternative to Scientometrics

Alternative uses of the Internet for really open access to scientific publication are possible and sustainable such as attested by the example of *Cybergeo, European Journal in Geography*. Choosing *Cybergeo* may appear anecdotal and immodest but is rather representative of the energy and ability of anticipation that are required to develop viable alternative solutions in the David against Goliath struggle of global publications. This pioneered entirely digital journal was launched as early as 1996. Without any cost to the author or the reader since its

[6]https://arxiv.org/.

[7]https://www.openaire.eu/.

creation, *Cybergeo* is a scientific journal open to all themes and currents in geography and related disciplines, with a multilingual content and a trilingual interface. It ensures a worldwide diffusion of research whose quality is guaranteed by several international reading committees. The doubly anonymised peer review system respects the highest scientific and ethical criteria. To promote cumulative, reproducible, and open geographic science, the journal launched a Model papers section in 2014 and a Data papers section in 2017. (To our knowledge, this is the only journal in geography to date that publishes this type of article). *Cybergeo* cares about international audiences with an interface in French and English, and all metadata (titles, abstracts and keywords) published in four languages (French, English, Spanish and Chinese). Regarding its economic model, *Cybergeo* joined the alternative *Freemium*[8] publishing model independent of commercial publishers and launched in 2012 by *OpenEdition*. The open access application *CybergeoNetworks* (Raimbault et al. 2021) reveals the journal's spatial relationships and impact through citation network and semantic content analysis.

Thus the Diamond Open Access model is an "authentic open access", without cost for readers and authors.[9] According to a recent survey,[10] there are 12,000 journals in all scientific disciplines in Diamond Open Access, whereas the DOAJ (Directory of Open Access Journals) itself announces (at the end of March 2021) that an identification number has been given to some 16,232 journals among which 11,674 without APCs, representing almost 6 million articles in 80 languages and 125 countries. Thus, Open access is widely diffused but not yet all over the world.

[8]For more information on Freemium: http://www.openedition.org/14043.

[9]https://www.coalition-s.org/diamond-unearthed-shining-light-on-community-driven-open-access-publishing/.

[10]Bosman et al. 2021: https://zenodo.org/record/4558704#.YEtA9tzjKUk.

8.4 Hubs and Gaps in Internationalisation

A striking feature accompanying the tremendous success of the Internet is how much inequalities it generates among the populations. A full programme of satellite launching is in preparation for filling the white zones in Internet access, especially to reach the most remote territories in Africa or South America. However, inequalities do exist as well in most developed countries, and one estimates for instance that in France the digital divide leaves behind 14% of the population. From a geographical view, the current image of the Internet connecting the whole world is hiding many holes in the World Wide Web. The visualisation tools used, for example, to depict the expansion of social networks on the surface of the globe has been the subject of debate (Joliveau 2011). Most of the "raw" cartographic documents representing the flows exchanged on social networks, or the broadcasts of messages on Twitter, do not so much represent the globalisation of exchanges, but first illustrate the location of the main masses of the connected population, that of the wealthiest countries. Moreover, internationalisation may appear now not so much between nations and states, but as directed by multinational firms. This is true in the activity of scientific publishing.

The JournalBase statistics speak alone: more than 70% of journals in social sciences and humanities are issued from the USA and UK. This absolute domination is ensured by the concentration of journals in an oligopolistic structure of publishing companies. Although we do not consider it as a reliable source of information, the *Scimago* web site—that represents the views of Elsevier Group via the list of journals they gathered in the Scopus database for competing the WOS, provides a ranked list of geographical journals that is supposedly worldwide. Some 728 journals are labelled in the category "geography, planning and development" of which only 177 are open journals, most of them probably with APCs. In that list, 463 journals (two thirds) are published in the UK,

North America and the Netherlands. The whole Asiatic continent has only 34 titles while all BRICS countries have 39. There are just five journals from China and two from Africa. The world of publishing in geography is definitely organised according to a centre-periphery model with a steep gradient of inequality. That strongly unbalanced internationalisation was consolidated in the last decades by the action of many of these major journals organising conferences and other marketing operations in Asiatic countries for collecting sources of new scientific articles, attracting local scholars and opening subsidiaries.

In this context of high domination and because of the adoption by a majority of scientists of English as a *lingua franca* for publication, one may conclude that internationalisation is in progress. Artificial intelligence and deep learning have boosted the immense progress of automated translation between a few world languages (Fan et al. 2020) although human reading remains necessary to avoid deep misunderstanding, such as coined in the still relevant Italian expression: *traduttore, traditore*. Nevertheless, for social sciences the question of language remains, more so than in other scientific disciplines, a difficult problem. The question is not so much in choosing vocabulary, even writing fluently and making an enjoyable use of gimmick, but in determining which questions are on the research agenda and how to formulate them. In human geography. Despite their supposed inspiration from the so-called French theory, or other Europeans such as Karl Marx, Henri Lefebvre or Michel Foucault, the new fashionable topics that successfully emerged during the last three decades were all, if not always fully invented, truly elaborated, fashioned and institutionalised in the USA: critical and radical geographies (Harvey 1973; Peet 1977; Johnston 2000), humanistic geography (Buttimer 1976; Tuan 1976; Ley and Samuels 2014), post-modern geography (Soja 1989; Smith 1992), gender studies (Monk and Hanson 1982; Pratt and Hanson 1994), cultural geographies revised as post-colonial geography (Blunt and McEwan 2003; Hart 2004). One could extend the enumeration as well as name dropping over many pages. The intensity of debates, the speed of fashion renewals and the vitality of communication processes together with the large size of a wealthy market are complementary factors explaining the leadership and ability to ensure a worldwide diffusion of the American scientific agenda in geography. Foreigners participating in AAG meetings frequently were extremely struck by the huge number of parallel sessions dedicated to minorities. Perhaps it is not random if the sense of ethics in geography is also made explicit and developed under a moral North-American inspiration (Smith 1997).

The inequalities and asymmetries in the internationalisation processes of geography in the digital age and Internet era are easily understandable with our common geographic knowledge. They are part of today's world economy and geopolitics. The IGU strategy fully supports the current counteracting trends in political economy that may bring about a more polycentric pattern in the next decades (Kolosov et al. 2017). Most of the internationalisation process has positive effects in encouraging knowledge creation and intellectual debates. A more careful attention may be paid to its possible negative effects when distorting some local and regional hierarchies of research agenda, and sometimes, worse but fortunately rarer, possibly inducing distortion of scientific discourses.

8.5 Conclusion: Still an Ambiguous Game

Technologies play an important role in social evolution. They transform individual practices and social interactions, with consequences that are partly beneficial and partly problematic. But in many ways they do not revolutionise the habits of economics and geopolitics. What has been conceptualised by Western imperialism as the "Gutenberg revolution" has accelerated the circulation of knowledge leading to the succession of two industrial and service revolutions, while affirming the construction of nation-states in colonialisms and conflicts, sometimes of global magnitude. The Internet revolution extends and accelerates the diffusion of information in a quasi-instantaneous way on the scale of the whole planet. But this expansion and the increased speed of globalisation processes, without having put an end to previous sources of conflict, are also taking place in the expanded confrontation between multinational companies such as GAFAM or BHATX and international regulatory institutions. For scientists, the Internet represents a formidable lever for knowledge production, methodological experimentation and the implementation of artificial intelligence tools. Geographers see the scope of their work potentially multiplied and enriched by the universals of georeferencing. They use massive data produced by sensors, satellite or personal, new institutions and all kinds of instruments of standardisation and validation. They are accessible in principle all over the world and participate in the enterprise of internationalisation of geographic knowledge which is one of the reasons for the existence of the IGU. Learned societies such as the IGU amplify their long-standing contribution to the internationalisation of knowledge. The movement towards open science is however not yet complete and requires their full support.

References

Agnew J (2009) Globalization and sovereignty. Rowman and Littlefield, Lanham

Amato F et al (2020) A novel framework for spatio-temporal prediction of environmental data using deep learning. Sci Rep 10(1):1–11

Arcaute E, Hatna E (2020) Scaling laws: insights and limitations. In: Pumain D (ed) Theories and models of urbanization. Springer, Cham, pp 45–66

Batty M (2013) The new science of cities. MIT Press, Cambridge, Mass

Bennett J (2010) OpenStreetMap: be your own cartographer. Packt Publications, Birmingham

Blunt A, McEwan C (eds) (2003) Postcolonial geographies. Bloomsbury Publishing, London

Bosman J et al (2021) The OA diamond journals study. Part 1: findings. http://doi.org/10.5281/zenodo.4558704

Boulos MNK, Geraghty EM (2020) Geographical tracking and mapping of coronavirus disease COVID-19/severe acute respiratory syndrome coronavirus 2 (SARS-CoV-2) epidemic and associated events around the world: how 21st century GIS technologies are supporting the global fight against outbreaks and epidemics. Int J Health Geogr 19(8)

Brenner N, Schmid C (2012) Planetary urbanization. In: Gandy M (ed) Urban constellations. Jovis, Berlin, pp 10–13

Burini F (2016) Cartografia Partecipativa: Mapping per la Governance Ambientale e Urbana. Franco Angeli, Milan

Burton I (1963) The quantitative revolution and theoretical geography. Can Geogr 7(4):151–162

Buttimer A (1976) Grasping the dynamism of lifeworld. Ann Assoc Am Geogr 66(2):277–292

Calabrese F et al (2010) The geography of taste: analyzing cell-phone mobility and social events. In: Floréen P, Krüger A, Spasojevic M (eds) Pervasive computing. Pervasive 2010. Lecture notes in computer science. Springer, Heidelberg, pp 22–37

Chavalarias D, Cointet JP (2013) Phylomemetic patterns in science evolution—the rise and fall of scientific fields. PLoS ONE 8(2):e54847

Cöltekin A, Janetzko H, Fabrikant SI (2018) Geovisualization. Geogr Inf Sci (Q2). http://doi.org/10.22224/gistbok/2018.2.6

Cottineau C et al (2017) Diverse cities or the systematic paradox of urban scaling laws. Comput Environ Urban Syst 63:80–94

Couclelis H et al (1987) Exploring the anchor-point hypothesis of spatial cognition. J Environ Psychol 7 (2):99–122

Cura R (2019) Model visualization. In: Pumain D (ed) Geographical modeling: cities and territories. Wiley, London, pp 151–191

Dassa M, Kosmopoulos C, Pumain D (2010) JournalBase—a comparative international study of scientific journal databases in the social sciences and the humanities (SSH). Cybergeo Eur J Geogr 484. http://doi.org/10.4000/cybergeo.22862

David PA, Foray D (2003) Economic fundamentals of the knowledge society. Policy Futures Educ 1(1):20–49

Denis E, Zérah MH (eds) (2017) Subaltern urbanisation in India: an introduction to the dynamics of ordinary towns. Springer India, New Delhi

Denis E et al (2020) Mapping the lockdown effects in India: how geographers can contribute to tackle Covid-19 diffusion. The Conversation. https://cybergeo.hypotheses.org/581

Dillman DA (2011) Mail and Internet surveys: the tailored design method—2007 update with new internet, visual, and mixed-mode guide. Wiley, New York

Dodge M, Kitchin R, Perkins R (2009) Rethinking maps: new frontiers in cartographic theory. Routledge, London

Duféal M, Noucher M (2017) Des TIC au TOC. Contribuer à OpenStreetMap: entre commun numérique et utopie cartographique. Netcom 31(1/2):77–98. http://doi.org/10.4000/netcom.2635

Emmer A (2018) Geographies and scientometrics of research on natural hazards. Geosciences 8(10):382. https://doi.org/10.3390/geosciences8100382

Fan A et al (2020) Beyond English-centric multilingual machine translation. https://arxiv.org/abs/2010.11125

Fen-Chong J (2012) Organisation spatio-temporelle des mobilités révélées par la téléphonie mobile en Ile-de-France. Université Panthéon-Sorbonne-Paris I, Unpublished Ph.D. thesis
Finance O, Swerts E (2020) Scaling laws in urban geography. Linkages with urban theories, challenges and limitations. In: Pumain D (ed) Theories and models of urbanization: geography, economics and computing sciences. Springer, Cham, pp 67–96
Garfield E (1963) New factors in the evaluation of scientific literature through citation indexing. Am Documentation 14(3):195–201. http://garfield.library.upenn.edu/essays/v6p492y1983.pdf
Giuliani G et al (2017) Bringing GEOSS services into practice: a capacity building resource on spatial data infrastructures (SDI). Trans GIS 21(4):811–824
Global SDG indicators database: https://unstats.un.org/sdgs/indicators/database/
González MC, Hidalgo CA, Barabási AL (2008) Understanding individual human mobility patterns. Nature 453(7196):779–782
Goodchild MF (2007) Citizens as sensors: the world of volunteered geography. GeoJournal 69(4):211–221
Goodchild MF (2016) GIS in the era of big data. Cybergeo Eur J Geogr. http://cybergeo.revues.org/27647
Graham M, Shelton T (2013) Geography and the future of big data, big data and the future of geography. Dialogues Human Geogr 3(3):255–261
Grekousis G (2019) Artificial neural networks and deep learning in urban geography: a systematic review and meta-analysis. Comput Environ Urban Syst 74:244–256
Haggett P (1990) The geographer's art. Blackwell, London
Harnad S (1991) Post-Gutenberg Galaxy: the fourth revolution in the means of production of knowledge. Public-Access Comput Syst Rev 2(1):39–53
Hart G (2004) Geography and development: critical ethnographies. Prog Hum Geogr 28(1):91–100
Harvey D (1973) Social justice and the city. Arnold, London
Johnston RJ (2000) Intellectual respectability and disciplinary transformation? Radical geography and the institutionalisation of geography in the USA since 1945. Environ Plan A 32(6):971–990
Joliveau T (2011) Le géoweb, un nouveau défi pour les bases de données géographiques. L'espace Géographique 40(2):154–163
Kanevski M, Pozdnukhov A, Timonin V (2008) Machine learning algorithms for geospatial data. Applications and software tools. In: International Congress on environmental modelling and software 53. https://scholarsarchive.byu.edu/iemssconference/2008/all/53
Kitchin R (2014) Big data, new epistemologies and paradigm shifts. Big Data Soc 1(1):2053951714528481
Kolosov V et al (2017) The Chinese initiative "the belt and road": a geographical perspective. Geogr Environ Sustain 10(1):5–20
Kosmopoulos C, Pumain D (2018) Humanities and social sciences (HSS) at the head of open access as demonstrated from Cybergeo's experience. Cybergeo Eur J Geogr 861. http://doi.org/10.4000/cybergeo.29251
Lane D et al (eds) (2009) Complexity perspectives on innovation and social change. Springer, Berlin
Langlais PC (2016) Critical study of the new ways of "editorialising" open access scientific journals. Bibliothèque Scientifique Numérique. https://hal.archives-ouvertes.fr/hal-01399286
Ley D, Samuels MS (eds) (2014) Humanistic geography: problems and prospects. Routledge, London
Liu C, Gui Q (2016) Mapping intellectual structures and dynamics of transport geography research: a scientometric overview from 1982 to 2014. Scientometrics 109(1):159–184
Liu X, Wang J (2015) The geography of Weibo. Environ Plan A 47(6):1231–1234
Longley PA et al (2005) Geographic information systems and science. Wiley, Chichester
Lopez L, Olvera-Lobo MD (2018) Public engagement in science via web 2.0 technologies. Evaluation criteria validated using the Delphi method. J Sci Commun 17 (2):A08
Lucchini F, Elissalde B, Sylviano-Freire-Diaz S (2011) From the event to the Ephemeral City. mesm 2011 conference. Amman, Jordan, pp 88–95
Maisonobe M et al (2017) The global geography of scientific visibility: a deconcentration process (1999–2011). Scientometrics 113(1):479–493
Mericskay B, Roche S (2011) Cartographie 2.0: le grand public, producteur de contenus et de savoirs géographiques avec le web 2.0. Cybergeo Eur J Geogr 552. http://doi.org/10.4000/cybergeo.24710
Mertens B et al (2012) Interactive Forest Atlas of Cameroon—Atlas Forestier Interactif du Cameroun (Version 3.0). http://cmr.forest-atlas.org
Miller HJ, Goodchild MF (2015) Data-driven geography. GeoJournal 80(4):449–461
Mitchell R (2018) Web scraping with Python: collecting more data from the modern web. O'Reilly, Beijing
Monk J, Hanson S (1982) On not excluding half of the human in human geography. Prof Geogr 34(1):11–23
OpenStreetMap. https://www.openstreetmap.org
Pearson A, Heffernan M (2015) Globalizing cartography? The international map of the world, the International Geographical Union, and the United Nations. Imago Mundi 67:58–80
Peet R (1977) The development of radical geography in the United States. Prog Hum Geogr 1(2):240–263
Peris A, Meijers E, van Ham M (2018) The evolution of the systems of cities literature since 1995: schools of thought and their interaction. Netw Spat Econ 18 (3):533–554
Pinchemel P (ed) (1972) La géographie à travers un siècle de congrès internationaux: Union Géographique Internationale, Commission Histoire de la Pensée Géographique. Caen

Pratt G, Hanson S (1994) Geography and the construction of difference. Gender Place Culture J Feminist Geogr 1(1):5–29
Pucci P, Manfredini F, Tagliolato P (2015) Mapping urban practices through mobile phone data. Springer, Heidelberg
Pumain D (ed) (2020) Theories and models of urbanization: geography, economics and computing sciences. Springer, Cham
Pumain D, Reuillon R (2017) Urban dynamics and simulation models. Springer, International. Lecture Notes in Morphogenesis, 123 p
Raimbault J (2019) Exploration of an interdisciplinary scientific landscape. Scientometrics 119(2):617–641
Raimbault J (2020) Indirect evidence of network effects in a system of cities. Environ Plann B Urban Anal City Sci 47(1):138–155
Raimbault J et al (2021) Empowering open science with reflexive and spatialised indicators. Environ Plann B Urban Anal City Sci 48(2):298–313
Ratti C et al (2010) Redrawing the map of Great Britain from a network of human interactions. PLoS ONE 5 (12):e14248. https://doi.org/10.1371/journal.pone.0014248
Reed C (2004) Internet law: text and materials. Cambridge University Press, Cambridge
Robic MC, Briend AM, Rössler M (eds) (1996) Géographes face au monde. L'union géographique internationale et les congrès internationaux de géographie. L'Harmattan, Paris, Montréal
Robic MC et al (2006) Couvrir le monde: un grand XXe siècle de géographie française. Association pour la diffusion de la pensée française, Paris
Rohat G et al (2018) Co-use of existing scenario sets to extend and quantify the shared socioeconomic pathways. Clim Change 151(3–4):619–636
Rozenblat C, Pumain D, Velasquez E (eds) (2018) International and transnational perspectives on urban systems. Springer, Singapore
Scott AJ, Storper M (2015) The nature of cities: the scope and limits of urban theory. Int J Urban Reg Res 39 (1):1–15
Sebastian Y, Chen C (2021) The boundary-spanning mechanisms of Nobel prize winning papers. PLoS ONE 16(8):e0254744. https://doi.org/10.1371/journal.pone.0254744
Sheppard E (2001) Quantitative geography: representations, practices, and possibilities. Environ Plann D Soc Space 9(5):535–554
Sheppard E, Leitner H, Maringanti A (2013) Provincializing global urbanism: a manifesto. Urban Geogr 34 (7):893–900
Sikka G, Yenneti K, Singh R (2021) IGU-YECG special issue: geospatiality and sustainable development goals. Geogr Environ Sustain 14(1):6–8
Smith N (1992) Geography, difference and the politics of scale. In: Postmodernism and the social sciences (pp 57–79). Palgrave Macmillan, London
Smith DM (1997) Geography and ethics: a moral turn? Prog Hum Geogr 21(4):583–590
Soja E (1989) Postmodern geographies: the reassertion of space in critical social theory. Verso, London
Takhteyev Y, Gruzd A, Wellman B (2012) Geography of Twitter networks. Soc Netw 34(1):73–81
Thrift N (2008) Non-representational theory: space, politics, affect. Routledge, London
Thrift N, French S (2002) The automatic production of space. Trans Inst Br Geogr 27(3):309–335
Thrift N, Leyshon A (1994) A phantom state? The detraditionalization of money, the international financial system and international financial centres. Polit Geogr 13(4):299–327
Tobler W (1959) Automation and cartography. Geogr Rev 49(4):526–534
Tobler W et al (1997) World population in a grid of spherical quadrilaterals. Int J Popul Geogr 3(3):203–225
Tuan YF (1976) Humanistic geography. Ann Assoc Am Geogr 66(2):266–276
Unwin D (1999) Euroquant at 21: "coming of age"? Cybergeo Eur J Geogr 114. http://doi.org/10.4000/cybergeo.563
van der Leeuw S (2020) Social sustainability, past and future: undoing unintended consequences for the Earth's survival. Cambridge University Press, Cambridge
Wu F (2020) Emerging cities and urban theories: a Chinese perspective. In: Pumain D (ed) Theories and models of urbanization: geography, economics and computing sciences. Springer, Cham, pp 171–182
Yeates M (ed) (1974) IGU commission on quantitative geography. McGill Queen's University Press, Montreal
Zhuang L, Ye C, Lieske SN (2020) Intertwining globality and locality: bibliometric analysis based on the top geography annual conferences in America and China. Scientometrics 122(2):1075–1096

Denise Pumain is Emeritus Professor at University Paris I Panthéon-Sorbonne, member of CNRS laboratory Géographie-cités. Her main scientific contribution is about building an evolutionary theory of urban systems and transferring concepts and models from self-organising complex systems towards social sciences. Founder of the research laboratory P.A.R.I.S. (1984), Director of the UMR Géographie-cités (CNRS 1992–2000), Chair of the Commission on Urban Development and Urban Life of the IGU (1992–2000), Director of the European Research Group S4 (Spatial Simulation for Social Sciences, 2006–2013), Founder and Director of *CybergeoEuropean Journal of Geography* (since 1996), PI of the ERC advanced grant GeoDiverCity (2011–2016).

Christine Kosmopoulos is Research Engineer at CNRS laboratory Géographie-cités, editorial manager of *Cybergeo European Journal of Geography*. Specializing in Open Science, she has made strategic and technological choices and pioneering developments in favour of open access, enhancing the journal's international impact. She acts at international level as an expert for scientific evaluation and digital publication: Expert Group for Program Monitoring and Evaluation (ERC 2015–2016); Revues 2.0 (Montreal University 2018); Expert with the Monitoring Committee for Digital Publishing (MESRI, 2017–2019). She directed the international CybergeoNet project aiming at pooling methods for metadata translation and testing machine translation.

9 To Be or Not to Be International: Geographic Knowledge, Globalization and the Question of Languages

Rindra Raharinjanahary, Nathalie Lemarchand, and Louis Dupont

Abstract

Even before geographers started to conceive the global world and the way it affected established worlds, their own scientific world has already changed. As IGU congresses show, starting in the 1960s English became the language of global science. However, this process did not go without questions and worries about languages from IGU prominent geographers; there has been recently a renewed interest among geographers. The authors focus here on the questions of production and transposition of knowledge, although the latter cannot be dissociated from its publication. The central question is: Can the global scale, which may offer a way out of, say, regional or national normative grips, have become itself normative. Say differently: does the acceptance of English as the international communication device ends all discussions on the hegemonic normative position it has on scientific productions? *Is going global the same as being international*? Does meeting the global scientific standards suffice for your research to be considered, "universal"? The authors do not pretend here to be able to answer all those questions completely and satisfactory, the merit of their contribution might simply be to shed a new light on some of them.

Keywords

Globalization · Language · Normativity · Knowledge

9.1 Introduction

In our era of global science, a geographer living somewhere on this planet called Earth may wonder what it takes to be 'international', and what it means to be an 'international geographer'? Does presenting a paper in English at an International Geographical Congress make you international? Are international journals of human geography really international, a question posed by Gutiérrez and Lopez-Nieva (2001), or are they simply 'Anglo American' as Bański and Ferenc (2013) suggest? A geographer living in, say, Madagascar, may wonder what the word 'international' means for native English speakers who participate in their national conferences. For that matter, British geographers felt the wind of

R. Raharinjanahary
Antananarivo University, Antananarivo, Madagascar

N. Lemarchand (✉)
University of Paris, Paris, France
e-mail: nlemarchand02@univ-paris8.fr

L. Dupont
Sorbonne Université, Paris, France
e-mail: Louis.Dupont@sorbonne-universite.fr

V. Kolosov et al. (eds.), *A Geographical Century*,
https://doi.org/10.1007/978-3-031-05419-8_9

globalization when they started to call their national conference, now located mostly in London, the 'RGS-IBG International Annual Conference' in 2003.[1] For French geographers, the act of crossing the Channel to attend the RGS International Annual Conference provide an international status in their university, whereas before 2002 they would merely have been attending an interesting conference in a friendly, if somehow exotic northern country. In the IGU, one can organize sessions and present papers in French. The vast majority of geographers, including some French-speaking colleagues, wonder why French is included, however, as they usually recognize that their linguistic hegemony has long since passed. We could call this a linguistic heritage, though some would say it's a vestige of the past! Why French and not Spanish? Or Chinese?

There is no use denying that geographers need to communicate on their research, make comparisons and debate their conclusions. These exchanges contribute to the advancement of science, and a better understanding of the worlds we inhabit. We have to make do with the global scale, and to reach this state of affairs we need to communicate in a common language. But does the acceptance of English as the international communication device end all discussions on the hegemonic normative position—on form and content—it has on scientific productions? We argue that it should not, that the matter remains open for discussions, especially in the social sciences and humanities, and even more in geography. Our general goal here is to look at what is left to be seen beyond the necessary horizon of English as the scientific globalized language of communication.

The material we went through confirms the impression we had when raising related questions on the subject at different venues: the more globalized geographic knowledge becomes, the less questions and discussions on the hegemonic place of English seem to matter.[2] Surprisingly, this was not the case at the beginning of the modern globalization of knowledge, circa 2000. Different issues and questions were raised by geographers who worried that a one-language science could impoverish the knowledge of the overlapping space, regions and places geographers seek to explore, analyse and try to understand. Why should we go back to those questions and issues? Maybe because twenty years later, we have a better grasp of globalization, and its impact on the scientific world. Some questions come to mind. First: Beyond English as a communication device, how does the so-called English-language hegemony affect the nature of geographic knowledge? Second: What is lost or gained in translation? How do we transpose one's world into someone else's words? Third: How have national and international scientific institutions reacted, adjusted to the existence of the global scale? Finally, we may wish to add epistemological questions: Is going global the same as being international? Does meeting the global scientific standards suffice for your research to be considered 'universal'? We do not pretend here to be able to answer all those questions completely and satisfactorily, the merit of our contribution might simply be to shed a new light on some of them.

The three authors have the French language in common, different language skills, and somehow an international profile (Africa, North America, Europe): one has to navigate through three languages in a post-colonial context, the other is quite familiar with the dominance of English, the last has to cope with a national language in a multilingual European. We do not bring issues

[1]From 1996 (when the Royal Geographical Society joined forces with the Institute of British Geographers, which represents the discipline in UK higher education) to 2002, the annual meeting of the joint Society-Institute was simply called the 'Annual Conference', with no explicit reference to its international reach and significance. URL: https://rgs.koha-ptfs.co.uk/cgi-bin/koha/opacdetail.pl?biblionumber=246022&query_desc=an%3A%2212369%22.

[2]Although the question is interesting, we cannot determine whether American or British English is hegemonic or dominant in writing or in presenting papers. It is a debate in itself that tends to put forward the concept of 'international English' as a common scientific language. See Paredes et al. (2002); Johnston and Sidaway (2004).

about language to go back to some sort of good old (French) time; this question is moot. Rather, we use this opportunity to reflect on the place of language, all languages, in the production of geographic knowledge in this modern global era.

9.2 Language and the Geographic Knowledge of Overlapping Worlds

The 2000/2010 decade saw many articles being published on the globalization of knowledge in geography, and the hegemonic/dominant position of English. They were for the greater part written in English, in Anglo American journals, and, for a lesser part, by English-speaking geographers. In the 2020s, we think that those worries are giving way to reconsidering languages in the production of geographic knowledge, and to looking at ways to have more exchanges in different languages. Said exchanges are seen as contributing to the betterment of geographic knowledge, and to the redefinition of what it is to be international.

9.2.1 English, the Modern Global Scientific Language

In an article published in the *Geographical Review* (91:4) by the American Geographical Society of New York, Harris (2001) looks at the evolution of geographic international organizations through the lens of languages. He provides numbers that clearly show the extent of this transformative process. He examined more than a hundred years of international geographical congresses (IGC), starting in 1871 in Antwerp where 76% of the communications were in French, up to the IGU-run international congress in Seoul, where 97% of the papers were in English. He notes that between 1871 and 1938, four languages, French, English, German and Italian, were in use in congresses, while Portuguese and Spanish were also officially present from 1938 to 1956. However, the 1952 IGU congress in Washington, where 70% of papers were in English compared to 28% in Lisbon in 1948, can be seen as the date when geography's scientific world started to be seen and expressed, decisively, in English on the international, not yet called global, stage. IGU's executive business, run mostly in French since its creation in 1921, shifted predominantly to English after the Second World War. Afterword, sessions in another language than English tend to be held in the language of the congress's host (including we expect IGU Paris 2022 congress), and attended mostly by 'local' geographers.

Likewise, American journals, which had already established a tight evaluating system as a way to show accountability and measure of excellence to their society, took precedence on the world stage. The UK and other English-speaking countries soon followed suit in adopting the same system of evaluation, thus contributing to a worldwide competition between Anglo American journals. In the non-English-speaking world, the national systems had to adjust to the open access to the world stage. With the aim of reaching the world stage, many European (national) journals started to be published in English.[3]

Reacting to this linguistic shift, Kitchin (2005) stated that the globalized knowledge economy is liable to intensify the dominance of English as science's *lingua franca*. Indeed, Paasi (2005) details how the academic world took, during that period, a capitalist turn. Congresses such as those associated with the Association of American Geographers, the Royal Geographical Society-Institute of British Geographers, and the IGU itself adopted a business model that provided the global structural foundation for the domination of English.

In a detailed analysis of international congresses in geography, Vollé (1996) explored the language policies and practices of the IGU. She distinguishes three periods, from an open multilingualism to a de facto monopoly of English, with an in-between period of French–English

[3]Of the 27,000 journals included in the *Web of Science* (WoS) indexes, only 9000 are published in a language other than English. According to Curry and Lillis (2013), most of these non-English journals are excluded from the most prestigious indexes.

'duopoly' during the 1960s. Seen from today, this change in language policy puts IGU in a paradoxical position: maintaining a bilingual French–English façade in its official written communication, while running congresses and businesses in English. The bilingual burden rests mostly on the French-speaking member(s) (Robic, Briend and Rössler 1996), on the good will of others, sometimes on automatic translation which is a less than perfect way to fill in the blanks.[4] Nevertheless, the presence of the French language as heritage, vestige or simply as a signal means that the IGU recognizes the importance of language in the production of geographic knowledge. In many ways, this differentiates the IGU from the AAG, where the majority of congress participants are not native speakers. In an interview published in *Chimera*, Anne Buttimer shared Chauncy Harris's concerns on the effect of the dominance of English has on geographic knowledge: "language provides an invaluable key into the lived situations of people in environment. There are so many instances of 'untranslatable taken-for-granteds' in people's everyday worlds. For the geographer of twenty-first century, language competence may be among the most important requirements" (Nally and Costello 2000, 20; see also Short et al. 2007).

9.2.2 Issues and Perspectives

We think these worries and positions on language(s) by two IGU historical figures still resonate today, as geographers are increasingly studying different interconnected worlds that coexist at different scales. How can we manage the interplay of language(s), national institutions and the globalization of science?

The first issue is the normative effect the dominance English imposes on geographic knowledge and practices. Kitchin (2005) worries that the dominant global trend could disregard certain geography, and create double standards in conferences, where, for instance, a normative way of presenting and questioning marginalizes non-native English speakers (even when they master English as a communication device). The Dutch Manuel Aalbers (2004), presenting his article as reflecting a 'non-Anglo American perspective', goes further in suggesting a "creative destruction" of this hegemony in publications. In both cases, the intent is somehow political, but they also point to the fact that this system affects how we conduct research and write about them.[5] One of the most telling examples is certainly the *casestudyification* of the local/national, as if every world we study becomes an exemplification of normative conceptions, theories or, even, vision of the world. Getting into trouble water, Desbiens and Ruddick (2006) wonder what type of universality this trend entails, and consequently, point out to its relativistic definition in the globalization of scientific knowledge.

Raising other issues, some geographers have moved away from a confrontational Anglo vs. non-Anglo perspective. Looking at those transformations at the national level, especially in Europe, Timár (2004) concludes that the central issue of concern for geography is not the Anglo-American linguistic hegemony but rather the wider western hegemony that sets the limits on the nature of geographical inquiry. This is certainly true of the IGU membership, including many IGU commissions where the problem is not that English is the dominant language but that themes and subjects identified for investigation reflect western intellectual priorities. Let say that there are cultural and epistemological differences, not to say political, that support Tilmar's point. Best (2009) suggests adopting a postcolonial, or decolonial critical perspective, as a way to appreciate the geography of the 'others,' and other geographies. As the editors acknowledge, this has an impact on the more reflexive

[4]In this case French speakers can easily detect it, and quickly go to the English version.

[5]It is worth mentioning that some native English speakers also have to adapt to this normative and conformist international scientific English. In some corners of geography, and in certain topics, this creates resentment which translates into an epistemological debate on the nature of geographical knowledge. See Schmitz (2013), Timmis (2002), Kong and Quang (2019).

approaches and innovative methodologies used in some articles published in the journal *Social and Cultural Geography* (Shaw et al. 2015).

Houssay-Holzschuch and Milhaud (2013) offer an interesting perspective in their essay "Geography after Babel: a view from the French province." Inspired by Fall and Minca (2012), they suggest using the geographical 'centre-periphery' model to better articulate the global system, with the peculiarities of other geographies. Following Wismann (2012: 51), they "point to some of the challenges and—crucially—advantages of practicing a multilingual geography and 'thinking in between languages'". A good example is provided by Fall and Rosière (2008) in their exchange "On the limits of dialogue between Francophone and Anglophone political geography."

To sum it up, the point is not anymore to challenge the dominance of English as a communication device at the international scale, but how we can translate and transpose the diversity of the many worlds we geographers study, into a scientific perspective. The question of language, and it is especially true in Europe, also challenges the ways the 'inter-national' can be achieved, in *thinking in between languages*. We would like to examine those issues and perspectives in the next two sections.

9.3 Navigating Through Languages as a Geographer in Madagascar

As a late decolonized country, Madagascar offers an interesting window to the complexity of languages in Africa, as such in their respective society and cultural area, but also in regard to the scientific production. In Madagascar, hesitation in language policy has had an impact on the choice of official language(s), and consequently on the adoption of the languages of communication and of education. Over the past fifteen years, the country has oscillated between two and three official languages, it is now down to two with French and Malagasy. In this way, Madagascar resembles the case of "most African countries [that] continue to use the language of the former colonizing country as the main language of education and government of the country" (Ouane and Glanz 2010: 5). This choice is above all political in the face of the lack of mastery of foreign languages by the population, and as a consequence of the degradation of the quality of education in general, even if French is taught in elementary school (6–10) and English starting in college (11–14). While every language has the power to say everything (da Silva Machado 2013), the contribution of the Malagasy language thus deserves to be questioned vis-à-vis research in geography since, to this day, published works in Malagasy are non-existent.

9.3.1 From French Heritage to English Necessity

Despite the Malagasyization programs undertaken in 1975 in basic education, the language used at the university has never been subject to this process because the Malagasy language was not solicited in scientific matters during the decades of colonization and decolonization—the period 1960–1972 during which France remained very present (Ranaivoson 2007). This is why in geography, the legacy of the founding fathers such as Gérald Donque, René Battistini, Hildebert Isnard, remains deeply rooted in the transmission of geographic knowledge, because the language of instruction was, and remains, French.

The first generation of Malagasy geographers who inherited the discipline maintained the language of Molière, all the more so since they continued to work with the French authorities until the 1990s.[6] This first generation taught in French and bequeathed this linguistic heritage as the language of teaching and research. They thus produced their work through translation.[7]

[6]Currently there are four departments of Geography in Madagascar, though the first teachers in the other provincial universities were trained by the same cooperates who taught in Antananarivo.

[7]The current teachers are of the second, third and fourth generation after the French who taught in the department.

However, today the use of '*vary amin'anana*',[8] the 'language of the place', or, literally, 'rice cooked in brèdes', in teaching beginners with explanations in French and Malagasy has become a regular practice. The use of the mixture of the two languages is seen by teachers as a means to better integrate new notions, and to provide a better understanding of the theoretical contexts, because "in a mixed discourse, the alternation of languages is de rigueur and is invested as a scientific mode of communication where the two linguistic systems subsist in a permanent face-to-face that does not hinder in any way those who practice them" (Rhaïb 1996, 153).

The appropriation of the country's geography is nevertheless evident through the publication of numerous works in Madagascar's *Revue de Géographie*, the country's only geography journal, where the issues it deals with are specific to the country, and are not those of the 'Westerners'. This reaction is similar to that express by Garcia-Ramon (2003, 2) when she states that: "Recent studies show that geographers from a large number of countries (some of which have a long and productive geographical tradition) are reluctant to publish their work in this type of journal, preferring instead to publish in their own countries. There are several reasons that might explain this situation". However, the language of publication remains French, which has never been the subject of debate in the editorial policy of the journal; we are convinced that writing in French allows us to be more widely read on a national and international scale, and yet, we know that reaching the level of language of international publications is quite problematic. Moreover, in Africa, no one doubts of the power of English and French, they become a *sine qua non* condition to have access to international colloquiums, organized both locally and internationally, and that they constitute a means to maintain this 'bridge' between the country and the international, the local and the global.

French or English are conceived as both a tool and a *lingua franca* in debates on the implementation of international projects for poverty alleviation, water management in urban and rural areas, landscape dynamics, biodiversity, and remote sensing. All these vocabularies are translatable into Malagasy but only the latter is used when interacting with the target population, as well as in national debates on programs such as the National Urban Forum or World Cities Days, where the audience is composed of mayors from all the communes of Madagascar. The '*vary amin'anana*' is then the language of exchange through a projection in French and a presentation in Malagasy, since the use of the official Malagasy language is strongly recommended (Ranaivoson 2007). In the field of geography research, the influence of foreign languages on Malagasy is far from diminishing, even if efforts have been made to enhance the value of this language at the national level. The national challenge is far from being met, whereas historically the geographical elements of the different Malagasy natural regions were described entirely in Malagasy by Street and Andrianaly (1870) during the period of the kingdoms.

9.3.2 Geography and Geographic Words

In geographical research, national languages such as Malagasy do not have their place and it is not an exception since African geographers use French or English in spite of the importance of the works on the continent since colonization until nowadays. However, some Malagasy words are known internationally, the impetus of a deterministic geography advocated at that time justifies these words related to geomorphological or hydrological phenomena such as *lavaka* by Riquier (1954), whose meaning is simply a 'hole', but takes another meaning and dimension in translation: "roughly ovoid excavations with steep walls, shaped in the alterites of crystalline and metamorphic rocks by runoff and underflow." (p. 169) The word tanety, which means high ground, hillside, hill (Webber 1853), was

[8]This is a way of cooking rice by adding lots of water as if to make soup and then mixing in several kinds of green leaves, like a salad.

also internationalized by Hallanger (1973), whose definition in geography is "ferralitic soils characterized by their acidity and low phosphorus availability." Finally, *baiboho*, spelled *baibo* in Webber's dictionary, is also widely used, its meaning is "a field, a plantation at the water's edge," whereas for geographers it means literally 'land flooded by floods' (Guebourg 1997, 246).

The 1990s saw a new turn, not towards a total questioning of the language(s) and a desire to write everything in Malagasy, but rather towards the selection deployment of a geographical vocabulary in Malagasy words are used without translating them, even if they are not internationally known. This coincided with various actions claiming an essential place for the Malagasy language. The Academy's Language Center, created in 1993, sought to provide regular scientific work in order to make it suitable for all the needs that the context imposes. The indirect objective was to consolidate, through the "heritage and the values it conveys", the culture and identity of Malagasy speakers (Ranaivoson 2007). Likewise, Rakoto-Ramiarantsoa (1995) adopted many Malagasy for ethno-geographic research, as a way of "valuing vernacular knowledge", illustrated by the title of his best-known work, *Flesh of the Earth*. This was a deliberate attempt to keep the meaning of words in Malagasy even if it does not call into question the language of writing. *Flesh of the Earth* means 'soil' in pedology, whereas the soil has no flesh, but only layers. Of course, there is never a direct translation but to speak of 'soil' would take away the true meaning for Malagasy farmers. Climate-related phenomena are also described in Malagasy because their "approach ... combines geo-climatological analysis with a reflection on local knowledge and cultural dimensions relating to atmospheric elements" (Rakoto-Ramiarantsoa and Peyrusaube 2007: 323), with notions such as *fahavaratra*, 'time of lightning', *mitohatra*, 'clouds that swell as they ascend', *Erika*, 'fine precipitation or drizzle', or *lohataona* 'head of the year'.

As such, using Malagasy words accounts for expressing a world according to the practices and habits of the inhabitants, such as *mpiantsoantso tavoahangy*, 'bottle shouters' or *mpitsindroka* 'crumb collectors' (Raharinjanahary 2011), or else *mpamatotra entana*, translated as 'the one who ties up luggage', that is, the person who helps the taxi-brousse driver in bus stations (Raharinjanahary and Rajoelina 2018). The first cited word can find its equivalence, but not its meaning, with 'ragpickers or biffins'. Keeping the meaning with Malagasy words is a way to stay with the social representations of the actors, as for *mpamatotra entana*, the fact of writing it in Malagasy contributes to the knowledge of the local practices. Authors such as Vololonirainy and Finoalahy (2017) also maintained *ramarasana*, which is 'the following of plots after three years of cultivation', a cultivation technique in the north-eastern part of Imerina described as 'an operation in a slash-and-burn system'.[9] It is a local variant in this region of *tavy*, or *hatsake*, vocabularies reflecting slash-and-burn practices in the east and south of the country, and now known in the world of research on Madagascar (Ramandraiarivony 2017).

It takes a real revolution for researchers to be convinced to write everything in Malagasy, but already these new expressions written in the national language are a way to decolonize geography through the prism of the daily life of the inhabitants and their values. This is a good perspective and even a challenge to be pursued, because Ramanambelina (2013) says during her PhD defence: "The way on the dissemination and publication of new terms deserves reflection if we really want to give a considerable place to the Malagasy language." The use of a language is always a political fact of "self-affirmation". Blanc-Pamard and Rakoto-Ramiarantsoa (2000) provided a Malagasy abstract of their work instead of an English version to demonstrate that more is possible.

[9]Imerina is part of the natural region of Madagascar's Central Highlands.

9.4 From the French Desk

The economic globalization of the twenty-first century is based, among other things, on what has been called the knowledge society, which Lesemann (2003) associates to the knowledge economy. In this context, governments tend to increase the pressure on academics working in universities and laboratories, private and public (Castells 1998). In most countries, the academic world has adapted to this neoliberal incentive, not always easily, not always happily. Whatever transformations that ensued, they all have to take into account the existence of the global scale. In this light, for the last twenty years, we have been able to observe this slow but assertive transformation of the French national system, where the global is mediated by the supranational, that is by Europe. As Best (2009) explains, in most European countries a new intellectual 'elite', professors who speak and write English, and who can afford travelling and attend international congresses, became in these nations the relay to the global system.

9.4.1 From National to International Conformism?

"Is there an 'Anglo American' domination in human geography. And, is it bad?" Rodríguez-Pose (2006) questions the extent of the English hegemonic position. He points out to the global scale liberating dimension from the "normative" national scale. Seen from the French desk, it appears this is an important issue that has not been well documented in European countries, including France. Were these European elites looking for ways to free themselves from the grip of a national normative system? Or put themselves above the pack, with the help of a global/international varnish? We do not have extensive factual information to answer this question adequately, but our experiences, observations and exchanges with scholars from other European countries tend to show that it was the case, most notably in subfields of geography in connection with the humanities and the social sciences.

The geography of gender, sexes and sexualities provides a good case in point in the French context. Although famous French scholars, including feminists, provided an important and essential contribution to the theorization and critic of modernity, and the power structure of liberal societies, the question of gender and sexualities did not make any official inroad within French universities before the 2010s. Up to that decade, they were often derided by self-declared serious scholars as an Anglo-American fad. Notwithstanding, a generation of geographers, influenced by the 'cultural turn' that went global, started to connect with the theoretical debates and practical research going in the geography of the Anglosphere. SciencePo, the famous French academic institution was the first to officially create in 2010 a research centre dedicated to gender and feminist research (PRESAGE); the CNRS follows suit in 2012 with LEGS (*Laboratoire d'études du genre et des sexualités*), and in 2014 with the *Institut du genre*, a multidisciplinary forum for researchers, teaching programs, and research centres. Inspired by this scientific advancement in the social sciences, French geographers, at ease with English and with the theoretical debate, put into place the '*biennale du genre*' in 2010. In so doing, they helped to shake up French geography's normative grip, thus contributing to the renewal of a well-established social geography, as well as the development of a more critical cultural geography. It was a period of many firsts: the first Ph.D. thesis on gender and geography; the first MA thesis on the question of sexuality and queer geography, etc.

In *Homo Academicus*, Bourdieu (1984) highlighted the social hierarchy that comes along with the academic hierarchy. The establishment of a scientific order prioritized by the importance attributed to the global/international dimension of scientific exchange has contributed in European countries to the advancement of a "global" class of scholars. In the French context, this change of guards, so to speak, did not come without raising some tensions and debates that were both epistemological and social, not to say political, in nature. The best piece of evidence is certainly the debate on postmodernism in geography that went on in the office of *L'Espace géographique*, at the invitation of Brunet (2004), its founder. For more than thirty pages, the debate, at times intense, has continued as Brunet's social and political geographical analysis associates those changes with American trends, not with the globalization of knowledge. Likewise, in a very interesting collective work, Chivallon et al. (1999) try to understand these changes by comparing French and British geographies in connection to their respective 'cultural contexts'. In a related but different register, Fall and Rosière (2008) try to understand divergences and convergences in Anglophone and Francophone political geography, in light of the globalization of knowledge as well as political events around the world.

Beyond these debates on the values of the global scale and its new 'international' class, French academic institutions, led by the national government, resolutely took a European turn, which means aligning to European standards and norms, themselves the product of an adaptation to the global sphere dominated by the English language (Berman 2007). For many scholars, these changes translated into meant a diminish national attraction: journals, conferences, regional studies took a hit. Consequently, some of the new French generation of geographers with an International-European profile, and networking inside the English-speaking network, ignore—do not know—the heritage of French geographers, for example, in Regional Geography! Therefore, some French geographers turned their back to what they saw as an international science disconnected to the social, cultural and political fabrics of society. But, ultimately, those two movements converged in reconsidering more positively the local/global interactions in the global scientific context. It can be advanced that the pressure of the international has favoured the construction of an internationalized researcher profile, becoming paradoxically more and more homogeneous. Faced with this trend, the national and international levels seem to be able to contribute to restoring meaning and dynamism to the International in infusing the global with diverse forms of geographic diversities.

9.4.2 Diversity of Knowledge and International Dynamism: Some Avenues to Explore

The world of research is made up of many people, spread across different public and private organizations, who are associated with the academic world. Becoming a member of this community requires a curriculum that includes, in addition to a high-quality thesis, elements that demonstrate the ability to do scientific work, to present and debate one's work and results by putting them up for discussion through communications, and increasingly to build them up in scientific networks. If we look at the different elements of a curriculum, we can see that it is ranked according to two criteria: interdisciplinary and international networking, usually implied outside the national language. This situation gradually leads to choices that can encourage the abandonment of scientific projects, and also of collaborations that are nevertheless of importance to one's society. The internationalization of exchanges can take place in many different ways, one of them being the setting up of binational and bilingual meetings (Müller M 2007).

(a) **Binational Meetings and Bilingualism**

We argue that a full recognition of the plurality of scientific expressions favours innovation. In geography, as in all human and social sciences, it is essential to maintain all scales of analysis and exchange. These feed each other and offer new perspectives. It was through observing his own society that Michel Foucault (1971) wrote *L'ordre du discours* in which he denounced the rigidity of a partitioned normative system in academia, especially the French educational system in which he was educated. His local observations have reached the global scale. Elsewhere, the British geographer David Harvey forged a new critical social geography by working within his national contexts, in the UK initially and later in the USA. Many geographers from various countries were inspired by these works. In many ways, it is difficult in geography to think about space without the interconnections of scales. There is a continuum, not a hierarchy.

Favouring internationalism while ensuring the diversity of its expressions must be based on meetings that are less complex to organize than the very large international congresses, such as binational meetings. Through the logistical and linguistic flexibility that they offer, they make it possible to promote the construction of scientific networks as well as to preserve or even encourage research on territories that may be gradually marginalized in research or simply become case studies. Indeed, in a competitive international scientific context where the curriculum is constructed through scientific communication in the dominant networks and in a common English language, the effort that the younger generations of geographers must make may lead them to privilege the study of the dominant cultural and linguistic areas, which brings together the time of study and communication in the same temporality. Moreover, the current scientific situation in geography, as in other disciplines, regularly leads to 'blind spots' in geographical knowledge. This situation is amplified through the linguistic bias when the recognition of conceptual and theoretical innovation is closely associated with dominant media of dissemination.

Setting up binational meetings can help maintain a specific and valued research in two different linguistic areas and feed the multilateral scale. Moreover, these meetings are more easily organized by a small team of geographers and therefore with reduced costs. These meetings can then be promoted in recognized national scientific journals, often with one or two issues devoted to them. The IGU can be an active promoter of this type of meeting. As we know, IGU's cornerstones are its commissions, and one of its actions is to promote—and rewards—joint sessions, where, for instance, the Cultural Approaches commission and the Urban Geography Commission propose sessions on 'urban cultures'. In the same way, it can set an incentive to the joint French-Spanish sessions on certain topics, where members of different commissions can intervene. They can be held before congress, as some commissions do, or during. This will ensure that we move away from local-language sessions in congresses where only native speakers participate.

(b) **Journals**

Scientific journals can be discussed in many ways: as commercial scientific publishers charging outrageous costs; by reference to the ranking that comes with the evaluation of journals dominated by Anglo American journals and which generally dominate the portfolios of commercial scientific publishers; as predatory platforms, facilitated by open-access science; or in terms of the publication timelines imposed on volunteer researchers. All these points have been widely debated in numerous articles and reports, including the International Science Council's 'The Future of Scientific Publishing'.[10]

Encouraging other international scientific publications while upgrading public academic journals can be done by exchanging articles already accepted by a journal to be published in a journal of another language within the framework of agreements. This practice would offer the advantage to journals of having articles ready for publication. The exchange of journal issues was a common practice when publications were made only in paper format. Agreements allowed libraries to make recent scientific production available to their readers, researchers, teachers, students, etc. Nowadays, the digitization of journals and the continuous on-line publication of articles on platforms dedicated to scientific publication have eliminated this practice. We admit that this sounds good in theory, while difficult to put in place in practice. Translation adds extra cost most journals cannot cover, and automatic translation, although improved, needs reviewing and, for a great deal, is as costly as direct translation. However, binational meetings

[10]URL: https://council.science/actionplan/future-of-scientific-publishing/

can help generate co-publication between two national journals (Germes M, Husseini de Araújo S 2016) .

Geographical journals continue to be produced in all countries in their national language, with editorial boards and reviewers, often volunteers, that defend this model of disseminating geographical knowledge. These journals are necessary because they are as close as possible to local systems of thought, models of innovation and knowledge production. They are able to measure the progress and emergence of new theories and new territorial concepts. As we have mentioned above, one solution for national journals is to decide to publish only in English. The goal is more to have local geographers being read by other geographers globally, although the journal remains peripheral to Anglo-American leading journals. However, their contribution might be to help geographers to have access to a diversity of works done in a language other than English, even if they cannot read it (and here automatic translation, imperfect as it is, can be useful). It can in principle enhances scientific knowledge in providing more international references. This might help to move away a little from normative conceptions, theories or, even, vision of the world.

Often underestimated by restrictive public research policies, these alternative forms of geographical dissemination must retain their place in geography's research landscape. The vitality and motivation of these non-English journals can be demonstrated in many ways and they would benefit from exchanging some of their articles with other academic journals to accelerate the dissemination of work internationally. For the authors who publish in these journals, reciprocal agreements of this kind would allow them to broaden their readership without being accused of self-plagiarism, an argument that could lead authors to restrict their support to journals that are already ranked internationally. The dissemination of results in several languages is therefore achieved by valorizing the work of translation, which allows a wider audience to access a more diverse production.

9.5 Conclusion

The IGU has a long history of engaging with the challenges posed by language. We think the IGU still has a role to play in a debate it helped to initiate in geography as science has grown steadily more global. The IGU's diverse membership, covering all cultural and linguistic parts of the world, means that it remains the principal international forum in which the relationship between geography and language can be explored for the betterment of geographical knowledge. As is widely acknowledged, not least by the IGU itself, language plays an important role in the production of geographical knowledge about the complex and overlapping worlds that geographers seek to understand and explain. Translation is not itself a problem in this respect, but transposing is. The relationship between language and geographical knowledge needs to be more carefully studied, and the IGU is well placed to encourage this. The question of conformism and the possible exclusions this can produce should also be investigated. We suggest creating a commission on languages and the production of geographic knowledge. As we have suggested above, encouraging bi-national meetings within and between commissions can also help. Being aware of the problems and keeping the discussions open should be, in this context, the best attitude.

Acknowledgements The authors wish to thank the editors and the two evaluators of their first draft for their insightful remarks and suggestions.

References

Aalbers M (2004) Creative destruction through the Anglo-American hegemony: a non-Anglo-American view on publications, referees and language, area, 36(3):319-322. URL: https://www.jstor.org/stable/20004397

Bański J, Ferenc M (2013) "International" or "Anglo American" journals of geography? Geoforum 45:285–295

Berman R (2007) The humanities, globalization, and the transformation of the university. Profession 2006:210–217. URL: http://www.jstor.org/stable/25595868

Best U (2009) The invented periphery: constructing Europe in debates about 'Anglo hegemony' in geography. Soc Geogr 4:83–91

Blanc-Pamard C, Rakoto-Ramiarantsoa H (2000) Le terroir et son double. Tsarahonenana 1966–1992 IRD Éditions, Marseille/Madagascar

Bourdieu P (1984) Homo Academicus. Éditions de Minuit, Paris

Brunet M (2004) Débat: le postmodernisme en géographie. L'espace Géographique 1(33):6–37

Castells M (1998) La société en réseaux. L'ère de l'information. Fayar, Paris

Chivallon C et al (eds) (1999) Discours scientifique et contextes culturels. Géographies françaises et britanniques à l'épreuve postmoderne. Maison des Sciences de l'Homme d'Aquitaine, Talence

Curry MJ, Lillis T (2013) A scholar's guide to getting published in English: critical choices and practical strategies. Multilingual Matters, Clevedon, UK

da Silva Machado J (2013) Le mythe de la langue franche dans les sciences : une idéologie objectivée. Revue Hermes 65:211–215. DOI: https://doi.org/10.4267/2042/51523

Desbiens C, Ruddick S (2006) Speaking of geography: language, power, and the spaces of Anglo-Saxon 'hegemony. Environ Plann D Soc Space 24(1):1-8. URL's: https://www.researchgate.net/journal/Environment-and-Planning-D-Society-and-Space-1472-3433, https://www.researchgate.net/publication/248881506

Fall JJ, Minca C (2012) Not a geography of what doesn't exist, but a counter-geography of what does: rereading Giuseppe Dematteis' *Le Metafore della Terra*. Prog Hum Geogr 37(4):542–563

Fall JJ, Rosière S (2008) On the limits of dialogue between Francophone and Anglophone political geography. Polit Geogr 27:713–716

Foucault M (1971) L'ordre du discours: leçon inaugurale au Collège de France prononcée le 2 décembre 1970. Gallimard-Collection Blanche, Paris

Garcia-Ramon MD (2003) Globalization and international geography: the questions of languages and scholarly traditions. Prog Hum Geogr 27(1):1–5

Germes M, Husseini de Araújo S (2016) Pour une pratique critique de la traduction en géographie. ACME Int E-Journal Crit Geogr (special issue on Transl Crit Pract) 15(1):1–14. URL: https://hal.archives-ouvertes.fr/halshs-01324093/

Guebourg J-L (1997) Analyse de la filière-tabac malgache: de l'étatisation pure à la société mixte. Les Cahiers D'outre-Mer (revue De Géographie De Bordeaux) 50(99):245–270

Gutiérrez J, Lopez-Nieva P (2001) Are international journals of human geography really international? Prog Human Geogr 25(1):53–69

Hallanger FS (1973) Diksionera Malagasy-Englisy. Antananarivo, Trano Printy Fiangonana Loterana Malagasy

Harris CD (2001) English as an international language in geography: development and limitations. Geogr Rev 91(4):675–689

Houssay-Holzschuch M, Milhaud O (2013) Geography after Babel: a view from the French province. Geogr Helv 68:51–55. URL: https://gh.copernicus.org/articles/68/51/2013/

Johnston RJ, Sidaway JD (2004) The trans-Atlantic connection: 'Anglo American' geography reconsidered. GeoJournal 59:15–22

Kitchin R (2005) Commentary: disrupting and destabilizing Anglo American and English-language hegemony in geography. Soc Cult Geogr 6(1):1–15

Kong L, Quang J (2019) Knowledge circulation in urban geography/urban studies, 1990–2010: testing the discourse of Anglo American hegemony through publication and citation patterns. Urban Stud 56:44–80

Lesemann F (2003) La société des savoirs et la gouvernance : la transformation des conditions de production de la recherche universitaire, Lien sociale et Politiques, 50:17–37. URL: https://www.erudit.org/fr/revues/lsp/2003-n50-lsp698/008277ar.pdf

Müller M (2007) What's in a word? Problematizing translation between languages. Area 39(2):206–213

Nally D, Costello B (eds) (2000) Interview with Professor Anne Buttimer. Chimera 15:16–21

Ouane A, Glanz C (2010) Pourquoi et comment l'Afrique doit investir dans les langues africaines et l'enseignement multilingue: note de sensibilisation et d'orientation étayée par les faits et fondée sur la pratique. Institut de l'UNESCO pour l'apprentissage tout au long de la vie, Paris

Paasi A (2005) Globalisation, academic capitalism, and the uneven geographies of international journal publishing spaces. Environ Plann A Econ Soc 37:769–789

Paredes XM, da Silva Mendes S (2002) The geography of languages: a strictly geopolitical issue? The case of 'international English'. Chimera 17:104–112. https://web.archive.org/web/20080414003235/, https://www.xoan.net/recursos/Geolinguistics.pdf

Raharinjanahary R (2011) Antananarivo et ses ordures: production, collecte, valorisation. Foi et Justice Édition, Antananarivo

Raharinjanahary R, Rajoelina S (2018) Les nuits, poumons de l'agglomération d'Antananarivo. Bollettino della Società Geografica Italiana 14(2):151–161. URL: www.fupress.com/bsgi

Rakoto-Ramiarantsoa H (1995) Chair de la terre, Œil de l'eau: paysanneries et recompositions de campagne en Imerina (Madagascar). IRD Édition, Marseille. URL: https://www.editions.ird.fr/auteur/49/Herve%20Rakoto%20Ramiarantsoa

Rakoto-Ramiarantsoa H, Peyrusaube D (2007) Nuages, vents et pluies: scruter le 'visage du firmament' en Imerina. Cahiers D'outre-Mer 240:319–340. DOI: https://doi.org/10.4000/com.2514

Ramanambelina H (2013) Problèmes d'aménagement linguistique à Madagascar: termes de géographie dans l'enseignement. Unpublished Ph.D. thesis, Inalco, Paris

Ramandraiarivony RJT (2017) Le périmètre irrigué de Dabara: un potentiel économique mal exploité. In: Proceedings of the colloquium on development,

spaces, territory and climate change, pp 86–94. URL: https://madarevues.recherches.gov.mg
Ranaivoson D (2007) La langue malagasy et les politiques linguistiques. Erudit: Études Littéraires Africaines 23:6–12. URL: http://doi.org/10.7202/1035446ar
Rhaïb D (1996) Le mélange de langues dans le discours d'enseignants marocains de français au Maroc, en dehors de la classe. Revue québécoise de linguistique 24(2):151–162. URL: https://www.erudit.org/fr/revues/rql/1996-v24-n2-rql2943/603118ar.pdf
Riquier J (1954) Etude sur les 'lavaka'. Mémoires de l'Institut Scientifique de Madagascar. Tananarive: Série D: Sciences de la Terre 169–189
Robic MC, Briend AM, Rössler M (eds) (1996) Géographes face au monde. L'Union géographique Internationale et les congrès internationaux de géographie. L'Harmattan, Paris
Rodríguez-Pose A (2006) Is there an 'Anglo American' domination in human geography? And, is it bad? Environ Plann A Econ Soc 38:603–610
Schmitz E (2013) The bilingual triangle: teaching content and language integrated geography. 1st edn (6 Mar 2008), GRIN Verlag, Franckfurt
Shaw W, Delyser D, Crang M (2015) Limited by imagination alone: research methods in cultural geography. Cult Geogr 22(2):211–215
Short JR et al (2007) Cultural globalization, global English, and geography journal. Prof Geogr 53:1–11
Street L, Andrianaly M (1870) Ny Geography vaovao, hianarana ny toetrany izao tontolo izao. London Missionary Society, Antananarivo
Timár J (2004) More than 'Anglo American', it is 'Western': hegemony in geography from a Hungarian perspective. Geoforum 35(5):533–538
Timmis I (2002) Native-speaker norms and international English: a classroom view. ELT J 56(3):240–249. DOI: https://doi.org/10.1093/elt/56.3.240
Vollé D (1996) L'universalité et ses limites. In: Robic MC, Briend AM, Rössler M (eds) Géographes face au monde. L'Union géographique Internationale et les congrès internationaux de géographie. L'Harmattan, Paris, pp 63–86
Vololonirainy R, Finoalahy F (2017) La mise en place de l'AP et la mutation du système agraire dans le terroir de Miadana sur le versant oriental du corridor forestier Anjozorobe-Angavo. In: Proceedings of the Colloquium on development, spaces, territory and climate change, pp 46–61. URL: https://madarevues.recherches.gov.mg
Webber J (1853) Dictionnaire malgache-français, rédigé selon l'ordre des racines par les missionnaires catholiques de Madagascar, et adapté aux dialectes de toutes les provinces. Établissement malgache de Notre-Dame de la Ressource, Ile Bourbon
Wismann H (2012) Penser entre les langues. Albin Michel, Paris

Rindra Raharinjanahary is an Associate Professor of geography at Antananarivo University (Madagascar). She is currently President of the *Union des Géographes* de Madagascar, and the editorial director of the on-line journal *Madagascar, revue de Géographie*. Her research interest is in the informal activities in cities, as well as in the waste process in Antananarivo. She is an active member of the research project "Towards a collaborative observatory of landscapes and their sustainability," where she assumed the responsibility of training and diffusion of knowledge.

Nathalie Lemarchand is a Full Professor of geography at the University of Paris 8 (France). She is currently Vice-President of the International Geographical Union, and member of the Council of French National Committee of Geography (CNFG). Her research interest is the geography of retailing and consumption, where she favours a cultural and social approach; her latest publication deals with local protest movements against global shopping mall projects. Very involved in academic institutions at all levels, she has been reflecting for several years on the languages of scientific production and diffusion.

Louis Dupont is a Full professor of geography at Sorbonne Université (France). He has been involved for more than a decade in the journal *Geography et Cultures,* of which he was the editor for 6 years; he is currently an elected official in the Research Commission of La Sorbonne. His research interest is in cultural and social geography, dealing with modern issues most notably in North American societies. His recent publications are on cultural diversity and tensions, as well as on the methodological (and reflexive) challenge posed by the study of culture.

Geography and International Education

10

Rafael de Miguel González
and Karl Donert

Abstract

International geographic education emerged at the beginning of the nineteenth century, as a consequence of the progressive implementation of modern educational systems in European countries. This process coincided with the consolidation of geography as a scientific discipline and with the creation of different national geographical societies. The foundation of the IGU in 1922 fostered the diffusion of the concept of international understanding in educational curricula, first at the various International Geographical Congresses, and later with UNESCO's initiatives for the international teaching of geography, which led to the creation of the IGU Commission on Geographical Education in 1952. Source Books for Geography Teaching (1965 and 1982), International Charters on Geographical Education (1992 and 2016), Symposia of Commission on Geographical education, collaboration with other associations (like EUROGEO, AAG, SEAGA), international projects and publications, International Geographical Olympiads, etc., have contributed to the internationalisation of geographical education. Thus, geography educators worldwide can network, share experiences in curriculum, pedagogies and assessment, collaborate in good practices and instructional resources like geospatial technologies, or promote global understanding and a planetary citizenship seeking a sustainable development.

Keywords

Geographical education · IGU-CGE · EUROGEO · International understanding · Curriculum · Source book · International charter

10.1 Introduction

In 2022, we celebrate the centenary of the International Geographical Union (IGU). Geography is, of course, a much older discipline, dating back to the time of ancient civilisations but it is consolidation as a science coincides with the establishment of the universal education systems for the entire school population implemented in European countries during the nineteenth century. Since then, geography education has been conceived from a national perspective and was codified by different national geographical societies and geographical education associations. This trend continued until the end of the first half of the twentieth century.

R. de Miguel González (✉)
EUROGEO & University of Zaragoza, Zaragoza, Spain
e-mail: rafaelmg@unizar.es

K. Donert
European Association of Geographers (EUROGEO), Wardamme, Belgium

V. Kolosov et al. (eds.), *A Geographical Century*,
https://doi.org/10.1007/978-3-031-05419-8_10

Despite the creation of International Geographical Union (IGU) in 1922 and the organisation of various International Geographical Congresses (IGCs), both before and after the IGU was established, it wasn't until the 17th IGC in Washington DC in 1952, when the Commission on Geographical Education (IGU-CGE) was formalised to promote international geographical education with the support of UNESCO. Since then, the internationalisation of geographical education has been promoted by the IGU-CGE in four ways: (i) by organising its own symposia and geographical education sessions at IGCs; (ii) by publishing various Source Books for geography teaching; (iii) by proclaiming different International Charters and Declarations; and (iv) by working with national and international geographical organisations. However, the IGU-CGE has focused on concepts such as international understanding, international cooperation, an international action plan on global understanding, as well as collaboration with other important associations dealing with geography education, such as the National Council for Geographic Education (NCGE), the European Association of Geographers (EUROGEO), the Geographical Association (GA) and the American Association of Geographers (AAG) (Bourke and Lane 2018; De Miguel 2020).

Initiatives like the International Network for Learning and Teaching (INLT) were promoted to strengthen links between geography higher education, education research and school education through teacher education and education research international networking (Bednarz et al. 2000). Other initiatives like the HERODOT project or the Center for Global Geography Education (CGGE) have implemented more collaborative and integrative international education development in networks, too (Shepherd et al. 2000; Donert et al. 2011; Higgitt et al. 2008; Klein and Solem 2008).

Authors such as Martin Haigh decisively contributed to the intellectual definition of internationalisation in geography education, described as having diverse forms as it is both complex and multi-layered (Haigh 2014) ranging from international collaborative learning and teaching, to curriculum creation in higher education (especially in the European Higher Education Area) and the development of international projects. Internationalisation has advantages like the dissemination of good practices and instructional resources thanks to new information and communication technologies, or the promotion of a planetary citizenship seeking a future sustainable in environmental, social and ethical terms. There are also disadvantages as internationalisation can be viewed as the offspring of economic globalisation seeking financial gains (Haigh 2008).

In contrast, and predominantly working at a different scale, the European Association of Geographers (EUROGEO) was established initially as an international open institution for all EU countries and with the support of the European Union. The association is thus an original, transnational and cooperative organisation producing relatively more integrated and balanced geography education outcomes among most European countries and with the inclusion of scholars, teachers, educators and other stakeholders from the wider world. This study examines, using a critical historiographical methodology, a century in the evolution of key ideas, paradigms, periods and milestones of international geographical education led by the aforementioned institutions, as well as their claims for increasing international collaboration among geography educators in a world today characterised by global trends like digitalisation or global challenges like sustainable development.

10.2 Early Steps Until 1922

Geography is a scientific discipline born in Europe. And the same is true of geography education. If Ancient Geography was created in Classical Greece, Modern Geography was developed first in Germany, the country of both Alexander Humboldt and Carl Ritter, the founders of Modern Geography. Humboldt's best-known publication, Kosmos (1845–1862), laid the groundwork for systematic geography. Ritter attained the first ever Chair of Geography in

1820 at the University of Berlin where he completed his main work Die Erdkunde (1822–1859). Both lived and wrote through the Industrial Revolution, where economic development required the constant search for raw materials and markets beyond national borders, and through an age of political revolutions when modern national identities were established, especially in Europe. Both transformations—economic and political—contributed to an increase in rivalry between European powers, and to the transformation of colonialism and imperialism. Furthermore, these political and economic changes had a strong social impact on the demographic transition in Europe, on the growth of cities and on the acquisition of essential political individual rights such as universal education in recognition of citizenship.

This was the time, during the first half of the nineteenth century, when geographical societies were created as institutions to promote the advancement of geographical knowledge, but also to support colonial exploration mainly in Africa and Asia as they had a strong support from the mercantile, diplomatic and military classes. The Societé de Geographie de Paris was founded in 1821, and subsequent institutions were created: the Geographical Society of Berlin (1828), the Geographical Society of London (1830) (later the Royal Geographical Society), the Russian Geographical Society (1845), the American Geographical Society (1851) (with the National Geographic Society following in 1888), the Austrian Geographical Society (1856), the Italian Geographical Society (1867), the Spanish Real Sociedad Geográfica (1876), the Association of American Geographers (1904), etc.

In 1819, the first geography textbook was published in London specifically for use in schools under the title A Grammar of General Geography. It was written by an English cleric, the Reverend J. Goldsmith (Fig. 10.1). By 1850, geography was widely extended as a school subject in primary and secondary education in several European countries. Furthermore, the results of the Franco-Prussian War accelerated an academic institutionalisation of geography as a scientific discipline and as a school subject. After 1870, political reasons related to a reaffirmation of national identity, supplemented by imperial expansionism, the advancement of natural and social sciences, and in particular by the new legal status of basic education led to the creation of national geographical curricula for schools in the resulting political regimes of the French Third Republic and German Empire.

In 1870, an Act to provide for public Elementary Education in England and Wales was set. In the USA, education was compulsory in Massachusetts after 1852 and District of Columbia after 1864, but eighteen States between 1870 and 1883. As expressed by Johnston (2018): 'Geography is one of the few academic disciplines, particularly in Europe, to have been established in universities as a result of pressure to produce people who could teach it in schools. As the demand for geographical information increased, more people required a foundation of geographical knowledge'.

Three main institutions supplied this demand for teachers of Geography: universities, geographical societies and associations for geography education. Thus, after many decades of Ritter's Chair in Berlin, a new Chair of Geography was established at the University of Leipzig in 1871, awarded to Oscar Peschel. In France, Paul Vidal de la Blache became Chair of Geography at the University of Nancy in 1872. The first Chair of Geography was created at the University of Amsterdam in 1877, and one decade later the University of Oxford created the first British Readership in Geography in 1887. In the USA, William Morris Davis taught geography at Harvard University from 1876 (Mamadouh 2005). In Spain, Eloy Bullón was awarded the first Chair of Geography in 1907 at the University of Madrid. Geographical societies began to play an important role in publishing textbooks and maps for teaching geography in primary and secondary education (Bosque 2004; Arroyo 2014). But in some countries, organisations specifically devoted to promoting geography education were created, including the Geographical Association (GA) in the UK (1893), the Association des Professeurs d'Histoire et de Géographie in France (1910), the

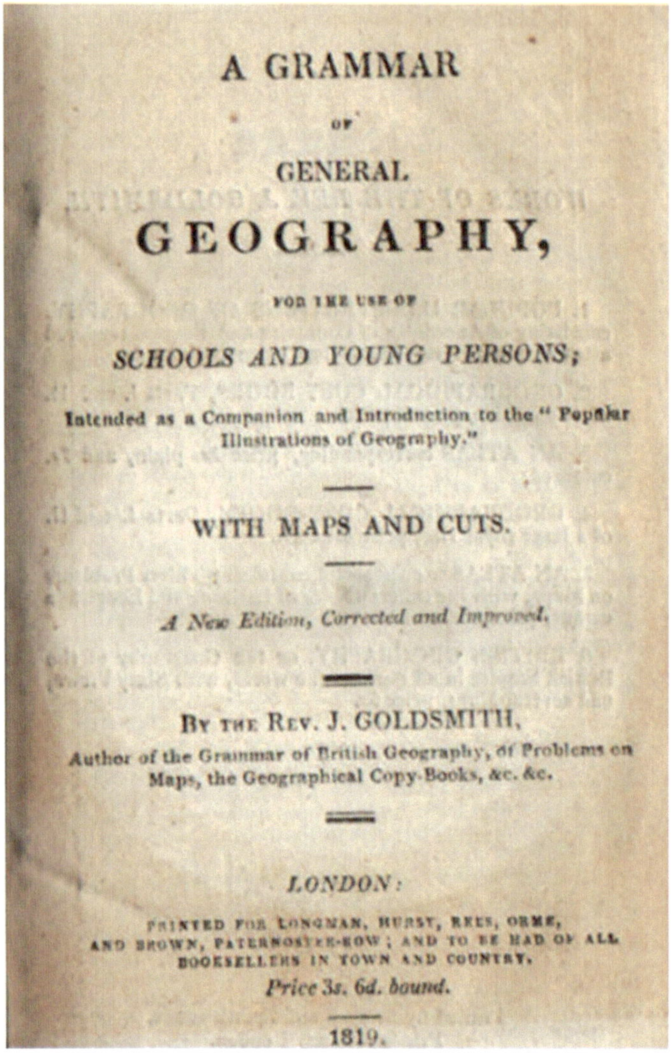

A GRAMMAR

OF

GENERAL

GEOGRAPHY,

FOR THE USE OF

SCHOOLS AND YOUNG PERSONS;

Intended as a Companion and Introduction to the "Popular Illustrations of Geography."

WITH MAPS AND CUTS.

A New Edition, Corrected and Improved.

BY THE REV. J. GOLDSMITH,

Author of the Grammar of British Geography, of Problems on Maps, the Geographical Copy-Books, &c. &c.

LONDON:

PRINTED FOR LONGMAN, HURST, REES, ORME, AND BROWN, PATERNOSTER-ROW; AND TO BE HAD OF ALL BOOKSELLERS IN TOWN AND COUNTRY.

Price 3s. 6d. bound.

1819.

Fig. 10.1 Cover of the reverend J. Goldsmith's *Grammar of General Geography* (1819)

Verband Deutscher Schulgeographen in Germany (1912) and the National Council for Geographic Education in the USA (1915).

One way or another, these three types of institutions were nationally based, so their outcomes in terms of what geography was taught or how to teach geography was influenced by their historical context, by the national identity reaffirmation and by the territorial ambitions of the imperial governments funding them. In other words, an international approach to geography education did not exist at this time. In spite of this, in 1871 the First International Geographical Congress was held in Antwerp. Until 1922 subsequent meetings had been held in Europe, except for the meeting in 1904 in Washington DC, but in the IGU archives, geography education is not referred to during this period.

10.3 The International Geographical Union, 1922–1952

The academic references on international geography education—and consequently the importance of teaching geography as a factor in the construction of peace between nations—goes back more than three decades before the creation of the IGU-CGE. The Treaty of Versailles, signed on 18th June 1919, included repeated references to cooperation between nations to guarantee peace and international security after the war. The first part of the Treaty (Articles 1 to 26) included the Covenant of the League of Nations based on the concept of international understanding. In education, these ideas directly inspired the World Federation of Education Associations (WFEA), created de facto in 1923 in San Francisco and de jure in 1925 at its first World Conference in Edinburgh (Smith 1944).

The first issue of the GA's journal, The Geographical Teacher, published in 1901, included some preliminary references to international geography education. In 1920, the GA was opened to geographers from outside the UK and members were welcomed from all continents in the world in an explicit attempt to internationalise geographical education on a British model. The geographer H.J. Fleure (1919) emphasised the potential role of geographical education in 'efforts for international understanding' just several weeks after the Treaty of Versailles was signed. Later, the GA's 1920 Annual Report recognised that the growth of overseas membership was helping education and international understanding as 'the continuous growth in importance of international questions makes our claim ever stronger: geographical education is one of the essentials for the development of international peace' (Fleure 1921). This idea was later supported by Atwood (1922) and Fairgrieve (1926) in their *Geography in School*, one of the most influential books on geography education (Thomas 1932).

The international institutionalisation of geography only formally occurred with the founding of the IGU in 1922 (Robic 1996; Schelhaas and Pietsch 2020). This was clear from reports published in the journals of the major geographical societies whose representatives had attended the inaugural meeting of the IGU in Brussels from 25–29 July 1922 (Fig. 10.2; Hinks 1922; Societé de Géographie de Paris 1922; Buen and Gómez Núñez 1923; Tur 1923). Few of these participants had been educated as geographers. Many were from the armed forces, and most were trained as engineers, surveyors, biologists or astronomers.

From the eight national geographical societies that took part in that inaugural meeting and the twenty-two founding delegates, only two of them had a significant scientific production in geography and geographical education (Buen 1909, 2003; Martonne 1911). Their participation reflects on the importance of renewing their scientific methods as well as engaging in less traditional learning and more outdoor learning. Emmanuel de Martonne was later President of the IGU between 1939 and 1949, at the important time to establish the most productive partnership between the IGU and UNESCO for the dissemination of international geography education. Odón de Buen (born and raised in Zuera, Zaragoza) inherited the powerful contributions to geographical education from other fellows at the Real Sociedad Geográfica like Martín Ferreiro, Rafael Torres and Ricardo Beltrán, and many years before the founding meeting of the IGU Odón de Buen had strongly advocated for geographical education, while he was Spanish Senator (Buen 1909).

The formal establishment of the IGU as a permanent organisation in 1922 was an important milestone. The IGU relaunched the series of IGCs, but it took a long time for geography education to be recognised. There is no record of a group of participants for geography education at either the IGCs in Cairo (1925), Cambridge (1928) or Paris (1931). The Warsaw IGC (1934) was the first to include an individual track on education. The fourth section in the Warsaw Congress was entitled 'Methods of teaching geography', with sixteen contributions. At the Amsterdam IGC (1938), twenty-nine papers were presented in its sixth section on methodology and didactics). One paper, related to international education by teaching the 'geography of human relations' based on the good understanding among the people of the planet (Orford 1938), was unfortunately written against the historical context of the Munich Agreement and confrontation prior to World War Two.

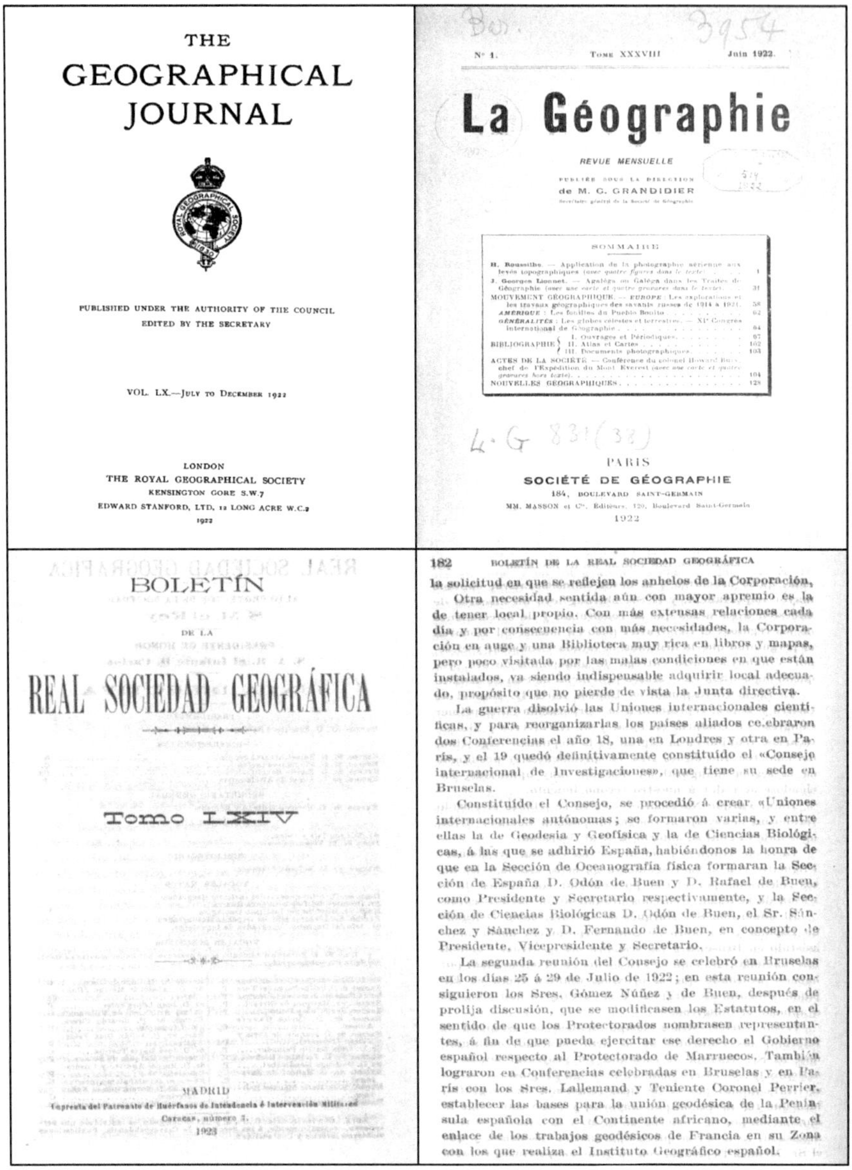
THE GEOGRAPHICAL JOURNAL

PUBLISHED UNDER THE AUTHORITY OF THE COUNCIL
EDITED BY THE SECRETARY

VOL. LX.—JULY TO DECEMBER 1922

LONDON
THE ROYAL GEOGRAPHICAL SOCIETY
KENSINGTON GORE S.W.7
EDWARD STANFORD, LTD, 12 LONG ACRE W.C.2
1922

N° 1. TOME XXXVIII Juin 1922.

La Géographie

REVUE MENSUELLE

de M. G. GRANDIDIER

SOMMAIRE

PARIS
SOCIÉTÉ DE GÉOGRAPHIE
184, BOULEVARD SAINT-GERMAIN
1922

BOLETÍN
DE LA
REAL SOCIEDAD GEOGRÁFICA
Tomo LXIV
MADRID
1923

182 BOLETÍN DE LA REAL SOCIEDAD GEOGRÁFICA

la solicitud en que se reflejen los anhelos de la Corporación.

Otra necesidad sentida aún con mayor apremio es la de tener local propio. Con más extensas relaciones cada día y por consecuencia con más necesidades, la Corporación en auge y una Biblioteca muy rica en libros y mapas, pero poco visitada por las malas condiciones en que están instalados, va siendo indispensable adquirir local adecuado, propósito que no pierde de vista la Junta directiva.

La guerra disolvió las Uniones internacionales científicas, y para reorganizarlas los países aliados celebraron dos Conferencias el año 18, una en Londres y otra en París, y el 19 quedó definitivamente constituído el «Consejo internacional de Investigaciones», que tiene su sede en Bruselas.

Constituído el Consejo, se procedió á crear «Uniones internacionales autónomas; se formaron varias, y entre ellas la de Geodesia y Geofísica y la de Ciencias Biológicas, á las que se adhirió España, habiéndonos la honra de que en la Sección de Oceanografía física formaran la Sección de España D. Odón de Buen y D. Rafael de Buen, como Presidente y Secretario respectivamente, y la Sección de Ciencias Biológicas D. Odón de Buen, el Sr. Sánchez y Sánchez y D. Fernando de Buen, en concepto de Presidente, Vicepresidente y Secretario.

La segunda reunión del Consejo se celebró en Bruselas en los días 25 á 29 de Julio de 1922; en esta reunión consiguieron los Sres. Gómez Núñez y de Buen, después de prolija discusión, que se modificasen los Estatutos, en el sentido de que los Protectorados nombrasen representantes, á fin de que pueda ejercitar ese derecho el Gobierno español respecto al Protectorado de Marruecos. También lograron en Conferencias celebradas en Bruselas y en París con los Sres. Lallemand y Teniente Coronel Perrier, establecer las bases para la unión geodésica de la Península española con el Continente africano, mediante el enlace de los trabajos geodésicos de Francia en su Zona con los que realiza el Instituto Geográfico español.

Fig. 10.2 Reports on the establishment of the IGU in British, French and Spanish journals from the geographical societies in those countries, 1922–1923

In 1939, the International Bureau of Education (IBE, later integrated into UNESCO) published the first report on 'The Teaching of Geography in Secondary Schools, according to the data provided by Ministries of Education'. This contributed to reinforce the international perception of geography education. The year also marked the end of the dream for international peace as World War Two started. After the War, two international organisations were established, the UN and UNESCO. The Constitution of UNESCO, reaffirming the main principles expressed in the Charter of San Francisco, established as its main purpose the promotion of mutual knowledge and understanding among nations as a means to maintain world peace and avoid a World War Three. As described by the second President of the Commission on Geographical Education (Brouillette and Vilá Valentí 1971), UNESCO was interested in geography, history and social sciences as strategic disciplines for the achievement of its objectives from the very beginning. Thus, in 1949, UNESCO fostered three initiatives that were key to understanding the later development of geography education worldwide (De Miguel 2020): a new programme on the Teaching of Geography and International Understanding (UNESCO 1949a), a new series of Handbooks for the improvement of textbooks and teaching materials (UNESCO 1949b) and regular reports containing 'Some Suggestions on the Teaching of Geography' (UNESCO 1949c).

Besides this, the 16th IGC was held in Lisbon in 1949, the first in eleven years after World War Two. A Geography Teaching Committee, an embryo of the future Commission, chaired by Neville Scarfe was organised in this Congress. Some months later, he became the counsellor of the Joint Seminar (Committee Teaching IGU-UNESCO) of Montreal in 1950 and used the booklet 'Some suggestions…' as documentation for the debates and studies. After the deliberations of the seminar, this document was reviewed by Scarfe himself and published with the new title of Handbook of Suggestions on the Teaching of Geography: Towards World Understanding (UNESCO 1951). It was widely disseminated by UNESCO and translated to several languages and was also the main reference for the 1965 source book.

10.4 The IGU Commission on Geographical Education and the Source Books for Geography Education, 1952–1992

The IGU Commission on Teaching Geography in Schools was formed at the 17th IGC held in Washington DC in 1952 but not without difficulties (Brouillette and Vilá Valentí 1971; Lidstone and Stoltman 2002) and despite a blurry precedent dating back to 1904 (Schneider 1972; Wise 1992). The first President was Neville Scarfe who continued his leadership until the 18th IGC held in 1956 in Rio de Janeiro, when Benoît Brouillette was elected as the new President of the Commission. He held that position until the 21st IGC held in 1968 in New Delhi, when the IGU-CGE got its current name. Subsequent Presidents have written a detailed history of IGU-CGE activities (Graves and Stoltman 2015), and in particular about the IGU-CGE Symposia, most of them ran in parallel to the International Geographical Congress and IGU Regional Conferences (Annex 1).

Brouillette reinforced the relationship between the IGU Commission and UNESCO thanks to the signatures of several research contracts from 1957 (Brouillette and Vilá Valentí 1971). Some of these research achievements were presented at the 20th IGC held in London in 1964. However, the final outcome was the publication of the UNESCO Source Book for Geography Teaching (UNESCO 1965), which had a great international diffusion not only as a result of its translation into eleven languages but also due to the Source Book becoming the reference manual for the training of geography teachers in many countries. Two supplementary books were published later as a regional localisation of the Source Book, one for Africa (Brouillette, Graves and Last 1974) and one for Latin America in Spanish (Brouillette and Vilá Valentí 1975).

The Source Book began with a declaration of principles—mutual understanding among people, global literacy and youth civic education—but it also included information on useful pedagogies such that the manual could be used in schools in all countries and contexts. The geography teaching approaches were based firstly on the local and regional spatial problems as they were more accessible to students, thus allowing later the change of scale to the national and international scales, as well as teaching international understanding and international solidarity.

The historical context of the revision of Source Book was very different to that produced two decades earlier. In the case of international geography, education was even more evident at the

23rd IGC was held in 1976 in Moscow, where political rivalries between the USA and USSR barely interfered in the advancement and consolidation of international collaboration in geography research. However, the new UNESCO Source Book for Geography Teaching (Graves 1982), written by the Chair of IGU Commission on Geography Education between 1972 and 1980, hardly included any reference to international collaboration in geography education or international understanding. This new Source Book was essentially oriented towards the postulates of new geography and quantitative approaches but also to the economic, social, ecological and spatial values of geography as a school discipline. It included new pedagogical language like curriculum organisation, instructional design, problem-solving and evaluating learning outcomes that are still valid today (Fig. 10.3).

10.5 Recent Work of the IGU Commission on Geographical Education and the International Charters on Geographical Education, 1992–2016

The Chair of the IGU Commission on Geography Education between 1980 and 1988 recognised the conceptual shift from international understanding to international education and global issues, particularly after the implementation of the Global Geography Project (Lidstone and Stoltman 2002; Stoltman 2006). The Chairs of the Commission between 1988 and 1996 (Haubrich 2006) and between 1996 and 2000 (Gerber 2003) agreed with the increasing global agenda for geography education in the last two decades of the twentieth century.

The 27th IGC held in Washington DC in 1992 proclaimed the International Charter on Geographical Education. This document was supported in the principles set out in the UN and UNESCO Charters, Constitutions and Declarations, and particularly the UNESCO Recommendation of 1974 concerning Education for International Understanding. However, a global approach was introduced in geographical education literature and the Charter proclaimed the 'right to high-quality geographical education', both in terms of a balanced regional and national identity and a commitment to international and global perspectives. Thus, many decades later, the foundational aim of geography education had changed from international understanding to international and global commitment. In other words, this recognised that the world had changed enormously and geography had become much more complex as a science. Moreover, geographical education needed further reflection to enable students to develop knowledge, understanding, skills and values regarding the increasing society–environment relationships, spatial interconnections, environmental impacts and scale interactions.

Fig. 10.3 UNESCO Source Book for Geography Teaching (1965, French version) and the New UNESCO Source Book for Geography Teaching (1982)

Based on the UNESCO Recommendation of 1974, the Charter for Geographic Education highlighted the importance of international geography education in promoting understanding, tolerance and friendship among all nations from an 'international dimension and a global perspective'. Thus, the Charter focused on International Cooperation to promote bilateral or multilateral exchanges for geography educators from all the countries. The consolidation of an international geography education workforce had been possible through the IGU-CGE, international geography associations (in particular EUROGEO, founded in 1979, whose previous name ESCGTA was included in the Charter Appendix 1), and national geography associations as active stakeholders who meet the national education policy makers to set curriculum (Stoltman 1997).

The International Charter was also pioneering as it introduced, thirty years ago, concepts for geography education that currently remain the focus of social debate like climate change and sustainable development. Other key issues in geography education like cultural diversity or the impact of geospatial technologies have been included in subsequent declarations (Fig. 10.4), namely the International Declaration on Geographical Education for Cultural Diversity (2000) and the Lucerne Declaration on Geographical Education for Sustainable Development (2007).

These two International Declarations have since been supplemented by others. The international community of geography educators has been even more actively engaged with global ethics in geography education so that a stronger global leadership in geography education is claimed to influence the processes of global education and globalisation (Haubrich 2009). Consequently, three important additional documents have been introduced to reinforce the international collaboration in geography education (De Miguel 2020): a Joint Declaration on Geographical Education in Europe was signed by IGU-CGE, EUROGEO and EUGEO in Rome 2013 (Fig. 10.5), an International Strategy for Geography Education was presented in 2014 at the EUROGEO Conference held in Malta and at the IGU Regional Conference held in Krakow and the International Declaration on Research in Geography Education which was proclaimed at the IGU Regional Conference held in Moscow in 2015. Two other important milestones in 2015 were the Paris Agreement within the United Nations Framework Convention on Climate Change and the Sustainable Development Goals set by the United Nations General Assembly, at which EUROGEO members participated as

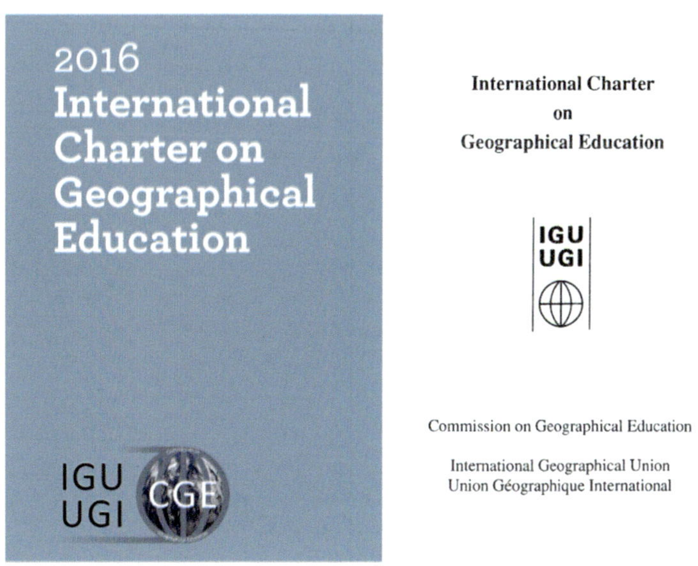

Fig. 10.4 IGU-CGE international charters on geographical education (1992 and 2016)

Rome Declaration on Geographical Education in Europe IV EUGEO Congress 2013

download PDF

Geographical education provides students with essential capabilities and competences needed to know and understand the world. Responsible and effective uses of geographical information are vital for the future of Europe. Therefore, all European citizens need to understand how to deal with it. Geographical education provides them with the knowledge and skills to do this. For example, an appropriate use of geospatial data and technologies is necessary to help analyse and address problems related to water, climate, energy, sustainable development, natural hazards, globalisation and urban growth. Most of these big issues also have a distinct European dimension. Geography also deals with the daily living environment of citizens where issues such as housing, employment, transportation, provision of services and green spaces are important. These must all be addressed but in an integrated way, which only the study of Geography provides. Geographical knowledge is indispensible for well informed citizens, successful businessmen and policy makers.

The representatives of the Italian Association of Geography Teachers (AIIG), the Association of Geographical Societies in Europe (EUGEO), the European Association of Geographers (EUROGEO) and the International Geographical Union (IGU), gathered for the congress session "Geography education's challenges in response to changing geographies". In this declaration, we underline clearly and strongly that the teaching of Geography in schools is fundamental for the future of Europe.

With this firm conviction, we are committed to take initiatives in the countries of Europe and with the relevant European institutions to provide standards and guidelines that will help authorities develop relevant syllabuses and school curricula, methods and approaches in Geography that:

- apply geographical knowledge, skills and understanding to the main issues linked with processes of change in society, nature and environment at local, national, European and global levels; and
- highlight the educational values and the role of geographical education in a changing world.

We urge those responsible in European governments and educational systems:

- to recognise the educational value afforded by the study of Geography as an essential school subject; and
- to acknowledge its strategic role for realising active citizenship and balanced social, economic and environmental development.

We therefore request that:

- sufficient time for the teaching of Geography is allocated in curricula for primary and secondary schools;
- the teaching of Geography is limited to teachers with a qualified training in Geography and Geography Education.

Gino De Vecchis, President of the Italian Association of Geography Teachers (AIIG)
Karl Donert, President of the European Association of Geographers (EUROGEO)
Vladimir Kolossov, President of the International Geographic Union (IGU)
Henk Ottens, President of the Association of Geographical Societies in Europe (EUGEO)
Joop van der Schee, Co-chair of the Commission on Geographical Education of the International Geographic Union (IGU-CGE).

Rome, September 5th 2013

Fig. 10.5 Joint Declaration on Geographical Education in Europe

members of the Council of Europe delegation and lobbied for action on human rights. These initiatives do not specifically belong to geographers; however, their real impact on the global agenda is so powerful that any kind of updated international geography education needs to incorporate them because of their spatial dimension.

Two co-chairs of the IGU Commission prepared the International Charter on Geographical Education 2016: Joop van der Schee and John Lidstone. Draft versions had been discussed with representatives of EUROGEO, EUGEO, AAG, SEAGA and others for international agreement on geography education. The 33rd IGC held in Beijing in 2016 endorsed the new Charter. This proposes and advocates for International Cooperation and an International Action Plan. In an age of globalisation, it is clear that an international geography education approach is needed even more than one century ago in IGU foundation times, it was important due to the contribution of geography education to international understanding for the development of international peace. Today, the significance of international understanding lies in the contribution of geography education to the understanding of global challenges.

10.6 Geographical Education in a Global Era

Even though the 2016 International Charter on Geographical Education does not contain the expression 'global understanding', the links between them are quite evident, as 2016 was proclaimed the International Year of Global Understanding (IYGU), by the IGU, with the support of leading international research organisations the International Council for Science (ICSU), the International Social Science Council (ISSC) and the International Council for Philosophy and Human Sciences (CIPHS). Global Understanding was one of the five key topics for the 33rd IGC, where the new Charter was approved. Global understanding has also become a key issue for IGU-CGE in recent Symposia and publications (Demirci, De Miguel and Bednarz 2018). Besides this, the principles of global understanding and the new 2016 Charter share a common approach to geography education: bottom-up processes for a quantitative and qualitative improvement of geography in education. Both principles underline the need for commitment from the international community of geography educators (Stoltman, Lidstone and Kidman 2017) as learning geography is an educational right for every person, and understanding sustainable development in the global context is a basic condition to protect the Earth's future.

Sustainable development is a crucial concept used in geography education for teaching in global times. Geography education is an essential contribution to the achievement of United Nations Agenda 2030 and Sustainable Development Goals (SDG). Despite the transversal approach of the SDG's, geography is the most important school subject addressing Education for Sustainable Development Goals, as stated by the incumbent IGU-CGE Chair (Chang, Kidman and Wi 2020), and also by those working in European education (De Miguel, Koutsopoulos and Donert 2019). More than this, the SDG's extend a spatial concern to the main challenges facing the world today, such as climate change, water, international migration, urban transformation and demographic growth, among others. In the first two cases, physical geography has traditionally been devoted to research into the biosphere and hydrosphere. In the remaining areas, human geography has studied people's places and how different societies settled over geographical space, politically organised can develop productive activities, as well as their impact on the environment (De Miguel 2019).

The third big concern for international geographical education is related to curriculum and pedagogies for spatial and geographical competences, as described by highlighted publications from the IGU-CGE flagship initiative book Series 'International Perspectives on Geographical Education': dealing with learning progressions, geographical thinking, powerful knowledge, assessment, etc. (Brooks et al. 2017; Muñiz et al. 2016). The American 'Road Map for 21st Century Geography Education' (Bednarz et al. 2013), the British 'Debates in Geography Education' (Jones and Lambert 2013) or 'Assessment in Geographical Education', a recent EUROGEO publication (Lane and Bourke 2021) are also basic references for current international research about methodologies for teaching geography.

Last, but not the least, geospatial technologies have dramatically changed the nature of geographical education in recent years thanks to desktop and online GIS, digital atlases, map viewers, virtual globes, mobile apps, GPS, story maps, geomedia, etc. (Milson, Demirci and Kerski 2012; De Miguel and Donert 2014; Muñiz, Demirci and Van der Schee 2015; De Miguel, Donert and Koutsopoulos 2019a, b). Undoubtedly, geography is among those school subjects that have benefited most from the implementation of new technologies in the classroom. Geospatial technologies can be used to develop inquiry-based learning pedagogies and help students to face the complexity and diversity of processes, systems and interconnections between human and natural environments by developing analytical thinking (data, visualisation), critical thinking (judgement, assessment) and lateral thinking (creativity, problem-solving) (De Miguel 2019). The potential of geospatial technologies for all levels of education was highlighted first in the USA (Esri 1998) and by EUROGEO (Donert and Charzyński 2005), then later by the IGU in the 2007 Lucerne Declaration and 2016 Charter.

This list of four key issues in current geographical education is not exhaustive, as increasing internationalisation has made a wide diversity of research lines possible, activities like International Geographical Olympiad (Fig. 10.7), projects and publications printed and online (books, journals, school books, instructional resources, etc.). A survey of 187 presentations at four recent IGU-CGE symposia (London 2015–Fig. 10.6-, Singapore 2016, Lisbon 2017 and Quebec 2018) shows that curriculum is today the main concern for geography educators (Fig. 10.8), but geopolitics is no longer a major topic in geography education, as it was in the early times, though it ought to be in global times.

Lastly, the international scope of geography education is reflected in one of the most active and participatory IGU Commissions, as the IGU-CGE had 958 registered people in 84 identified countries as of December 2018 (Fig. 10.9). Nevertheless, IGU-CGE has been dominated by Eurocentrism and Anglocentrism in its activities, membership, but particularly through its leadership. There is strong evidence for this, as since its creation, 50% of the fourteen IGU-CGE Chairs (or co-Chairs) have been European, 52% of IGU-CGE Symposia were held in Europe (see Annex 1) and 70% of IGU-CGE Chairs belong to Commonwealth countries or the USA. This is a situation mirrored in international geography research publications. Studies quoted by Johnston and Sidaway (2016) showed that about 80% of the papers published in top geography journals have been written by scholars established in these cultural areas. This presents a real challenge of inclusion to the global community of geography educators. Meanwhile, the IGU itself has had 26 Presidents in this century, of which 16 have been from Europe (62%). Besides this, Europe has held most of the International Geographical Congresses, 26 out of 34 (76%).

Fig. 10.6 IGU-CGE conference, London 2015. Photo: Rafael de Miguel González

Fig. 10.7 International geographical Olympiad, 2018 Quebec City. Photo: Rafael de Miguel González

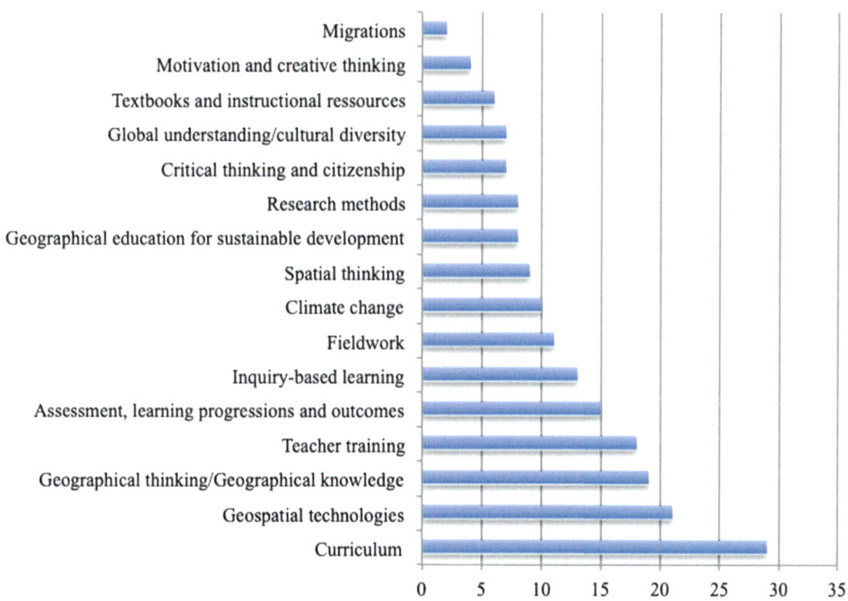

Fig. 10.8 Subjects of IGU-CGE paper presentations in the last four IGU-CGE conferences

Beside this, the European Association of Geographers (EUROGEO) has reinforced the internationalisation of geographical education from a European approach, due to the implementation of the European Higher Education Area in 1999 and the European Education Area

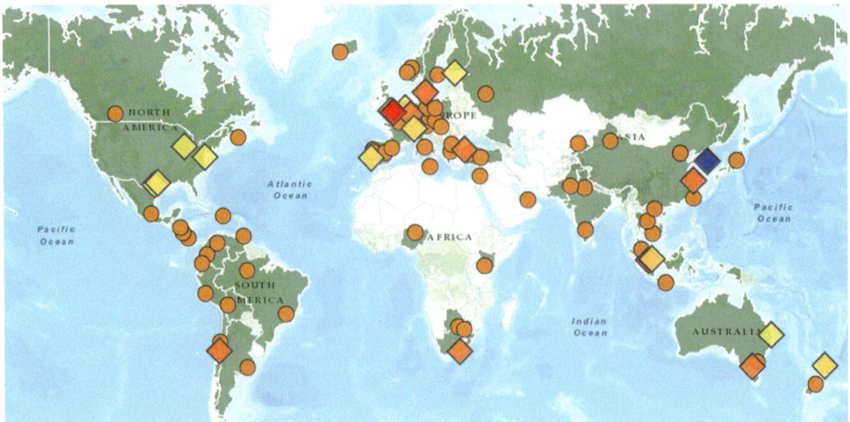

Fig. 10.9 IGU-CGE steering committee members and regional contacts

in 2017. International Conferences (Annex 2), international networks and international projects like HERODOT, Digital-earth.eu, GI-Learner, Geocapabilities, GeoDem and many more (De Miguel 2021) have made possible a genuine process of international collaboration in geographical education through a direct exchange of hundreds of geographers and geography educators in teaching, learning and research, coming from more than 30 European countries and beyond. Among the EUROGEO, outcomes are the contribution to promote a better geography education from a European perspective and values, and the definition and experimentation of specific geographical competences for lifelong learning, involving school education, higher education, adult education and non-formal education. In addition to that, EUROGEO projects have fostered research practices to define and disseminate six key cutting-edge concepts in geographical education: European education, digital geographical education, spatial thinking, spatial citizenship, geocapabilities and geographical education for European sustainable development.

These experiences have allowed the signature of a memorandum of understanding in 2005 between EUROGEO and the IGU to increase synergies between both organisations in international geographical education. Some of the outcomes are: 2013 Joint Declaration, consultation for 2016 International Charter, and 2016 IYGU Story Map Competition (https://collections.storymaps.esri.com/global-understanding/). In this way, EUROGEO is contributing to strengthening international geographical and geospatial education, digital and data literacy, sustainable development, or European values and citizenship. In the twenty-first century, global trends will intensify, but places, spaces and environments will remain more diverse than ever and geography education more essential than ever. Thus, geography education must be taken into consideration in a wider sense than the basics of the discipline—curriculum, pedagogy and assessment (Chang and Kidman 2019) for the implementation of the European Education Area and for the enhancement and international cohesion of a European geographical education research, innovation, good practices and dissemination, guided by five principles (De Miguel 2021):

- Employability
- Enhancing personal development and social inclusion
- Empowering young people through participation
- Enabling for digital skills
- Engagement with democracy and European citizenship.

10.7 Conclusions: A Road Map for International Collaboration in Geography Education

During the century of its existence, the International Geographical Union has been an essential body for the establishment and strengthening of international geographical education. Despite the undisputed initiative and leadership of European, North American and, more recently, Asia–Pacific geographers, challenges of an increasingly and interconnected global world must be faced for the second century of the IGU. So, geographic education must play a basic role in global understanding in the new geopolitical scenario. If in 1922 or in the second half of the twentieth century, geographic education was a factor of international understanding in the face of bilateral confrontations, in the coming decades geographic education for global understanding must continue to be an essential element in the construction of world citizenship in a context of multilateralism.

In this way, we reframe the Road Map for twenty-first-century geography education (Bednarz et al. 2013) and the Road Map to empower geography education for global understanding (Bednarz et al.2018) into a new Road Map for International Collaboration in Geographic Education.

We agree that geography educators must move from charters, declarations and Road Maps to actions—the 'all hands on deck imperative' (Solem and Bohem 2018), hence the importance of scientific research networks. Under the leadership of IGU-CGE, other associations like EUROGEO, NRCGE, SEAGA or REDLAGEO are playing an important role for international collaboration in geography education in Europe, North America, South East Asia and Latin America, respectively.

It is strategic to take advantage of the fact that current IGU-CGE Chairs are, at the same time, Director of NRCGE and President of SEAGA, but President of EUROGEO (Spanish native speaker) collaborates strongly with REDLAGEO. Beside this, EUROGEO and SEAGA Presidents have close relations with Chinese Universities. In 1922, the world population was two billion. It is currently almost eight billion and likely to exceed 10 billion by 2050 according to UN demographic projections. In this new world, totally different to when the IGU was founded in 1922, more than 2 billion inhabitants have as a native language English, Spanish or Chinese, but 4 billion (almost one out of two people in the world) speak fluently these three languages or two others similar to Spanish (as Romance languages), like Portuguese and French. Some words missing. Otherwise, English is the official language for 59 countries in the world, but Spanish–French-Portuguese is in 40 countries more, this is, 99 countries in the world from 193 UN State members.

Not forgetting other important languages (Arabic, Hindi, Russian, Turkish, German or Japanese), the three people mentioned above are able to reach and understand half the world, not only higher education professors, but also school teachers who may only speak their native language. This is the power of geography, the understanding of cultural diversity across the spaces in the world in contrast to monolingual approaches, which identify the future of geography education with a single country and a single language (Gilbert Grosvenor Center for Geographic Education 2020). Geography educators involved in international collaboration must acquire a sense of spatial empathy, they must understand that intercultural, and intercomprehension approaches are vital to achieve international understanding as founding principles of IGU, and later of IGU-CGE.

Geography is an old scientific discipline and a traditional school subject with an established body of knowledge, but it faces the challenges of a changing world, a technological revolution, mass media impact and the teaching of rapid political, economic, social and cultural transformations that have an impact on space, countries and regions of the world. Some of this body of knowledge and the means to teach *it* will remain invariant, as we have known in previous decades, as a common legacy and heritage of geography

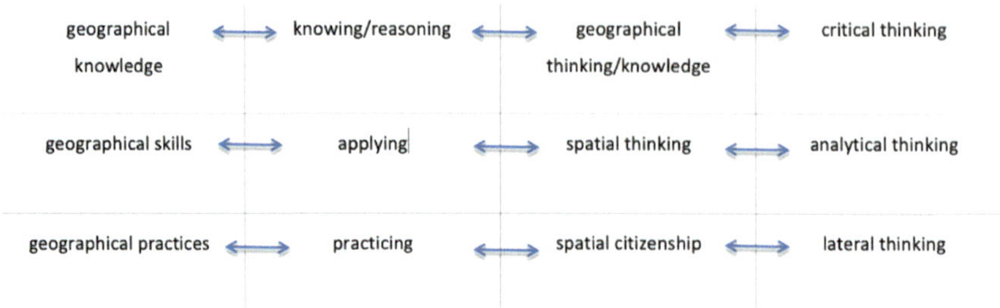

Fig. 10.10 Integrated framework of geographical competences for curriculum and assessment

educators. But geographical education is committed to innovation and practices responding to the twenty-first century, and particularly in the two most important aspects of international collaboration: curriculum and assessment. Fortunately, the two authors of this chapter have contributed to an international curriculum from a European Higher Education Area approach (Donert 2009) and by defining seven geographical competences from the European lifelong learning approach (De Miguel 2021). In addition to that, two chairs of IGU-CGE have also contributed to a framework for key themes of geography curriculum (Chang et al 2019) or a framework for international assessment in geography education (Solem et al. 2018).

In this framework, geography education worldwide would be available as a shared framework (Fig. 10.10) to know, apply and reason and to the acquisition of spatial thinking, geographical thinking and spatial citizenship by the students, in order to develop analytical, critical and lateral thinking and to improve geography achievement results (De Miguel 2021).

The constant evolution of contemporary societies, their needs and their impact on the planet earth and the natural and human landscape, the growing human interactions with the environment, the challenge of global sustainability, etc., are reasons that compel a greater and more complex geographic education, also involved in geoethics. In the twenty-first century, we need to teach more and better geography, so an international framework for curriculum and assessment is needed. Other aspects of geographic education are no less important: instructional materials, geospatial technologies and digital geographical education, pedagogies, teaching and learning strategies, fieldwork, teacher training and professional development.

Finally, we must increase international collaboration in geography education to get a higher impact for its research, to keep up with debates on big topics in geography education, to be more inclusive, but also to support young geography educators and teachers, in particular from disadvantaged countries and contexts. In 1971, Benoit Brouillette expressed his desire for the widest possible international participation in international education, with Africa, Asia and Latin America in particular in mind. Half a century after his words, a century after the founding of the IGU, it still remains the main objective of this Road Map for the twenty-first century.

Annex 1: IGU-CGE Symposia 1952–2022

Year	City	Related to IGU event
1952	Washington	17 IGU International Geographical Congress
1956	Rio de Janeiro	18 IGU International Geographical Congress
1960	Stockholm	19 IGU International Geographical Congress
1964	London	20 IGU International Geographical Congress
1968	Madras	21 IGU International Geographical Congress
1972	Quebec City	22 IGU International Geographical Congress
1974	Palmerston	IGU Regional Conference

(continued)

Year	City	Related to IGU event
1976	Moscow	23 IGU International Geographical Congress
1978	Lagos	IGU Regional Conference
1980	Tsukuba	24 IGU International Geographical Congress
1982	Curitiba	IGU Regional Conference
1984	Freiburg	25 IGU International Geographical Congress
1986	Barcelona	IGU Regional Conference
1988	Brisbane	26 IGU International Geographical Congress
1990	Hong Kong	IGU Regional Conference
1992	Boulder	27 IGU International Geographical Congress
1994	Berlin	IGU Regional Conference
1996	The Hague	28 IGU International Geographical Congress
1998	Porto	IGU Regional Conference
2000	Gyeonggi	29 IGU International Geographical Congress
2002	Durban	IGU Regional Conference
2004	Glasgow	30 IGU International Geographical Congress
2006	Brisbane	IGU Regional Conference
2007	Lucerne	
2008	Tunis	31 IGU International Geographical Congress
2010	Istanbul	IGU Regional Conference
2011	Santiago	IGU Regional Conference
2011	London	
2012	Freiburg	32 IGU International Geographical Congress
2014	Krakow	IGU Regional Conference
2015	London	
2015	Moscow	IGU Regional Conference
2016	Singapore	33 IGU International Geographical Congress
2017	Lisbon	
2018	Quebec City	IGU Regional Conference
2019	London	
2021	Prague	34 IGU International Geographical Congress
2022	Rennes	IGU Centennial Conference

Annex 2: EUROGEO Conferences 1979–2022

Year	City	Co-organiser	Conference Theme
1980	Brussels		
1982	Brussels		
1984	Paris		
1986	Brussels		
1988	Brussels		
1990	Brussels		
1992	Brussels		
1994	Brussels		
1996	Salzburg		
1998	Luxemburg	Luxembourg Geography Association	*Geography in Europe*
2001	Liverpool	Liverpool Hope University	*Teaching geography, skills and curricula*
2002	Funchal	APG, Association of Geography Teachers in Portugal	*Geography teaching in an enlarging Europe*
2004	Bled	University of Ljubljana, Slovenian Geography Teacher Association	*New developments in Geography*
2005	Torun	Institute of Geography, Nicolaus Copernicus University	*Changing Horizons in Geography Education*
2007	Stockholm	Stockholm University	*Geography for Society: Putting Bologna into Action*
2008	Liverpool	Liverpool Hope University	*Future Prospects in Geography*
2009	Ayvalik	Balikesir University	*Celebrating Geographical Diversity*
2010	Prague	Charles University	*Sustainable Geographies*

(continued)

Year	City	Co-organiser	Conference Theme
2011	Athens	National Technical University of Athens	*Geography: Your world – A European Perspective*
2012	Dublin	St. Patrick's University College	*Geography and Global Understanding: Connecting the Sciences*
2013	Bruges	Ghent University	*Geography: Linking Tradition and Future*
2014	Valletta	University of Malta	*The Power of Geography and the Role of Spatial Information*
2015	Ankara	Turkish Association of Geographers, Gazi University	*Communicating Geography: Serving our world*
2016	Málaga	University of Málaga, Royal Geographical Society, AGE	*Geographic Information: For a better world*
2017	Amsterdam	Utrecht University	*Challenges for geographical education*
2018	Cologne	University of Cologne	*Geography for all*
2019	Paris		*Teaching Geography in challenging times*
2019	Ljubljana	University of Ljubljana, Slovenian Geography Teacher Association	*Hidden geographies*
2021	Madrid	National University of Distance Education, Royal Geographical Society	*Sustainable Development Goals for all*

(continued)

Year	City	Co-organiser	Conference Theme
2022	Mytilene	University of Aegean	*Re-visioning geography for sustainability in the post-COVID era*

References

Arroyo F (2014) La educación geográfica en la Sociedad Geográfica de Madrid. In: La Real Sociedad Geográfica en la Biblioteca Nacional de España. Biblioteca Nacional de España, pp 25–30

Atwood W (1922) Geography and world relations. J Geogr 21(3):81–89

Bednarz S et al (2000) The international network for learning and teaching geography: developing links with school education. J Geogr High Educ 24(2):277–284

Bednarz S, Heffron S, Huynh N (eds) (2013) A road map for 21st century geography education. National Geographic Society, Washington DC

Bednarz S, De Miguel R, Demirci A (2018) A Road Map to empower geography education for global understanding. In: Demirci A, De Miguel R, Bednarz S (eds) Geography education for global understanding. Springer, Cham, pp 247–252

Bosque J (2004) La Real Sociedad Geográfica. 125 años de existencia. In La Geografía Española ante los retos de la sociedad actual. IGU Spanish Committee, pp 209–237

Bourke T, Lane R (2018) A comparison of the international charters on geographical education. J Geogr 117 (4):174–180

Brouillette B, Graves N, Last G (1974) African geography for schools: a handbook for teachers. Longman, London

Brouillette B, Vilá Valentí J (1971) Las comisiones La Enseñanza de la Geografía y la Geografía en la Educación de la Unión Geográfica Internacional. Revista De Geografía 5:85–107

Brouillette B, and Vilá Valentí J (Eds) (1975) Geografía de América Latina: Métodos y temas monográ. Teide.

Buen O (1909) Discurso inaugural leído en la solemne apertura del curso académico de 1909 a 1910 ante el claustro de la Universidad de Barcelona. University of Barcelone.

Buen O, Gómez Núñez S (1923) La Asamblea Extraordinaria de la Unión Internacional de Investigaciones celebrada en Bruselas. Constitución de la Unión Geográfica Internacional. Revista de Geografía Colonial y Mercantil 20 29.

Buen O (2003) Mis memorias (Zuera 1863—Toulouse 1939). Institución Fernando el Católico

Brooks C, Butt G, Fargher M (eds) (2017) The power of geographical thinking. Springer
Chang C, Kidman G, Wi A (2020) Issues in teaching and learning of education for sustainability. Routledge
Chang C, Kidman G (2019) Curriculum, pedagogy and assessment in geographical education—for whom and for what purpose? Int Res Geogr Environ Educ 28(1):1–4
Chang C et al (2019) Advancing a framework to describe school geography curricula around the world. Res Geogr Educ 21(2):55–72
De Miguel R (2021) Geographical and geospatial competences from school education to higher education: the contribution of EUROGEO in the international projects. J-READING: J Res Didactics Geogr 2 (10):57–65
De Miguel R (2020) From international to global understanding: toward a century of international geography education. Int Res Geogr Environ Educ 30(3):202–217
De Miguel R (2019) Europe in a global context: EUROGEO and the role of geography and European geographers. Eur J Geogr 10(4):160–176
De Miguel R, Donert K, Koutsopoulos K (eds) (2019a) Geospatial technologies in geography education. Springer, Dordrecht
De Miguel R, Koustopoulos K, Donert K (2019b) Key challenges in geography research with geospatial technologies. In: Koustopoulos K, De Miguel R, Donert K (eds) Geospatial Challenges in the 21st Century. Springer, Dordrecht, pp 1–10
De Miguel R, Donert K (2014) Innovative learning geography in Europe. New Challenges for the 21st Century. Cambridge Scholars Publishing, Cambridge
Demirci A, de Miguel R, Bednarz S (eds) (2018) Geography education for global understanding. Springer, Dordrecht
Donert K, Hay I, Theobald R, Valiunaite V, Wakefield K (2011) International collaboration in organizations promoting geography education: Exploring success and acknowledging limitations. J Geogr High Educ 35 (3):445–455
Donert K, Charzyński P (2005) Changing horizons in geography education. Herodot Network.
Donert K (2009) Geography: the world at your fingertips. EUROGEO, Wardamme
ESRI (1998) GIS in K–12 Education: An ESRI White Paper. ESRI.
Fairgrieve J (1926) Geography in schools. University of London Press, London
Fleure HJ (1919) Editorial. Geogr Teach 10(3):77–79
Fleure HJ (1921) Annual report 1920. Geogr Teach 11 (1):1–2
Gerber R (2003) Globalization and geographical education. In: Gerber R (ed) International Handbook on Geographical Education. Springer, Dordrecht, pp 21–33
Gilbert Grosvenor Center for Geographic Education (2020) A Report on a Round Table of the Future of Geography Education. Research in Geography Education 22(1):7–16
Graves N, Stoltman J (2015) The commission on geographical education of the international geographical Union 1952–2012. Manuscript review paper. Retrieved from http://www.igu-cge.org/wp-content/uploads/2018/02/The-IGU-Commission-on-Geog-Education-modified-0915.pdf
Graves N (1982) New UNESCO Source Book for Geography Teaching. Longman, London
Haubrich H (2006) Changing philosophies in geographical education from the 1970s to 2005: an international perspective. In: Lidstone J, Williams M (eds) Geographical education in a changing world: past experience, current trends and future challenges. Springer, Doredrecht, pp 39–53
Haigh M (2008) Internationalisation, planetary citizenship and Higher Education Inc. Compare 38(4):427–440
Haigh M (2014) From internationalisation to education for global citizenship: a multi-layered history. High Educ Q 68(1):6–27
Haubrich H (2009) Global leadership and global responsibility for geographical education. Int Res Geogr Environ Educ 18(2):79–81
Higgitt D et al (2008) Developing and enhancing international collaborative learning. J Geogr High Educ 32(1):121–133
Hinks AR (1922) The international geographical Union. Geogr J 60(4):291–294
Johnston RJ (2018) Geography. Encyclopaedia Britannica. https://www.britannica.com/science/geography
Johnston RJ, Sidaway J (2016) Geography and geographers: Anglo-American human geography since 1945. Routledge, London
Jones M, Lambert D (eds) (2013) Debates in geography education. Routledge, London
Klein P, Solem M (2008) Evaluating the impact of international collaboration on geography learning. J Geogr High Educ 32(2):245–267
Lane R, Bourke T (eds) (2021) Assessment in geographical education. Springer, Dordrecht
Lidstone J, Stoltman J (2002) International understanding and geographical education. Int Res Geogr Environ Educ 11(4):309–312
Martonne E de (1911) Sur le rôle des excursions à long parcours dans l'enseignement de la géographie. In: Comptes rendus du 9e Congrès International de Géographie: Genève, Tome III, pp. 321–323
Mamadouh V (2005) Geography and war, geographers and peace. In: Flint C (ed) The geography of war and peace: from death camps to diplomats. Oxford University Press, Oxford, pp 26–60
Milson A, Demirci A, Kerski J (eds) (2012) International perspectives on teaching and learning with GIS in secondary schools. Springer, Dordrecht
Muñiz O, Demirci A, van der Schee J (eds) (2015) Geospatial technologies and geography education in a changing world. Springer, Dordrecht
Muñiz O, Solem M, Boehm R (eds) (2016) Learning Progressions in Geography Education. Springer, Dordrecht

Robic MC (1996) La naissance de l'Union Géographique Internationale. In: Robic MC, Briend AM, Rössler M (eds) Géographes face au monde. L'Union Géographique Internationale et les congrès internationaux de géographie. L'Harmattan, Paris, pp 23–39
Schelhaas B, Pietsch S (2020) (Re-)Writing the history of IGU? A report from the archives. In: Schelhaas B et al (eds) Decolonising and internationalising geography: essays in the history of contested science. Springer, Cham, pp 127–140
Shepherd ID, Monk JJ, Fortuijn JD (2000) Internationalising geography in higher education: towards a conceptual framework. J Geogr High Educ 24 (2):285–298
Schneider J (1972) Les Commissions. In: IGU (ed.) Geography through a century of international commissions. International Geographical Union, Brussels
Smith H (1944) The world federation of education associations. Annals Am Acad Political Soc Sci 235 (1):107–112
Societé de Géographie de Paris (1922) Nouvelles géographiques: le Conseil International de Recherches Scientifiques 38: 256–257
Solem M, Boehm R (2018) Research in geography education: moving from declarations and road maps to actions. Int Res Geogr Environ Educ 27(3):191–198
Solem M et al. (2018) An assessment framework and methodology for trends in international geography assessment study (TIGAS) geographical education 31:7–15
Stoltman J, Lidstone J, Kidman G (2017) The 2016 international charter on geographical education. Int Res Geogr Environ Educ 26(1):1–2
Stoltman J (2006) Turning points in geographic education. In: Lidstone J, Williams M (eds) Geographical education in a changing world: past experience, current trends and future challenges. Springer, Dordrecht, pp 23–37
Stoltman J (1997) The international charter on geographical education: setting the curriculum standard. J Geogr 96(1):32–32
Thomas A (1932) International understanding through the teaching of geography. Education 52(5)
Tur L (1923) Tareas de la sociedad. Boletín De La Real Sociedad Geográfica 64:167–185
UNESCO (1949a) The teaching of geography and international understanding. UNESCO, Paris
UNESCO (1949b) A handbook for the improvement of textbooks and teaching materials as aids to international understanding. UNESCO, Paris
UNESCO (1949c) Some suggestions on the teaching of geography. UNESCO, Paris
UNESCO (1951) A handbook of suggestions on the teaching of geography: towards world understanding. UNESCO, Paris
UNESCO (1965) UNESCO source book for geography teaching. Longman, London
Wise MJ (1992) International geography: the IGU Commission on Education. In: Naish M (ed) Geography and education: national and international perspectives. University of London, London, pp 233–246

Rafael de Miguel González is professor of geographical education at the University of Zaragoza, where he is also Vice Dean. Ph.D. in urban geography and planning and Ed.D. in geographical education. Top scholar on geography education in Spanish language and an elected member of Academia Europea, Rafael is the Director of the Book Series Key Challenges in Geography (Springer) and Editorial Board of International Research in Geographical and Environmental Education. Board member of the Real Sociedad Geográfica, regional contact for Spain of the IGU-CGE, he is the incumbent President of EUROGEO. Rafael represents EUROGEO and lobbies for geography and geographical education at EU institutions and at the United Nations.

Karl Donert is a senior academic. Former President and currently Vice President of EUROGEO. He is a UK National Teaching Fellow, consultant and project manager. Karl is an expert to the European Commission in Brussels, advocate for geography and education at the Council of Europe, and an elected member of Academia Europea. He researches and regularly publishes widely on many aspects of geographical education, citizenship, the use of interactive media and new technologies.

The ‘North-South’ Problem in Geography

11

Pascal Clerc and André Reyes Novaes

Abstract

The chapter aims to analyze the ‘North-South’ problem in geography considering how representations strongly participate in the organization of spaces and the geopolitics of knowledge production. In the first part, we study the way in which these representations produce and reflect a relationship of domination of the ‘North’ over the ‘South’. The second part takes another perspective and shows that some so-called ‘southern’ countries can also turn this imposed identity into a means of contestation. Finally, we deal with the situation of the IGU which, with its 120 or so national committees, has, since its creation, been constantly confronted with this way of thinking about the world. As an important institution for the reproduction and contestation of geographic practices around the world, the International Geographical Union will continue to be a privileged stage to identify continuities and changes in ‘North-South’ relations.

Keywords

North-South · Imaginative geographies · Knowledge production · Means of contestation · International Geographical Union

11.1 Introduction

The need to classify and name is necessary for us to access and account for what surrounds us (Vignaux 1999). This need operates in all fields: social categories, animal species, landforms… We have to go through this to think about the world.

Classifying and naming also means characterizing, differentiating, and setting limits, ordering according to qualitative or quantitative criteria and finally ranking. These thought-based organizations are sometimes binary. This is the case with the ‘North-South’ division of the world. This binarity refers to an essential scheme of social life: distinguishing ‘us’ (or ‘self’) from ‘others’ (see Said 1978; Todorov 1989).

The purpose of this text is to analyze the representations of this ‘North-South’ division. We hypothesize that representations strongly participate in the organization of spaces. In the first part, we study the way in which these representations produce and reflect a relationship of domination of the ‘North’ over the ‘South’. The second part takes another perspective and shows

P. Clerc
Cergy Paris University, Paris, France

A. R. Novaes (✉)
Universidade do Estado do Rio de Janeiro, Rio de Janeiro, Brazil
e-mail: andrereyesnovaes@gmail.com

V. Kolosov et al. (eds.), *A Geographical Century*,
https://doi.org/10.1007/978-3-031-05419-8_11

that some so-called 'southern' countries can also turn this imposed identity into a means of contestation. Finally, we deal with the situation of the UGI which, with its 120 or so national committees, has, since its creation, been constantly confronted with this way of thinking about the world. It is both a mirror of the logic of domination and a place of action for a fair and united world.

11.2 North-South: A Recent Construction?

There are different ways to consider the division of the world between 'North' and 'South': as is an idea, as words and as spaces. The idea presents a binary and relational World. The words take place in the long story of the different ways to designate spatial entities of this World. Spaces show a divided world with a very stable line of separation on the planispheres. These different points of view are of varying temporalities. However, the idea, the words and the spaces constantly converge to signify the existence of two stable, unequal entities, defined in opposition to each other and in a relation to domination of one over the other.

Most authors who have attempted to retrace the history of the 'North-South' division propose as a starting point *The Report of the Independent Commission on International Development issues* published in 1980, more colloquially 'Report North-South' or simply the Brandt Report (see for example Dickenson 1983). In January 1977, to relaunch exchanges and cooperation that seemed to be deadlocked after the 1973 oil crisis, Robert McNamara, president of the World Bank, proposed the creation of a study group to facilitate 'North-South dialogue. He entrusted Willy Brandt, the former socialist Chancellor of the Federal Republic of Germany, to lead a group of twenty prominent personalities representing both rich and poor countries. In February 1980, the group submitted a report to the United Nations Secretary-General, Kurt Waldheim.

We shall return to the Brandt Report in due course but should emphasize at the outset that the world that late 20th-century regionalization reflected centuries of earlier attempts to identify meaningful global distinctions dating back to Antiquity and the Aristotelian zonal division of the Earth (Besse 2003). In this early period, the objective was to deduce the definition of the ecumene to a northern temperate zone that was habitable and inhabited, separate from a South intemperate zone that was inaccessible and inhabited only by another species that did not qualify as human. This was the first attempt to challenge an earlier unitary vision of the globe and raise the question of the relationship between the earth's diversity and the universal nature of humanity (Cosgrove 2001).

Later, the question of otherness would be reinterpreted through the different waves of colonization. With the first transoceanic travel, the thought of the world—always through the dominant gaze of westerners—opposes the European 'we' to the 'others' from the rest of the world. Gradually, a new dual representation of the world was created, more accurate and less geometric than zonation. From the nineteenth century, this division opposed Europe and North America from the tropical world. The contemporary divisions that separate a rich and dominant world from a poor and dominated world are born from these situations of colonial domination.

After World War Two, during the process of decolonization, the economic question, through the concept of development, gradually produced a new pattern of global division translated into quantitative data, measured by the Gross National Product, that gave an appearance of neutrality and objectivity to a relation of domination.

The notion of an 'underdeveloped country' was famously mentioned in 1949 by US President Harry S Truman. This terminology already situated the binary organization of the world in relation to a form of Western ethnocentrism: the development of Western countries as the standard against which the rest of the world was benchmarked. An underdeveloped country was defined by what it lacked. The subsequent creation of the expression 'developing countries' served only to raise unrealistic hopes among poor countries and allow richer countries to convince themselves that they were not responsible for a developmental process in which they were always the winners.

But very quickly, at the end of the war, another division of the world shook up these economic referents. It is the start of the Cold War. It establishes another binary pattern, a political one, West-East, within the dominant countries and temporarily makes forget the economic organization of the World. But this new pattern does not cover the whole planet. In this context, the French demographer Alfred Sauvy proposed the expression 'Third World' in 1952. With this expression, a tripartite division is established which aims to give consideration to those whom the division into two blocks had provisionally disqualified: 'We are happy to talk about the two worlds in presence, their possible war, their coexistence, etc., forgetting too much often that there is a third one, the most important, and actually, the first in the chronology. It is the set of those that are called, in United Nations style, the underdeveloped countries'.[1] It was also with reference to the third estate of the French Revolution that Sauvy had created this expression, and therefore from a more militant perspective than a simple practice of appointment.

This 'Third World' sought to establish a new balance of power. This new terminology deliberately dodges the question of development in favor of the idea of political independence from the two blocs. But almost all the states represented in Belgrade are part of the so-called 'underdeveloped countries', the future 'South'. There are Latin American, African, Middle Eastern, and Asian states. The 'non-aligned countries' want to redirect global geopolitical concerns around a South-North axis.

In 1964, U Thant, the Burmese politician, and UN secretary-general, prophetically declared that the tensions between the North and the South were as serious as the tensions between the East and the West. This announced a new way of thinking about organization of the world that Cold War détente would facilitate: from then on, the world was increasingly viewed as divided between the rich countries located for the most part in the northern part of the globe and the poor countries located mostly in the southern part that the World will be organized.

The words 'North' and 'South' and even the expression 'North-South' had appeared in political and media discourse a few years before U Thant's speech. The first occurrence of the term 'North-South' was probably in a report by a British diplomat, then president of Lloyds Bank, Oliver Franks in 1959 (Capdepuy 2018). Franks spoke of the 'North-South problem' and argues that the countries affected by this problem are located in relation to each other along a North-South axis.

But it was only in the mid-1970s, when the East-West opposition weakened, that the expression 'North-South' gained widespread popularity (Clerc 2020). The global axis of tension and the organization of the world were changing direction. The 1973 oil crisis was the trigger and opened a new chapter in the 'North-South dialogue'. The new dialogue was not motivated by a desire on the part of oil-producing countries to reduce the considerable differences in wealth in the world and the debate was essentially about how rich countries could maintain their supply of raw materials without altering the fundamental balance of power within the world economy.

It was in relation to these successive contexts that Willy Brandt and his colleagues published their report. Let us consider again the three dimensions stated at the beginning of this section: ideas, words, and spaces. Despite the report's fairly clear directions in favor of more balanced relations between the two major parts of the world, and despite positions that one might even qualify as Third-World perspectives, the floating terminology and the cartographic choices produced an ambivalent or even contradictory discourse.

In the introduction, Brandt used all the terminology available to describe the two entities

[1] Alfred Sauvy, *L'Observateur* 118, 14th August 1952, 14.

that the report describes. All these words seemed to be equivalent as if 'to name (was not) to express a point of view on the named object' (Dufour 2007, 32). Why did the members of the commission finally choose 'North-South' as the title of their report? There are no clear explanations in the published text. However, we can think that for Brandt and his colleagues, talking about 'North' and 'South' is a way to neutralize the semantic load of references to development. By using terms that refer to cardinal points, by mobilizing a descriptive geography that would report the World 'as it is', they undoubtedly thought they were escaping stigmatizing representations. As we will see later, the result was the opposite.

The third point of view concerns the map and the spatial division. We were able to consult nine editions of the report in seven different languages (German, English, Spanish, French, Italian, Norwegian, and Portuguese). The analysis of the cover pages makes it possible to distinguish those that used the map of the original edition in English and those that selected another iconography. The latter favored the idea of solidarity (an edition published in Mexico in 1981 showed two hands clasping each other, but one at the top, to the 'North' and one at the bottom, to the 'South') or the idea of emergency (one edition used a reproduction of Géricault's painting, *Le Radeau de la Méduse* (1818–1819) which shows men in distress on a makeshift lifeboat at sea after a shipwreck). A third non-cartographic representation can however be linked in part to what the cartographic discourse expressed. This was an image published in Colombia in 1980, in which the division of the world was indicated by photographs at the top (the 'North') of the skyscrapers of Manhattan and below (the 'South') showing faces, sad and closed, of men and women, presumably Colombians. Other editions used a planisphere for the cover page on which the world appears to be divided into two parts, separated by a very deep line (Fig. 11.1).

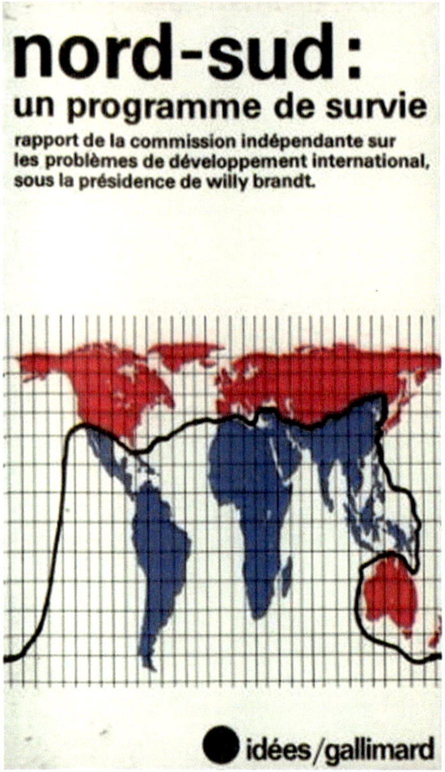

Fig. 11.1 Cover image from a French edition of the Brandt Report, entitled 'Commission indépendante sur les problèmes de développement international, 'Nord-sud': un program de survie' 1980, Gallimard, Paris

The maps used in Brandt Report covers generally used the cartographic projection known as 'Gall-Peters', popularized by the German historian and geographer Arno Peters in 1974.[2] The choice of this projection was political. Unlike the dominant Mercator projection, it respected the size of surfaces. Therefore, it was possible to visualize the importance of the intertropical zone, traditionally atrophied on the common planisphere. This projection is frequently used today by various groups to support a militant discourse against capitalist globalization. Under these conditions, this cartographic

[2] There were two exceptions. A Norwegian edition (1980) used a Winkel-Triple projection, a compromise projection that minimises surface and shape distortions and which allows the whole world to be correctly represented. A Portuguese edition (1981) showed a succession of six representations of the globe to signify the earth's rotation, based on a Mercator projection and with the two parts of the world identified by different colors without a dividing line.

choice cannot be without significance. Moreover, it was clearly assumed in the preamble to the report: 'This projection marks an important advance compared to the conception which attributed a preponderant world role to Europe, geographically as well as culturally.'

The map of a world was divided into two parts, separated by a thick and clearly visible line that has become an iconic image.[3] Its simplicity, and its obviousness, were reinforced by words that seemed to confirm what could be read cartographically, and facilitated its entry into the scholar culture of many countries through textbooks and probably in a 'world-culture' through the media. This image survived over the years and the upheavals of the last 40 years almost without modifications apart from the attachment of Israel and Singapore to the 'North', and the integration in the 'South' of the states of the Caucasus and Asia Central. This stability was reinforced by the choice of vocabulary. Development terminology introduced a time dimension and a possible evolution. With the concept of a 'North-South' divide, time has been replaced by space and a location on the world map.

By using these words, the authors of the Brandt Report were very probably trying quite sincerely to challenge Eurocentrism and the associated references to progress and the positive or negative values of developmental terminology. In reality, however, the effect was the reverse. These images, and this language, have naturalized a binary representation of the world by making the question of development into a cultural, even a natural construction in which the spatial organization of the world seems to be obvious. A country can have relative control over its development, not over its spatial location. The 'North-South' model affirms an intangible world order. It indicates where each 'other' is located and shows the relation to 'other' regions. It still prioritizes the 'North' which remains firmly at the top of the image. If the terminology of development can give hope to all the countries, the 'North-South' couple exists by its final binarity: the countries will never all be of the 'North'. The 'North' needs the 'South' to be the 'North'. The iconic map of the Brandt Report is one of the strongest metageography (Lewis and Wigen 1997) of the world order. It is a continuation of a very old situation of domination and the 'North-South' limit is only its updated form.

11.3 South-North: Inverting Directions?

Although the globe's division into North-South can be considered a contemporary form of imagining the world in a Eurocentric and colonialist way, it is essential to recognize that the narratives associated with the 'Third World' or the 'Global South' were also appropriated by other groups of actors. If the Conference on International Economic Cooperation (1976) is a landmark in the popularization and naturalization of the world division between 'North-South', the event was also a turning point in the use of the 'South' as 'an axis of articulation and a discursive device' for peripheral and subaltern claims in arts, sciences, and politics (Moura 2015, 24). Considering engaged uses of the South as a concept and discourse, this section explores the appropriation of this terminology by inhabitants and countries to build a counter-hegemonic identity and make another proposition for the world's organization.

Cartographic language, which actively influences our perception of the world, is often associated with a 'science of princes' because maps were primarily appropriated by ruling classes and colonialist interests (Harley 1988). However, it is crucial to recognize that maps' language was also often appropriated by non-hegemonic groups, developing what could be called a 'counter-cartography'. Artistic maps were frequently used in these narratives, and in addition to 'stripping the mask from the map' (Wood 2010, 189), these images often portray

[3] However, the report does not seem to have made a lot of noise when it was released, mainly because of current events and the entry of Soviet troops into Afghanistan. Then its distribution was limited. Despite about 20 translations, the report is said to have sold only around 350,000 copies worldwide, half in the United Kingdom, and few in Germany, France and the United States.

anti-colonialist arguments. One of the best-known pieces of map art, the 1929 Surrealist map of the world, published in the review *Variétés*, is a vibrantly anticolonial map. The unknown artist, probably Éluard (Wood 2010, p. 198), portrayed a world erasing the United States and most of Europe and exaggerated the South Sea islands' size to disrupt European hegemony.

Although the surrealist map is now well known, illustrating T-shirts and coffee mugs, other artistic maps probably have similar influence in many countries in the Global South. This is the case of *The Inverted Map of South America*, a famous image of Latin America created by Uruguayan artist Joaquin Torres Garcia (Fig. 11.2).[4] Initially printed in his manifesto *Escola del Sur* (1943),[5] this image is still reproduced today to endorse counter-hegemonic policies and 'southern' identities (Sales 2015). *The Inverted Map of South America* is a simple image with high communication power and was one of the first explicit attempts in twentieth-century art to use cartography to denaturalize the world with the 'North' at the top. By proposing a break with previous representational patterns, Torres Garcia allied himself with other modernist Latin-American artists who sought to avoid mechanical reproductions of European models (Ades 1997). In opposition to behavior that he called 'exiles from Europe', the artist suggested a focus on Latin-American motives and aesthetics, neither rejecting nor privileging external influences.

Torres Garcia's pioneering claims demonstrate how, before the idea of a 'North-South dialogue' became popular during the 1970s, the 'South' has a specific counter-hegemonic history and was already used as an anti-colonialist narrative circulating outside the hegemonic centers. The aim to 'invert' the map and 'correct' power relations

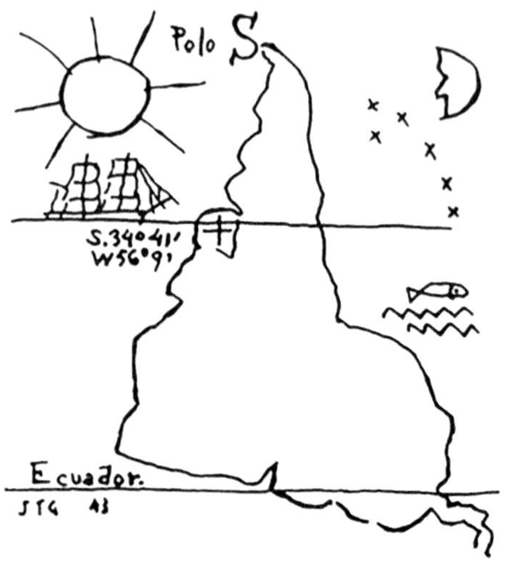

Fig. 11.2 Joaquín Torres García, Mapa Invertido da América do Sul (Inverted map of South America) 1943. Printed with permission © Museo Torres García. www.torresgarcia.org.uy

appears in many other artists' work throughout the twentieth-century. Another famous example is the *Corrective Map of the World*, made by the Australian artist Stuart McArthur in 1979. Inverting the Mercator projection and placing Australia in the center of the map, the artist used irony to claim a new world order: 'This map, a subtle but definite first step, corrects the situation. No longer will the South wallow in a pit of insignificance, carrying the North on its shoulders for little or no recognition of her efforts. Finally, the South emerges on top'. During a period of many diplomatic and economic initiatives to develop North-South and South-South relations, a narrative about a southern identity seems to increase both in politics, arts, and sciences.

Despite earlier uses of the word, the South was not explored as a central category in academic scholarship before the late 1990s and early 2000s. As runoff from the legacy of postcolonialism and other critical theories, such as subaltern, feminist, and Third-World studies, the 'South' emerged as a conceptual schema to discuss power relation and knowledge production in the colonial period and the contemporary world (Moura 2015). From the 2000s onwards, the

[4] Due to the knowledge and experience of the authors, we use a Latin American example here. But it is important to recognise that many other examples of counter-cartography can be found in Africa and Asia.

[5] Torres Garcia published a previous version of the map in 1936, using other cartographic elements such as a coordinate grid.

attempts to explore 'southern epistemologies' (Santos 2014) and 'southern theories' (Comaroff and Comaroff 2012) become commonplace, discussed by authors associated with different approaches in de-colonial scholarship (Mignolo 2006; Huish 2006; Maldonado-Torres 2006). In contrast with the static and fixed map that classified the countries between the 'North' and the 'South', some scholars conceived the 'South' as a metaphor, a field of 'epistemic challenges that seek to repair the damage and impacts historically caused by capitalism and its colonial relationship' (Santos and Meneses 2010, 12).

As a relationship and a place of enunciation, the 'South' only partially overlaps with its geographical delimitation on the map. Of course, it is possible to identify nations with similar histories of violence, conflict, and political and cultural subordination to European colonialism. However, the overlap between the metaphor or a region is not complete because in the geographical North's interior many classes and social groups (workers, women, indigenous people, Afro-descendants, Muslims) were subjected to colonial domination. Besides, there were always 'small Europes' in the interior of the geographical 'South' where local elites benefited from colonial domination (Santos and Meneses 2010). The line of demarcation between 'North' and 'South', between zones of prosperity and zones of underdevelopment, is not drawn stably, and it is often porous, broken, and illegible (Balibar 2004; Comaroff and Comaroff 2012, 84). The South has expanded globally, configuring a map of historical and political experiences that has refused to keep within hemispheric lines (Moura 2015, 15).

Considering this metaphorical and relational definition, the 'South' can group a series of subaltern knowledge produced in different scales and places. By stressing the diversity and the 'pluriversality' (Grosfoguel 2008) of modern science, this set of approaches seeks to challenge narratives focused on 'core' areas, which identified the peripheries always by the lack of development, science, theories, or epistemologies. During the last decades, a wide range of scholars related to postcolonial and decolonial scholarship developed a body of writing that attempts to shift the dominant ways in which the circulation of knowledge between 'North' and 'South' is viewed (Sharp 2009). To suggest new cartographies of the global power, these scholars challenge the creation of European modernity as isolated phenomenon that could be understood without coloniality (Mignolo 2000). Represented as the cradle of modernity, the 'North' had its knowledge often associated with the 'international' or the 'universal', a 'view from nowhere' (Haraway 1988) formulated by the collection of data from different parts of the world.

In contrast, southern knowledge is understood as 'local', contextual and empirical, ignoring that much of what is understood as 'western science' was practiced and carried out outside Europe, with agents and actors located in different parts of the world (Raj 2006). Showing how the categories and reasonings of social history were not, in any simple way, translated from Europe to colonies, all these categories and studies become powerful tools to demonstrate how Western science should be understood as a mixture of practices that escape the limits of Europe. The traditional view of science as a product conceived in Europe and exported to the rest of the world, reproducing a unidirectional North-South circulation, has been progressively substituted by an understanding that knowledge is often 'contaminated' and 'hybrid', produced during the encounters between multiple actors in different places.

By recognizing other agencies, these approaches highlighted a wide range of powerful social thought from the colonized and postcolonial world and encouraged new meanings of 'South' as a social category in scholarship (Connell 2007). Through these new approaches, the South would no longer be subject to a colonial relationship in knowledge production, supplying empirical raw materials, and importing theories. Instead of fitting external models into their realities at any cost, southern scholars have been recognizing their agency in knowledge production and identifying the circulation and appropriation of their knowledge in many other places and realities.

It remains to be seen how these narratives about the 'South' in art and scholarship influenced the balance of power in geography's international community. Beyond a simple internationalization and spread of institutions such as the International Geographic Union (IGU), which since the World War II has sought intensively to include the most significant number of members in different countries, it is crucial to address how did the dialogues between geographers from different areas of the world have been taking place. How can the prominent positions in the IGU and the places where the conferences take place could show new actors and agents in the international geographic community? What efforts have been made to discover 'other geographic traditions' (Ferretti 2019) and tell the discipline's history from different gazes and points of view? Moreover, which practices in making and communicating science guarantee the maintenance and persistence of colonialist positions in knowledge production? These questions will remain open but will be addressed shortly in the third section of this chapter.

11.4 North-South at the IGU

During the first decades of its existence, the IGU was a western union and specifically European union. Apart from Japan, the seven countries that founded the IGU in 1922 were all European. This is both a legacy of its history, which began half a century before the official establishment of the IGU at the Antwerp International Geographical Congress (IGC) in 1871 and a reflection of the balance of power within international science. The IGU still largely reproduces, even today, an established world scientific organization through its member countries, congress venues, and the origins of presidents and secretaries-general.

This organization is spatial, with Europe and North America at the center and more or less integrated peripheries surrounding the core. It was not until the Washington IGC in 1952 that an IGU congress left European or European colonized space and four more years to involve a country from the 'South' (Rio de Janeiro Congress in 1956).[6] Two-thirds of the 24 geographers who have served as IGU President were from Europe and only two come from 'Southern' countries: Shiba P. Chatterjee, from India, between 1964 and 1968, and Akin L. Mabogunje, from Nigeria, between 1980 and 1984. The balance is similar as regards the secretaries-general and it is only in 2010 that the vital center of the IGU, its official address, is in a country of the 'South': South Africa.

Nevertheless, this specialization of the IGU is not set in stone. Over time, the IGU has developed and concerned itself more and more with countries traditionally located in the 'South'. Let us take two important milestones in the history of the IGU, two IGCs in Amsterdam in 1938 and Rio de Janeiro in 1956. The Amsterdam congress has often been described as the triumph of colonial geography. Above all, it is clearly colonialist. Jacques Leclerc wrote of the 'racist paradigm that organizes and instructs the discourse of the colonial geography section' (Leclerc 1989, 93) and showed that very few speakers contested this orientation of the debates. This is not surprising. In this pre-war period, very few European geographers criticized colonial activities. But this congress reminds us above all that the relationship of domination between European countries and their North American extensions on the one hand, and the colonized regions of the world on the other, shapes within the IGU the place of countries and their scholars. The representatives of the colonized and/or dominated countries are intellectually discredited due to their 'race'.

Barely twenty years later, the situation had changed considerably. After Washington in 1952, the congress was once again held outside Europe and for the first time in a country considered to be 'developing'. This first (and still only) Latin-American congress promoted an important movement to internationalize the IGU, but also, paradoxically, marked the beginning of tightening around the English language (Lamego

[6] The Cairo Congress in 1925, discussed in Chap. 2, by Heffernan, was a geopolitical choice that reflected European considerations and colonial calculations.

2020). The issue of working languages is important. Dominique Volle (1996) has shown that various languages, overwhelmingly European, were used at congresses until the middle of the twentieth-century. The choice of two dominant languages, French and English, was officially encouraged as early as the Lisbon Congress in 1949 (Volle 1996) and became evident from the 1960s before English became almost the exclusive language. This development can be analyzed as the result of domination but it also makes it possible, if we consider English as a language of international communication, to open up the IGU to countries traditionally on the fringes.

A number of African, Asian and Latin-American countries joined the IGU in this period and the congress of Rio de Janeiro itself is described by Mariana Lamego as a 'tower of Babel' (2020, 124). The location effect played a strong role: 67% of the congress participants are South Americans and it is hoped that this opening will have a knock-on effect for the years to come. Above all, structural decisions have been taken with the creation of the status of 'associate member'.

It is an important choice. The IGU is an organization whose members are states.[7] During the first decades of the IGU's existence, they were all considered in an equal manner. But this equality within the structures affects spatial entities with very diverse characteristics. Some countries have the infrastructure, the financial, and human resources to organize congresses; geography is a structured science within universities and research organizations; the scholarly community of geographers is large and diverse; in various ways, these are the countries whose geographical knowledge is the most internationalized. Other countries do not have the opportunity to organize congresses and they do not have a sufficiently large and long-standing scholarly community to play a major role. The international situation of the community of geographers is very unequal and for a long time, egalitarian treatment within the IGU has contributed to the reproduction of inequalities.

But, for more than half a century, IGU policy has taken into account the unequal situation of geographical science in the world. The status of 'associate member' allows countries with few financial resources and a small number of geographers to join the IGU. The creation of regional congresses is also worth mentioning, enabling countries with limited resources to organize smaller congresses and, above all, enabling geographers from neighboring countries to participate in the congress at a lower cost.[8] In the current statutes, member countries are divided into fifteen categories and pay their membership according to their category of membership. In addition, the IGU has established a fund to develop geographical science and assist geographers from the 'less equipped countries'. The creation of working groups, lighter in structure than the commissions, which also contribute to a broadening of international participation in the activities of the IGU, is also worth mentioning.

Was the Rio de Janeiro Congress a turning point in the history of the IGU and its internationalization? Should we see the glass half empty or half full? The road traveled or the road still to be traveled? After 1956, in addition to the accession of new countries, the location of congresses changed considerably. Less than half of them were held in Europe. Moreover, a third of the member countries are countries of the 'South' and more than a third of the committee chairs are geographers from these countries.[9] This development was also linked to the decolonization

[7] It should nevertheless be noted that for a short period after the Rio de Janeiro congress, British dependencies such as Sudan or Kenya, and French dependencies such as French West Africa (AOF) were integrated into the IGU.

[8] However, this policy of equity is fragile. At the 1988 Sydney Congress, the desire to clarify the financial situation of the IGU led to the suspension of members who had not paid their dues for more than three years. This decision had considerable effects. Twenty-one countries disappeared from the list of IGU members: 5 Asian countries, 4 Latin American countries and 12 African countries, leading to the virtual disappearance of African representation in the IGU (Robic et al. 1996, 59).

[9] Between 1949 and 1996, 37 committee chairmanships were awarded to France, 30 to the United States, and only 5 to so-called 'southern' countries (Robic et al. 1996).

movement that led many countries to gain sovereignty and played a decisive role in the internationalization of the IGU.

However, this internationalization masks a persistent under-representation of the countries of the 'South': of the eight congresses held outside Europe since the Rio de Janeiro Congress, only for has been held in countries conventionally recognized as 'Southern countries'—the New Delhi IGC in 1968, the Tunis IGC in 2008, the Beijing IGC in 2016 and the Istanbul IGC, which was held virtually in 2021 due to the coronavirus pandemic. In its century of existence, the IGU has organized only one congress in Latin America and only one in Africa. From another point of view, it can be noted that these congresses of the 'South' now follow one another at a faster pace. It is essential. A congress like that of Tunis allows openings not only to the geographers of the organizing country but to those of many Arab and Muslim countries. The Beijing Congress impressed many participants with its scale. Here too, the expansion of the scientific community is notable. These developments mark a major turning point, perhaps the beginning of a new equilibrium in the geopolitics of geographic science.

Since its creation, the IGU has been faced with a geoeconomic and geopolitical situation with considerable inequalities and relations of domination between regions and states of the world, based on major divisions that oppose—to simplify things considerably—rich and poor countries. It is confronted with this situation, this world order, like other international organizations. But within an international organization that is concerned with geography, these problems are amplified and made very sensitive insofar as geography is a science that studies in particular these inequalities and relations of domination.

11.5 Conclusion

If the division and qualification of spaces are essential practices for assigning meaning to the world around us (Vignaux 1999), it is essential to recognize that these classifications are dynamic. Although the division of the world between 'North' and 'South' may suggest an immobile space stability, endorsing the continuation of a very old colonial condition, the appropriation of the idea of 'South' by a series of counter-hegemonic claims points to alternative balances in the global geopolitics of knowledge. Just as much as a space-bound essences, 'North' and 'South' express an asymmetrical relationship, which can be challenged by practices of circulation and cross-cultural exchanges. As an important institution for the reproduction and contestation of geographic practices around the world, the International Geographical Union will continue to be a privileged stage to identify continuities and changes in 'North-South' relations in the production of geographical knowledge.

References

Ades D (1997) Arte na América Latina: a era moderna 1820–1980. Cosac e Naify Edições, São Paulo

Balibar E (2004) We, the People from Europe? Princeton University Press, Princeton, Reflections on Transnational Citizenship

Besse J-M (2003) Les Grandeurs de la Terre: Aspects du Savoir Géographique à la Renaissance. ENS Editions, Lyon

Capdepuy V (2018) 50 histoires de mondialisation. De Néandertal à Wikipédia. Alma, Paris

Clerc P (2020) Do not cross the 'North/South' divide: a means of domination? In: Schelhaas B et al (eds) Decolonising and Internationalising Geography: Essays in the History of Contested Science. Springer, Cham, pp 47–56

Comaroff J, Comaroff J (2012) Theory from the South, or, How Euro-America is Evolving towards Africa. Routledge, London and New York

Connell R (2007) Southern theory: the global dynamics of knowledge in social science. Allen and Unwin, Sydney

Cosgrove D (2001) A cartographic genealogy of the earth in the western imagination. Johns Hopkins University Press, Baltimore

Dickenson JP et al (1983) A Geography of the Third World. Methuen, London

Dufour F (2007) Dire le 'Sud': quand l'autre catégorise le monde. Autrepart 41:27–39

Ferretti F (2019) Rediscovering other geographical traditions. Geography. Compass 13(3):1–15

Grosfoguel R (2008) Para descolonizar os estudos de economia política e os estudos pós-coloniais:

Transmodernidade, pensamento de fronteira e colonialidade global. Revista Crítica de Ciências Sociais. 80 (março)
Haraway D (1988) Situated knowledges: the science question in feminism and the privilege of partial perspective. Fem Stud 14(3):575–599
Harley JB (1988) Maps, knowledge and power. In: Cosgrove DE, Daniels, S (eds) The iconography of landscape. University of Cambridge Press, Cambridge
Huish R (2006) Logos a Thing of the Past? Not so fast, World Social Forum! Antipode 38(1):1–6
Lamego M (2020) How international was the International Geographical Congress in Rio de Janeiro 1956? On location and language politics. In: Schelhaas B et al (eds) Decolonising and Internationalising Geography: Essays in the History of Contested Science. Springer, Cham, pp 113–126
Leclerc J (1989) Amsterdam 1938: un tropique bien blanc sinon rien. In: Bruneau, M, Dory D (eds) Les enjeux de la tropicalité. Masson, Paris
Lewis MW, Wigen KE (1997) The myth of continents. A Critique of Metageography University of California Press, Berkeley and Los Angeles
Maldonado-Torres N (2006) Post-continental philosophy: its definition, contours, and fundamental sources. Worlds and Knowledges Otherwise 1(3)
Mignolo W (2000) Local histories/global designs: coloniality. Princeton University Press, Princeton, Subaltern Knowledges and Border Thinking
Mignolo W (2006) Citizenship, knowledge, and the limits of humanity. Am Lit Hist 18(2):312–331
Moura S (2015) Panoramas do Sul: Leituras para outras geografias do pensamento. Sesc, São Paulo
Raj K (2006) Relocating modern science: circulation and the construction of scientific knowledge in South Asia and Europe, 1650–1900. Palgrave Macmillan, Basingstoke, UK
Robic M-C, Briend A-M, Rössler M (eds) (1996) Géographes face au monde. L'Union Géographique Internationale et les congrès internationaux de géographie, L'Harmattan
Said E (1978) Orientalism. Pantheon Books, New York
Sales C (2015) Mapa invertido da América do Sul: e as visões críticas sobre o Sul global. Novas edições acadêmicas, Saarbrucken
Santos B (2014) Epistemologies of the South. Paradigm Publishers, New York
Santos BS, Meneses MP (eds) (2010) Epistemologias do Sul. Editora Cortez, São Paulo
Sharp J (2009) Geographies of postcolonialism. Sage, London
Todorov T (1989) Nous et les autres: la réflexion française sur la diversité humaine. Le Seuil, Paris
Vignaux G (1999) Le démon du classement: penser et organiser. Le Seuil, Paris
Volle D (1996) L'Universalité et ses limites. In: Robic M, Briend A, Rössler M (eds) Géographes face au monde. L'Union Géographique Internationale et les Congrès Internationaux de Géographie. Éd L'Harmattan, Paris, pp 63–82
Wood D (2010) Rethinking the power of maps. Guilford, New York and London

Pascal Clerc is full professor of geography at Cergy Paris University and member of the research team EMA (École—Mutations—Apprentissages). He is also associated researcher in the team EHGO (Épistémologie et histoire de la géographie). He works mainly about epistemology of geography, teaching of geography microgeography of schools.

André Reyes Novaes is an Associate Professor at the Universidade do Estado do Rio de Janeiro, Brazil. He is currently an Honorary Research Associate in the Department of Geography at Royal Holloway University of London and a member of the commission on the history of geography at the International Geographical Union. He works mainly about map histories, cross-cultural knowledge exchange and geopolitical imaginations.

12 Gender and International Geography

Mireia Baylina, Maria Dolors Garcia-Ramon, and Janice Monk

Abstract

The Commission on Gender and Geography of the International Geographical Union (IGU) has made great leaps ahead in the field of gender and geography. It has challenged conventional methodologies and epistemologies and has been a major tool for making gender geography more inclusive and truly 'international'. This chapter looks at the role played by the Commission in international geography and gives an overview of the worldwide differences in the development of feminist geography from its beginnings. The authors highlight the role played by agents, networks, and scales as well as the transformative role of situated gender geography in different world contexts. The Gender and Geography IGU Commission—which has been a real school for many geographers—has now the challenge of bringing together and promoting new generations in a framework of professional precariousness and instabilities within the neoliberal academic institutions.

Keywords

Commission on Gender and Geography · Agents · Networks · Challenges · Feminist geography

12.1 Introduction

The international women's political and social movement during the 1970s had an impact on academia and led to attention being given to gender in geographical research within the Anglophone world at that time. A little more than a decade later, in 1988, the International Geographical Union's Gender and Geography Commission was established. Since then, this Commission has made tremendous strides ahead in this discipline as it has challenged conventional methodologies and epistemologies and has been a major tool (probably the most important) for making gender geography more inclusive and really 'international'.[1]

This chapter has two sections. The first deals with the development of the Commission and the

M. Baylina (✉) · M. D. Garcia-Ramon
Universitat Autònoma de Barcelona (UAB), Barcelona, Spain
e-mail: mireia.baylina@uab.cat

J. Monk
SIROW, University of Arizona, Tucson, USA

[1]We use gender geography and feminist geography as synonymous, understood as a 'geography which explicitly takes into account the socially created gender structure of society, and in which a commitment both towards the alleviation of gender inequality in the short term and towards its removal through social changes towards real equality, in the longer term, is expressed' (WGSG-IBG 1984: 21).

V. Kolosov et al. (eds.), *A Geographical Century*,
https://doi.org/10.1007/978-3-031-05419-8_12

role it has played in advancing feminist teaching and research in geography. The second section provides a general overview of worldwide differences in the development of feminist geography, taking into account the roles played by agents, networks, and scales as well as the transformative role of situated gender geography in different contexts of the world. Finally, we present some possible ways forward for creating more inclusive gender geographies at the international level.

12.2 The IGU Gender and Geography Commission

The Commission of Gender and Geography has been the most important agent for internationalizing gender geography and counterbalancing the Anglophone hegemony since the very beginning. At the IGU congress of Paris in 1984, Jan Monk sent a written request (that was never answered) to the local organizers for a meeting time and space to discuss the possibility of creating a group. Nevertheless, a 'spontaneous' meeting took place and an informal list of interested people was created. In the following IGU Regional Conference, in Barcelona in 1986, a very well attended Round Table was organized within the program of the conference hosted by the University of Barcelona. Important steps toward initiating the Commission were taken in informal conversations between Jan Monk, Janet Momsen, and Maria Dolors Garcia Ramon (who would become vice-president, president, and secretary of the group) in December 1987 at a conference hosted by Garcia Ramon in Barcelona on 'Agriculture, Gender, and Space'. These were soon followed by discussions at British meetings, after which Janet Momsen took the lead in preparing an application and shepherding it through the IGU Executive Committee, drawing on her long-standing networks. At the Sydney IGU Conference in 1988, Jan Monk (Australian-born) was invited to give one of the plenary lectures on 'Encompassing Gender: Progress and Challenges in Gender Research'. And the Gender and Geography Study Group was approved with Janet Momsen as the founding chair, Jan Monk as vice-president, and Maria Dolors Garcia Ramon as secretary. Jan Monk was also named Chair when the Study Group was 'upgraded' to Commission status at the conference in Washington in 1992 with the same team. The creation and consolidation of the Commission made it possible to move from 'the Ladies' program to the feminist session', as expressed by Rössler (1996), and to contextualize the role of women in the IGU (Robic and Rössler 1996; see also Chap. 7 in this book, by Droogleever Fortuijn).

Since the beginning, the commission has had a Steering Committee that has paid a lot of attention to representing a wide range of regions of the world (Huang et al. 2017). Currently, the Commission has about 740 correspondent members and it holds sessions at the IGU Conferences and specialized meetings in other locations, usually hosted by members of its international steering committee. To date, its events have taken place in more than 20 countries. Work from these has been published in outlets beyond the dominant Anglophone sites in order to reach other audiences and, in some cases, to use languages other than English. The Commission has also collaborated in joint programs with other IGU Commissions, such as those in political geography, geographical education, health and development, history of geography, and economic geography. The advantage of working separately is that it is possible to have in-depth discussions, while collaboration makes it possible to bring gender perspectives to other audiences.

An early example of the Commission's potential to disseminate research by scholars from multiple regions of the world was the book series International Studies of Women and Place. It was edited by Janet Momsen and Jan Monk and the series is published until the present by Routledge. Janet Momsen (the first Chair of the gender group) hosted the Group's first formal event: a workshop on gender and development held at the University of Newcastle-upon-Tyne in 1989. This workshop was the basis of one of the first books of the series, which also developed from Momsen's initiative with the publisher. The book *Different Places: Different Voices* (Momsen and Kinnaird 1993), brought studies on gender and development in Africa, Asia, and Latin America to a wide audience, and gave a voice to scholars from around the world. The series has been very successful, with more than 35 volumes published so far.

However, in terms of inclusivity, the most important publication is the Newsletter which is published twice a year in English. It began in 1988, and in November 2020 reached issue number 64. Since 2006 it has also been circulated to all members in Spanish and French to widen the audience in those regions of the world where English is not so common in academia. Jan Monk was the editor for more than 30 years (up to 2018 and for 60 issues), and she has played a vital role in showcasing the strengths and connecting feminist geographers through the Newsletter (Huang et al. 2017). Joos Drooglever Fortuijn is currently the editor and was former chair of the Commission and first vice-president of the IGU until August 2020. The Newsletter reports on Commission activities and publishes academic and scientific news of feminist geographers as well as gender-related activities around the world. It gives particular emphasis to publications on gender (books and articles) and aims to cover not only the Anglophone world but also publications in languages other than English. Some of the English publications also include work by geographers from countries where English is not the native language.

In an analysis carried out by the Gender Group of the Autonomous University of Barcelona of the first 53 issues of the Newsletter, the authors divide their results into four sections. The first section is 'News around the World', which includes references to activities that have taken place in 49 different countries and only 50.11% belong to Anglophone areas (continental European countries represent 33.04%). There is a growing percentage of news from certain South Asian countries, contrasting with a decrease in news from African countries. The section devoted to 'Monographic issues of journals' includes 102 items, and 81 are completely written in English. However, a large number of issues are in other languages (in total 21 languages including Portuguese, Swedish, French, Spanish, Catalan, German, Italian, Hebrew, Chinese, Norwegian, Dutch, Polish, Turkish, and Hungarian). In the section referring to 'Published books' the percentage of books in English is a little higher (82.41%) but a quite considerable percentage (17.59%) of books are in languages other than English, in particular European languages. In the section 'Articles and Chapters of books', of a total of 3769 items 84% are in English and 16% in other languages, a much higher percentage than we can find in many international lists of publications. In more recent years, this percentage has increased significantly, thanks to the role played by multi-lingual journals such as ACME and *Revista Latino-Americana de Geografia ê Genero*. In summary, the efforts that editors of the Newsletter have made to include gender research outside the Anglophone world have been successful, although we need to move further in this direction.

12.3 Differences in the Development of Feminist Geography Around the World

In 2021 feminist geography now has a fifty-year trajectory in the founding countries and a history of several decades in other areas. Geographers have worked constantly to make this trajectory visible by documenting the state of gender geography in the world, providing updates on its progress, and, most importantly, making new recruits visible and creating support networks among researchers. At the end of the eighties, Bowlby and Peake (1989) confirmed the existence of diverse feminist geographies with different velocities in 21 countries and geographical areas of the world. This review follows others made from different contexts (Garcia Ramon 1989; Garcia Ramon and Monk 2007; Monk and Garcia Ramon 2013; Monk 2015; Ybarra and Escamilla 2016, among others). The latest update on the state of feminist geography in the world is that collected in the journal *Gender, Place and Culture* in the issue that celebrated its 25 years of publication. The monograph documents 39 countries, although only three are in Africa (Blidon and Zaragocín 2019).

The international health of feminist geography can be considered to be good in general if we look at its scientific progress in many places and its clear success in some of them, mainly English-speaking places. However, it is still non-existent in many countries and has only a very fragile presence in others. Paradoxically, it is also in a situation of unstable equilibrium in those places where it is most consolidated.

12.3.1 Agents, Networks, and Scales

The history of feminist geography is linked to specific people and the specific institutions that have supported them. For example, the formal creation of the Women's Geography Study Group of the Institute of British Geographers in 1982, with the aim of promoting the study and consideration of gender in geography and establishing networks to work collectively with and for women, has played a pivotal role in establishing feminist geography in the UK. The researchers' foundational works, including the manual *Geography and Gender* (WGSG of IBG 1984, have opened up many issues and influenced research in this country and many others. Some of these researchers participate in international geography conferences, including those of the International Geographical Union, the Royal Geographical Society-Institute of British Geographers, and the Annual Conference of the Association of American Geographers, where they establish contacts that become international networks of collaboration and mentoring. The researchers are professionals from different countries, with consolidated university jobs, with financial support to participate in international conferences and for research, and with a vocation for leadership in promoting feminist geography in their workplaces. International relations for the exchange of knowledge, strategies, and mutual aid are the engine for carrying out research and being legitimized in their own countries (Baylina and Rodó de Zárate 2019). Some geographers point out how in patriarchal and parochial institutional contexts, academic control can be mitigated with active networking, horizontal connections with feminists from other places, and with formal and informal mentoring (Datta 2019).

Leadership has also involved the direct implication in networking, such as management positions and membership in the international commission on geography and gender, the organization of international conferences and seminars in researchers' countries of origin and teaching stays in gender geography abroad. One of the clearest examples of mutual international and intergenerational learning is that of the Erasmus Network on Gender and Geography, which for ten years (1989–1998) allowed teachers and students from five countries (Denmark, Greece, the Netherlands, Spain, and the United Kingdom), six universities and six languages of the European Union to meet for one week each year in the different countries to carry out research seminars. Some of the teachers in this network still recognize the importance of these productive and innovative meetings (Droogleever 2008) for promoting teamwork when academia was already becoming increasingly individualistic and competitive (Vaiou 2019).

Various international conferences and seminars had already been held during the 1980s and 1990s, forging alliances between feminist geographers from different geographical areas. In Spain, the Research Group on Geography and Gender, founded in 1987 by Maria Dolors Garcia Ramon, organized international seminars and conferences on geography and gender on a regular basis (Garcia Ramon 1989). According to the Greek scholar, Dina Vaiou, this research group has been an inspiration for other geographers both for the development of feminist approaches grounded in Southern European experiences and intellectual traditions and for promoting group work within an increasingly individualistic and competitive academe (Vaiou 2019). International mentoring work has been carried out with geographers of different generations (Lan 2016). In Italy, as early as 1993, feminist geographers Maria Luisa Gentileschi and Gisella Cortesi organized the pre-conference of the IGU Population Geography Commission in Cagliari (Sardinia), which addressed the role of women in territorial organizations (Cortesi and Gentileschi 1996).

On occasions, to overcome national limitations, the Anglophone cultural hegemony, and the hierarchy of forms of knowledge production, researchers in other linguistic communities have joined international networks such as the Iberian and Latin American Network of Geography, Gender, and Sexuality (REGGSILA), to disseminate and make the research visible in Spanish-speaking contexts (Ferreira 2019). Similarly, since 1989 German-speaking countries have organized, through the Working Group Geography and Gender, interdisciplinary debates on feminist theory and attempts to re-theorize space, place, and gender within human geography in their linguistic area. Several activities such as workshops and panels at German-speaking conferences are added to a communication platform, the digital newsletter Feministisches Geo-Rundmail. This publication is a medium to discuss new strands of research and advance feminist knowledge production in these areas (Bauriedl et al. 2019).

At the national, regional, and local levels, researchers have been committed to gender geography in different fields using various strategies, including leading research projects, promoting teaching, organizing scientific meetings, directing scientific journals (a potential means of disseminating research), and presiding over geographic institutions. These strategies have been empowering although they require high personal dedication to create a study environment, as well as strong personal resilience in the face of much opposition and quite a bit of indifference from the scientific community. Researchers have also worked at various geographical scales, from personal to global (Silva and Ornat 2019), and have underlined the importance of publishing in different languages (Ulloa 2019) due to the potential influence of language in various geographical areas of the world.

Today, the founding Anglophone countries define the current situation of feminist geography as vibrant (United States, United Kingdom, Canada, Australia). However, Evans and Maddrell (2019) warn, in the case of the United Kingdom, of the risk that the clear normalization of gender in geography leads to invisibility and oblivion. This is not the case in most countries, which, recognizing international support, request more collaboration in order to resist in their countries, as in the case of Ghana (Wrigley-Assante and Ardayfio-Schandorf 2019). Even in the Netherlands, where feminist geography is accepted, it is recognized that integration still depends on the feminist commitment of teachers and researchers and that we must be vigilant to preserve what has been achieved so far (Droogleever Fortuijn 2019).

Developing spaces for collaboration and solidarity between countries is an essential task in the current agenda, especially between North and South (Gahman and Collins 2019), but also between neighboring geographic areas (Germany, Austria, Hungary, Poland) (Pitonak and Kingorova 2019) and between areas that speak the same language (Germany, Austria, Switzerland) (Schurr 2016). Dialogue is also needed between different linguistic communities within the same country, such as in Canada (Martin and Latendresse 2019) or in Switzerland (Duplan 2019), cases that show clear cultural hegemonies.

12.3.2 Gender as a Transformative Approach in Geography

Gender has become a transformative approach to the geographic discipline. It has transformed disciplinary analyses of landscape, place, and space, problematizing the regulations associated with people, politics, and places. Early works explore the spatial constraints that women face as well as the relationship between women, capitalism and urban landscapes. The increased number of feminists works and themes in the eighties in English-speaking countries gave an account of the material realities of women's lives. There were many, increasingly sophisticated, theoretical interventions on gender as an instrumental force and as an explanatory category of the discipline (Nelson 2016). Monk and Hanson (1982) explained in depth the construction of masculinist geography, and poststructuralist theory provides the conceptual language to bring feminist criticism to an epistemological level. Rose (1993) shows that the arguments about universality in geographic theory are based on a male producer of knowledge. Furthermore, the separation of the 'researcher' and 'researched' also reproduces existing hierarchies and social exclusions.

Researchers today recognize that many topics in geography can be rethought using theoretical frameworks that address patriarchy, queer theory, cultural studies, postcolonialism, and postmodernism (Colombara 2019). In Brazil, Silva and Ornat (2019) affirm that feminist geography has renewed Brazilian geography and has been crucial in confronting the different forms of oppression that mark the national territory: environmental justice, racial policies, and the decolonization of knowledge production. Trans and queer scholarship have pushed, mainly in the Anglophone world, for a new and innovative understanding of the spatiality of gender and the creation of gender through socio-spatial relations, but also the challenges and resistances trans people experience in the spaces they use, create, and reject (Browne et al. 2010). Indeed, feminist geography continues to expand the boundaries of the discipline. Currently, the greatest challenges, many of them championed by social movements (Me Too, Black Lives Matter, Fridays for Future among others) and those derived from the current global pandemic, question how geography can and should represent a multiplicity of communities and common interests.

Perhaps because of this transformative potential, feminist geography has encountered and encounters many obstacles to becoming established in many parts of the world. To begin with, there have always been problems with terminology. The concept 'feminist' is rejected in many places, as it suggests activism and political commitment. Even the researchers themselves have not called themselves feminist geographers for fear of being devalued and excluded, for example in France (Blidon 2019; Louargant 2019). In Japan, gender geography is mainly researched by men, who use life stories to relate the lived space of women in their daily lives, without adopting gender theory or using this concept (Yoshida 2019). In Muslim societies, such as Iran, feminist geography carries negative connotations and serious adverse consequences. Geography is an apolitical science and 'feminism' is a sensitive word, as are intersectionality, diversity, activism, and social change. These are political terms, and therefore, related to national security (Bagheri 2019). In Albania, there are also hostile prejudices against feminism, although there are some works on gender geography within the context of the social sciences (Danaj et al. 2019).

The power structures and organizations of many universities are strongly patriarchal. Researchers face the sticky floor and the glass ceiling and develop various mechanisms (such as informal mentoring) to survive and challenge, if they can, the context of masculinism and elitism of the neoliberal university (Duplan 2019). In Taiwan, geography is also dominated by men, who are insensitive to gender and relatively conservative in relation to other disciplines (Chiang and Stephenson 2019). In some places, such as India, despite a long history of work and internationalization of feminist geographers, there is still an established academic misogyny, which avoids taking new directions in the discipline by applying different mechanisms, such as promoting physical geography (Datta 2019). Similarly, in Israel, where a few geographers carry out powerful feminist geography in a conservative, hegemonic and patriarchal Jewish context, there is currently a renewed emphasis on physical geography to the detriment of human geography and quantitative methods versus qualitative methods (Fenster and Misgav 2019).

Added to the threat of seeing the status quo of patriarchal power in academia weakened, is the rejection of the epistemological orientation of feminist studies, which challenges the neutrality of scientific knowledge and the concepts of objectivity and neutrality of traditional empirical science. Consequently, quantitative methods are still privileged and little or no room is left to question androcentrism and to enter into critical social theory. Thus, in some post-socialist countries, such as Hungary, we see regressive movements such as the elimination, after two decades of emergence, of feminist geography and of master's degrees in gender, questioning their usefulness and qualifying these studies as ideology rather than science (Timar 2019).

In some countries, human geography is not powerful enough, which could explain why the gender approach is having difficulties becoming established. In Italy, feminist geography has had a lot of cultural and academic resistance and the current vision of this approach, from the geography discipline, is that it is about the geography of women carried out by women, despite the participation of researchers on an international scale and the constitution in 2005 of the geography and gender group in the Association of Italian Geographers (Schmidt di Friedberg and Pecorelli 2019). In Ecuador, feminist geography has recently penetrated social science departments, although there are also works by geographers from Anglo-American institutions on feminist, decolonial, and political ecology geography (Zaragocín 2019).

12.3.3 A Situated Feminist Geography, Young Generations, and Future Directions

Feminist geography must be diverse. Grounded in the everyday, feminist research recognizes that all knowledge is situated: it develops from a particular social and geographic location and therefore pays attention to the power dynamics of knowledge production and legitimation (Kern 2019). Thus, the direction it takes should reflect the interests and concerns of the researchers in order to be useful to the populations with which they work.

Very diverse topics have been studied over all these decades and in the different places, such as the gendered lives of women in the global North and South, urban mobility, gender and employment, migration studies, women's experiences in rural areas, gender and fear, women's experiences in public spaces, mother's care practices, use of transport, women's use of online spaces, men, masculinities and place, LGBT experiences, gender, empire and postcolonialism, body and space, fieldwork, lesbian geographies, women in informal economies, feminist geographies of children and youth, emotional geographies, participatory geographies, and so on. In foundational places, feminist geography has swung from an interest in the geographies of women to feminist analyses of spaces, the geographies of sexualities, relationships with other structural inequalities (intersectionality), and decolonial thinking.

The appearance of queer theory in feminist studies has led to the development of the geographies of sexualities. The first studies show that public spaces are constructed as heterosexual spaces through the repetition of certain practices. The research also reveals the heteronormative character of these spaces (Bell and Valentine 1995). These studies lay the foundations for understanding the sexualities constructed through places and in places and make visible that spaces are constructed and organized according to specific ways of understanding sex, gender, and sexuality. Although many feminist geographers embrace these new research initiatives, this topic has undoubtedly attracted a new generation of researchers (see, for example, Brown and Browne 2016) who study the experiences of white gay men in urban settings to discuss the exclusions within the LGBT collective itself (Rodó and Baylina 2019). It is important to see how feminist geography appears or expands in some countries of the world based on the geographies of sexualities and issues related to gender identity and expression, as is the case in Brazil, France, Chile, and South Africa (Silva and Ornat 2019; Blidon 2019; Astudillo 2019; Oldfield and Tucker 2019). Simultaneously, in the Anglophone context, scholars continue debating the state of sexualities education in geography curricula for over 20 years (Zebraki and Hall 2020)!

It is very enriching to see how in most countries research reflects local/international problems or interests: in Latin America (Ecuador, Colombia) 'territory' has been the priority spatial concept of identity. Therefore, feminist debates on territory follow this trend and deal with issues such as gender, conflict and corporal and territorial violence, water and territory, extraction processes, and gender relations (Zaragocín 2019; Ulloa 2019). In Mexico, research focuses on violence against women, migration, and sex tourism (Ybarra and Escamilla 2016). The disproportionate interest in urban versus rural studies contrasts with scientific production in Australia, where there is a large body of work on women's experiences in rural sectors, and also on the relationship between rurality and sexuality (Gorman-Murray and Baganz 2019). In the Mediterranean region, there is a general focus on women's work and daily life (Baylina 2019). In Southeast Asia, interests revolve mainly around migration, mobility, and transnational work, for example, the case of domestic workers (Huang and Ramdas 2019). All these social issues and concerns truly set the international agenda in geography and gender, beyond the thematic and chronological direction led by Anglo-American feminist geography.

The countries that have had the most funding have carried out collaborative, interdisciplinary, multicultural, and transnational research in order to better understand the consequences of neoliberalism and globalization processes (Sircar 2019; Lund et al. 2019; Myrdahl 2019). These countries are perhaps the ones that are now rethinking their research and teaching most from a decolonial and intersectional perspective. Researchers have entered the debates on the content of feminist geography, who practices it, what are its boundaries, and who gets to speak and for whom. In this regard, Ramdas (2016) offers a way to go beyond the standpoint theory with an emphasis on 'experiential knowledge' (see Hooks 1994) that allows us to overcome the ethical and political dilemmas of conducting research in distant social contexts. According to Ramdas, it is possible to observe and talk about an area from multiple perspectives, from near and far, and from feminism, with a commitment to minimizing the distance between the self and the other. Furthermore, North-South, East-West constructions can be very simple when researchers occupy multiple and complex positions. Along the lines of this debate, it is enriching to observe emerging feminist theorizations from the global South and about the global South (Huang and Ramdas 2019) and the feminisms of the Pacific (Underhill-Sem 2016; WGGRN 2019).

Many of these works, as well as those of the aforementioned geographies of sexualities and LGBT communities, are carried out by people beginning their careers, and therefore indicate future paths. These are not the only topics that interest the new generation of researchers, other topics include care work, reproductive justice, gender, and intersectional activism, indigeneity, gender and sexuality in disasters, digital activism, environmental activism, and climate justice, youth geographies, new techno feminisms, and the performative constitution of identities, among others. In the countries with the longest experience, these researchers are the third generation of feminist geographers. Some of these researchers are examples of brain drain, who come to countries with greater opportunities, others are in continuous training waiting for a job opportunity in a neoliberal institution that makes their situation precarious, and many of them are in close dialogue with social reality through political activism. The future of feminist geography is in the hands of national policies in different countries, and whether these policies prioritize the stability of new generations in academia so that they can continue the work of their predecessors: to integrate the feminist perspective into the core of teaching and researching mainstream geography.

12.4 A Way Forward

Feminist geography is in a time of transition. The founding geographers, key figures who have left a historic legacy in the discipline, are leaving, and the new, younger generations face numerous obstacles in establishing themselves in the neoliberal institutions. This means that there are few established academics who can promote research or teaching initiatives. In this situation, the role of the Gender and Geography IGU Commission as a real school for many geographers is more important than ever. New representatives have the challenge of bringing together and promoting new generations, whatever their professional situation. It is also an interesting moment due to the effervescence of the debates in feminism on the conceptualization of the gender category and other social constructions that form part of the dialogue with the discipline of geography. A stronger feminist geography will surely emerge from these debates.

Fostering inclusive perspectives is essential for challenging Anglo-American hegemonies in gender and geography and for international enrichment of our work (Garcia Ramon 2003). One important strategy is to diversify contacts and networks and thereby challenge unidirectional flows, facilitated by developing collaborations within international structures. These 'Other' to 'Other' communications, exchanges, and networking across national boundaries include not only communication between 'cores' and 'peripheries' but also between different 'peripheries', that is, not only North-South but also South-South and South-North, so that different languages, paradigms, and methodologies are accepted. This effort will provide a valuable counterbalance to Anglophone dominance. The Commission of Gender and Geography of the IGU is an excellent example of the creation of such an international space for gender scholars.

Acknowledgements Our warm thanks are extended to the reviewers and editors for their helpful comments on this chapter.

References

Astudillo P (2019) The study of homosexual space: an example of feminist geographies' limits in Chilean academia? Gend Place Cult 26(7–9):1013–1020

Bagheri N (2019) Avoiding the "F" word: feminist geography in Iran. Gend Place Cult 26(7–9):1128–1136

Bauriedl S et al (2019) Celebrating 30 years of feminist geographies in the German-speaking countries Germany, Switzerland and Austria. Gend Place Cult 26 (7–9):1049–1063

Baylina M (2019) Feminist interventions: Southern Europe/Mediterranean. In Orum A (ed) The Wiley Blackwell encyclopedia of urban and regional studies. Wiley, London

Baylina M, Rodó de Zárate M (2019) Mapping feminist geography in Spain. Gend Place Cult 26(7–9):1253–1260

Bell D, Valentine G (eds) (1995) Mapping desire: geographies of sexualities. Routledge, London

Blidon M (2019) Still a long way to go: gender and feminist geographies in France. Gend Place Cult 26(7–9):1039–1048

Blidon M, Zaragocín S (eds) (2019) Mapping gender and feminist geographies in the global context. Gend Place Cult 26(7–9):915–925 (Special issue on gender and geography worldwide)

Bowlby S, Peake L (1989) The challenge of feminist geography. J Geogr High Educ 13(1):90–92

Brown G, Browne K (2016) The Routledge research companion to geographies of sex and sexualities. Routledge, Oxford

Browne K, Nash C, Hines S (2010) Introduction: towards trans geographies. Gend Place Cult 17(5):573–577

Chiang L-HN, Stephenson R (2019) The challenges of feminist geography in Taiwan. Gend Place Cult 26(7–9):1280–1287

Colombara M (2019) Gender geography in Argentina: a brief overview. Gend Place Cult 26(7–9):935–944

Cortesi G, Gentileschi M (1996) Donne e geografia: studi, richerche, problema. Franco Angeli, Milan

Danaj E, Lame E, Kalaja D (2019) Gender and feminist studies in Albania -a brief state of the art. Gend Place Cult 26(7–9):926–934

Datta A (2019) But this is not geography…! Of ontological circumcisions and writing feminist geographies from India. Gend Place Cult 26(7–9):103–1110

Droogleever Fortuijn J (2008) Balancing the margin and the mainstream. In: Moss P, Falconer-al Hindi K (eds) Feminisms in geography. Rowman and Littlefield, Lanha, pp 75–83

Droogleever Fortuijn J (2019) Gender and geography in the Netherlands: from separation to integration. Gend Place Cult 26(7–9):1297–1303

Duplan K (2019) A feminist geographer in a strange land: building bridges through informal mentoring in Switzerland. Gend Place Cult 26(7–9):1271–1279

Evans S, Maddrell A (2019) Feminist geographies in the UK: the dialectics of women-gender-feminism-intersectionality and praxis. Gend Place Cult 26(7–9):1304–1313
Fenster T, Misgav C (2019) Israeli feminist geography: women, gender and queer geographies. Gend Place Cult 26(7–9):1119–1127
Ferreira E (2019) Women's, gender and feminist studies in Portugal: researchers' resilience vs institutional resistance. Gend Place Cult 26(7–9):1223–1232
Gahman L, Collins T (2019) Recognizing and undisciplining feminist geography in the Caribbean. Gend Place Cult 26(7–9):988–1000
Garcia Ramon MD (1989) Geography and gender in Spain: new lines of research and teaching. J Geogr High Educ 13(1):110–112
Garcia Ramon MD (2003) Globalization and international Geography: the questions of languages and scholarly traditions. Prog Hum Geogr 27(1):1–5
Garcia Ramon MD, Monk J (2007) Gender and geography: world views and practices. Belgeo 3:1-3-
Gorman-Murray A, Baganz E (2019) Geographies of gender and sexuality in Australia. Gend Place Cult 26 (7–9):945–955
Hooks B (1994) Teaching to transgress: education as the practice of freedom. Routledge, New York
Huang S et al (2017) A continuing agenda for gender: the role of the IGU Commission on Gender and Geography. Gend Place Cult 24(7):919–938
Huang S, Ramdas K (2019) Generative spaces of gender and feminist geography in Singapore: entanglements of the personal and political. Gend Place Cult 26(7–9):1233–1242
Kern L (2019) Anglophone world. In Orum A (ed.) The Wiley Blackwell encyclopedia of urban and regional studies. Wiley, London
Lan D (2016) Los estudios de género en la geografía argentina. In Ybarra V, Escamilla I (eds) Geografías feministas de diversas latitudes. Orígenes, desarrollo y temáticas contemporáneas. Universidad Nacional Autónoma de México, México D F, pp 55–70
Louargant S (2019) Feminist interventions: French connections. In: Orum A (ed) The Wiley Blackwell encyclopedia of urban and regional studies. Wiley, London
Lund R et al (2019) Feminist geographies in Norway from the turn of the millennium. Gend Place Cult 26(7–9):1198–1214
Martin P, Latendresse A (2019) Feministes tant qu'il le faudra! The multiple possibilities of Francophone feminist geographies au Québec/Canada. Gend Place Cult 26(7–9):976–987
Momsen J, Kinnaird V (1993) Different places, different voices: gender and development in Africa, Asia, and Latin America. Routledge, London and New York
Monk J (2015) Gender and feminist studies in geography. In: International encyclopedia of the social and behavioral sciences, vol 9. Elsevier, London, pp 689–693
Monk J, Hanson S (1982) On not excluding half of the human in human geography. Prof Geogr 34:11–23
Monk J, Garcia Ramon MD (2013) Bridges and barriers: some cartographies of 'international' practice in gender studies. Querelles, Jahrbuch für Frauen und Geschleterforschung 16
Myrdahl TM (2019) Canadian feminist geography in the 21st century. Gend Place Cult 26(7–9):964–975
Nelson L (2016) Geografía feminista anglosajona: reflexiones hacia una geografía global. In: Ybarra V, Escamilla I (eds) Geografías feministas de diversas latitudes. Orígenes, desarrollo y temáticas contemporáneas. Universidad Nacional Autónoma de México, México D F, pp 21–54
Oldfield S, Tucker A (2019) Persistent pasts, present struggles, imagined futures: gender geographies in South Africa after apartheid. Gend Place Cult 26(7–9):1243–1252
Pitonak M, Kingorova K (2019) Development of Czech feminist and queer geographies: identifying barriers, seeking progress. Gend Place Cult 26(7–9):1001–1012
Ramdas K (2016) Feminist care ethics, becoming area. Environ Plan D: Soc Space 34(5):843–849
Rose G (1993) Feminism and geography: the limits of geographical knowledge. University of Minnesota Press, Minneapolis
Rodó M, Baylina M (2019) Presentació: Geografies de les Sexualitats. Documents D'anàlisi Geogràfica 65 (3):441–449
Rössler M (1996) From the Ladies' program to the feminist session. Robic M-C. Géographes face au monde. L'Harmattan, Paris, pp 259–267
Robic M-C, Rössler M (1996) Sirens within the IGU—an analysis of the role of women at International Geographical Congresses (1871–1996). Cybergeo. https://doi.org/10.4000/cybergeo.5257. Accessed 3 Sept 2021
Schmidt di Friedberg M, Pecorelli V (2019) Gender and geography in Italy. Gend Place Cult 26(7–9):1137–1148
Schurr C (2016) ¡¿Otras geografías son posibles?!. Geografías feministas en Suiza, Austria y Alemania. In: Ybarra V, Escamilla I (eds) Geografías feministas de diversas latitudes. Orígenes, desarrollo y temáticas contemporáneas. Universidad Nacional Autónoma de México, México D F, pp 153–188
Silva J, Ornat M (2019) Feminist geographies: fight and achievement of a place in the Brazilian scientific production. Gend Place Cult 26(7–9):956–963
Sircar S (2019) Feminism and intersectionality in Swedish geography. Gend Place Cult 26(7–9):261–1270
Timar J (2019) Hungarian feminist geography in a curved space? Gend Place Cult 26(7–9):1094–1102
Ulloa A (2019) Gender and feminist geography in Colombia. Gend Place Cult 26(7–9):1021–1031
Underhill-Sem Y (2016) Gender research in the Pacific 1994–2014: beginnings. Pacific Women Support Unit, Suva. http://www.pacificwomen.org/

Vaiou D (2019) Feminist geographies in Greece: a personal trajectory between 'local' and 'international.' Gend Place Cult 26(7–9):1073–1080

WGSG-IBG (1984) Geography of gender. Hutchinson, London

WGGRN (Women and Gender Geographies Research Network of Aotearoa New Zealand) (2019) Feminist geographies in Aotearoa New Zealand: cultural, social and political moments. Gend Place Cult 26(7–9):1182–1197

Wrigley-Assante C, Ardayfio-Schandorf E (2019) The emergence and institutionalization of feminist geography in Ghana. Gend Place Cult 26(7–9):1064–1072

Ybarra V, Escamilla I (eds) (2016) Geografías feministas de diversas latitudes. Orígenes, desarrollo y temáticas contemporáneas. Universidad Nacional Autónoma de México, México D F

Yoshida Y (2019) Gender geography in Japan: the trajectory, fruits of research and future challenges. Gend Place Cult 26(7–9):1149–1158

Zaragocín S (2019) Feminist geography in Ecuador. Gend Place Cult 26(7–9):1032–1038

Zebraki M, Hall J (2020) Teaching geographies of sexualities: 20 years on. J Geogr High Educ 44 (2):179–187

Mireia Baylina is senior lecturer in Geography at the Universitat Autònoma de Barcelona (UAB). She has done research on social and economic issues from a gender perspective. Her interests of research and publications are related to women's work and daily life in rural areas, representation of gender and rurality, and geographies of children and youth from a gender and intersectional perspective. Her teaching has been mainly related to regional geography and gender geography. She has been editor of the journals *Documents d'Anàlisi Geogràfica* and *European Urban and Regional Studies,* Coordinator of the *Interuniversity Doctorate Program of Gender Studies: Cultures, Societies and Policies* for the UAB, and board member of the Commission of Gender and Geography of the International Geographical Union.

Maria Dolors Garcia-Ramon is Emerita Professor of Geography at the Autonomous University of Barcelona and has been professor/research fellow at several universities in the US, UK and Argentina. She has published several books and a large number of articles from a gender perspective on the fields of rural geography, urban public spaces and the history of geography from a critical perspective. She has been the coordinator of many international and national research projects on gendered social and economic activities: She has been a pioneering figure in Spain on gender studies as well as member of the editing board of several international journals. She has been the secretary of the Gender and Geography Commission and *Laureat d'Honneur* (2016) at the International Geographical Union. She is also member of the Academia Europea since 1998.

Janice Monk has been Research Social Scientist Emerita at the SIROW (University of Arizona) and Executive Director for more than 25 years. She is Emerita Professor of Geography and Development at the same university. She was the third woman president of the AAG in 2001. Among others, her fields of interest are social and cultural geography, feminist geography, history of women in geography and international collaboration. She has several books and a very large number of articles and chapters of books in international publications. She has been a sustained supporter of international collaboration and a well-recognized pioneering figure —as well as mentor—in gender and geography internationally. She was a co-founder of the Gender and Geography Commission of the IGU and her vice-president. She was awarded the *Laureat d'Honneur* at the IGU in 2012.

Part III

International Geography in the 21st Century: A Dream Discipline

Interactions of Geography with Other Natural and Social Sciences and the Humanities

13

Ruishan Chen, Annah Zhu, Yingjie Li, Pengfei Li, Chao Ye, and Michael E. Meadows

Abstract

Geography has historically enjoyed strong interactions with other disciplines in addressing major challenges related to social, economic, and environmental issues and in contributing overall to sustainability. More especially in the Anthropocene, issues such as climate change, biodiversity loss, terrorism, poverty, refugees, environmental hazards, and pandemics have emerged that require an improved understanding of spatial and temporal patterns, processes, and impacts. Consideration of scale and place-based perspectives are essential in helping to resolve such complex issues. The chapter highlight five arenas of interaction between geography and other disciplines, viz. the natural sciences, socioeconomic sciences, humanities, human-environment relationships, and sustainability science. The International Geographic Union (IGU) provides a platform to unite geographers globally to share ideas, promote communication, and advance the interaction of geography with other disciplines, and also with different stakeholders from NGOs, governmental agencies, and international organizations. At this critical juncture, Geography must continue to develop through its vibrant connections with other fields and geographers should continue to exhibit interdisciplinary leadership by embracing different perspectives, by supporting institutional arrangements that foster interdisciplinary activity, and by seeking the knowledge and techniques that other fields can contribute to geographic perspectives, approaches, and insights to the collective effort. The IGU continues to play an important role in facilitating knowledge development and sharing, and in encouraging transformational actions that promote a just, peaceful, and sustainable planet.

R. Chen (✉)
Department of Landscape Architecture, School of Design, Shanghai Jiao Tong University, Shanghai 200240, China
e-mail: rschen@sjtu.edu.cn

A. Zhu
Environmental Policy Group, Wageningen University, Hollandseweg 1, 6706 KN Wageningen, The Netherlands

Y. Li
Department of Fisheries and Wildlife, Center for Systems Integration and Sustainability, Michigan State University, East Lansing, MI, USA

P. Li
Haskayne School of Business, University of Calgary, 2500 University Dr NW, Calgary, AB T2N 1N4, Canada

C. Ye
School of Geographic Sciences, East Cina Normal University, Shanghai 200241, China

M. E. Meadows
Department of Environmental & Geographical Science, University of Cape Town, Private Bag X01, Rondebosch 7701, South Africa

V. Kolosov et al. (eds.), *A Geographical Century*,
https://doi.org/10.1007/978-3-031-05419-8_13

Keywords

Interdisciplinarity · Sustainability · (SDGs) · Convergence · Pattern and process · Human-environment relations

13.1 Introduction

Social, economic, environmental, and related health issues are receiving increased attention globally and have been brought into particularly sharp focus by the coronavirus pandemic. The continued spread of COVID-19 is an illustration of the increasingly interconnected nature of the world. At the same time, the pandemic is a timely reminder, not only of the importance of distance and geographic space but also of our vulnerability as a species. From its 'rediscovery' (National Research Council 1997) to its 'secret powers to save the world' (Bednarz 2019), there has never been a more relevant, or important, time for the discipline of geography, sitting astride—as indeed it does—the social and natural sciences.

The origin and evolution of geography have been shaped by its interactions with other disciplines, and its response to emerging global issues. In early times, astronomy, geometry, physics, anthropology, and history, among others, provided questions, perspectives, methods, and tools for geography as geographers accumulated knowledge of different places and countries in what was essentially exploration, discovery, and cartography (Baker 1931). The development of an identifiable academic discipline (Stoddart 1986) and its broader subdisciplines of physical geography and human geography in the last century benefited from interaction with other disciplines. Geographers have played active roles in multi-disciplinary, cross-disciplinary, interdisciplinary, and transdisciplinary research because the subject is itself inherently interdisciplinary. As Baerwald (2010) claims, geography has built on its fundamental element of spatial analysis to explore human-environment interactions and place-based and regional analyses to encourage communication and interaction with a myriad of other disciplines. The active pursuit of inquiry related to space and place, and to the dynamic interactions within and between spaces and places, has encouraged geographers to range far from its traditional core and explore the peripheral realms where geographic perspectives and insights intersect with those from other fields (Baerwald 2010). The International Geographical Union itself has a record of fostering such interactions, not least in the number of its Commissions that are interdisciplinary in nature (e.g., Geography for Future Earth, Sustainability of Rural Systems). Indeed two IGU Commissions are formally co-hosted with other major scientific organizations (Geomorphology and Society, with the International Association of Geomorphologists, and Toponymy, with the International Association of Cartography).

As a discipline, geography endured a particularly difficult time in the second half of the twentieth-century, as many universities in the United States, notably including Harvard, Columbia, and Michigan closed their geography departments (Fink 1979). Such challenges were not confined to the US and became further intensified as geography, even in the universities where it did survive, began to lose its identity as departmental names changed and incorporated other disciplinary labels (see Hall et al. 2015). More recently, the development of Geographic Information Systems (GIS), Remote Sensing (RS), and Global Positioning System (GPS) has reinvigorated, perhaps even revolutionized, geography. Several IGU Commissions, most notably the Commission on Geographical Information Science and Geography of Information, Innovation and Technology, have been established in response. The COVID-19 maps developed by Johns Hopkins University (https://gisanddata.maps.arcgis.com/apps/opsdashboard/index) are universally consulted and have certainly assisted in tracking the spread of the virus such that the significance of spatial analysis has been brought into the global spotlight. What this serves to emphasize is the importance of geography in its relationships with other disciplines. The history of the discipline is a mirror for the future in highlighting that geography has to be flexible and responsive, not only to developments in technology but also to the needs of society at large.

Currently, the study of high profile global issues such as climate change, loss of biodiversity, human migration, land degradation, refugees, poverty, terrorism, and pandemics all enjoy a legitimate place in the field of geography, but it is only through the relationship of geography with other disciplines, including atmospheric science, ecology, demography, political science, psychology, and medical science among others, that these complex issues can be comprehensively understood and addressed. Therefore, the interactions with other sciences, natural, social, human, and economic, have played their part in shaping the modern discipline and it is through such relationships that the future of geography will take shape. In this chapter, therefore, we highlight five arenas of interaction between geography and other disciplines, viz. the natural sciences, socioeconomic sciences, humanities, human-environment relationships, and sustainability science.

13.2 Geography and the Natural Sciences

Geography has deep roots as a natural science and, in the form of the sub-discipline of physical geography, typically uses many methods that evolved within the natural and physical sciences. The sciences, in general, have been, and remain, highly influential, although two concepts, in particular, have played an enormous role in the development of physical geography, viz. Hutton's 1795 uniformitarianism and Darwin's theory of evolution (Gregory 2000). Uniformitarianism, amplified notably by the geologist Charles Lyell as its 'high priest' (Chorley et al. 1964), was a counter to both scientific catastrophism and biblical fundamentalism and led to the prospect of using 'the present as a key to the past' to explain landscape formation. How can we imagine the contemporary science of geomorphology, in its various manifestations, without recourse to this seminal principle that allows us to interpret past features in terms of contemporary patterns and processes? The influence of Charles Darwin has been similarly pervasive and, as Stoddart (1966) argues, the idea of change through time strongly impacted the development of physical geography.

Physical geography has obvious and important relationships with the fundamental sciences, including Mathematics, Physics, Chemistry, Biology, and Geology. Indeed, for many students of physical geography to this day, these disciplines are essential (and often compulsory) building blocks for their curriculum. The physical sciences offer what Richards (2008) refers to as a 'role model' for physical geography which has borrowed from them the basic idea of testing theory through experimentation to develop systematic and formulated knowledge. Ultimately, these foundations led to the development of the recognizable branches of the subject, perhaps the earliest example of which is Huxley's (1877) *Physiography*, but which diversified through the first part of the twentieth-century into soil science, biogeography, climatology, hydrology and, of course, geomorphology. In respect of the latter, the work of W. M. Davis has been enormously influential, as the 'cycle of erosion' concept, rooted strongly in the concept of evolution and change over time, characterized physical geography up to 1950 (Gregory 2000). The cycle of erosion concept is at once elegant in its simplicity and convincing due to its perceived widespread applicability. Chorley (1973) went so far as to suggest that this notion essentially established geomorphology as a mature science and as an academic discipline (albeit one that is usually taught in US universities in geology departments). The second half of the twentieth-century witnessed a number of developments, among them a shift from a focus on longer-term landscape evolutionary processes to a consideration of shorter time scales. The Quaternary, with its dramatic shifts from glacial to interglacial conditions, proved an attractive line of inquiry for physical geographers that strengthened the linkages in particular with geology and biology, among other sciences. Studies of contemporary physical environmental processes were also to become a prominent theme. Before 2000, for example, physical geography research in China was dominated by the study of sediment sequences on the loess plateau, ice cores on the Qinghai-Tibet plateau, and speleothems in karst caves that record changes during the Quaternary through to the late Holocene, including the historical period. There has, however, been a broadening of emphasis to include landscape processes and the effects of recent climate change.

Physical processes are fundamental to all features of the earth system, but there is more to physical geography than this since it is concerned with 'phenomena that unfold in an unconstrained social and environmental space, across a wide range of scales' (Richards 2008: 24). Moreover, the methods by which such complex processes are observed, measured, and interpreted have also changed. Thus, physical geography engaged not only with the trend toward quantification and positivism, but ultimately also modeling, detailed investigations of processes, and a concern with human impact on the environment including, for example, studies of global climate change, land degradation, hazards, and risks, among others—themes that are the focus of several IGU Commissions. Developments were also prompted by novel techniques and technologies. Advances in remote sensing, geographical information systems, and information technology have at once prompted highly detailed small-scale analyses but also facilitated global-scale analysis. A fresh impetus has been given by the emergence of a more culturally-based approach throughout many branches of physical geography. A series of issues can be identified including the increasingly holistic trend prompted by a greater awareness of global environmental problems, the development of earth system science, and of the timely opportunities which can arise from closer links with human geography and with other disciplines. Extrapolating present trends suggests a bright future for physical geography involving a more integrated approach, even greater concern for environmental futures, and closer links to human geography, along with other disciplines. As physical geography evolves and reflects these trends, it seems likely that pluralist approaches will increasingly feature and that university departments may take on distinctive flavors according to the expertise of their staff and the nature of the research relationships they develop with other disciplines.

13.3 Geography and the Social and Economic Sciences

As Johnston (2009: 58) puts it: 'Geography came late to the social sciences' and its relationship with economics, political science, and sociology, the 'core' social science disciplines, is essentially a post-World War Two phenomenon. Prior to this time, geographers' concern with regions, distinctive areas with characteristic landscape features that were different from neighboring areas, meant that there was little, if any, orientation toward the social sciences. Johnston argues that while geographers were concerned with mapping economic activity and relating this to particular environments, they appear to have made few connections with economists prior to the 1950s and other social sciences were connected only very tangentially at best. This was to change, however, as human geography successfully remodeled itself and gained recognition as a social science by the 1970s, drawing on a wide range of additional traditions, such as Marxism, feminism, and postmodernism (Johnston and Sidaway 2004). The ideas of key social theorists, in particular, Anthony Giddens in revolutionizing social theory and Manuel Castells in considering the nature of 'the network society' were certainly instrumental in reshaping human geography (Giddens 1971; Castells 1996). Although, as Johnston (2009) suggests, geographers have been largely 'net importers' from the social sciences, we focus here on developments in economic geography by way of illustrating that interactions between disciplines may lead to mutual strengthening of both theory and practice.

In recent decades, intellectual exchange and collaboration between economic geography and economics and management have been flourishing, a trend that has been given further impetus due to the rise of globalization. Economic geographers adopt a spatial perspective to analyze activities, such as entrepreneurship, manufacturing, and innovation, which exhibit a highly uneven pattern, and try to answer questions such as why do some industries agglomerate in specific areas, e.g., high-tech in Silicon Valley, cellphones in Shenzhen, and finance in Wall Street? Why do some regions develop steadily while others decline rapidly, for example, Detroit? These questions attract geographers as well as economists and management scholars. Trained in their own disciplines, economic geographers, economists and management scholars tackle the same phenomenon from different perspectives using different methods. However, given that any economy, at whatever scale, is multifaceted, a single disciplinary investigation often neglects other important dimensions of the research subject, generating a biased understanding. For example, the assumption of firms with the same preference to maximize profits in economics rules out their heterogeneity in managerial practices and choices, which can be place-specific (Clark 2018; Storper 2013).

The significance of location to economic theory, previously largely neglected by economists, was emphasized by Paul Krugman (1991). In economics, economy was traditionally modeled in an abstract location-free market, an assumption challenged by Krugman who argued for the formal integration of location into economic models (see Krugman 1993). Together with others, Krugman introduced geography into mainstream economics and was indeed awarded a Nobel prize in 2009 for his contribution (see also Fujita et al. 1999). Although critiqued by geographers, his original contribution laid out the intellectual basis for dialogue between the two disciplines (Martin and Sunley 1996). A formal marriage between geography and economics was signified by the establishment of the *Journal of Economic Geography* in 2001 by Oxford University Press, edited by leading scholars split equally across geography and economics, with the aim 'to redefine and reinvigorate the intersection between economics and geography.' Geographers and economists have been prompted to discuss the strengths and weaknesses of their research methods and explore how one can learn from the other (Overman 2004; Storper 2013). The mathematical modeling approaches deployed by economists to identify and quantify mechanisms and geographers' more qualitative analysis to identify patterns and processes complement each other and deepen our understanding of spatial economies. Interestingly, related to their different preferences of research methods, geographers and economists implicitly developed a division of labor when examining the same phenomenon. For example, in relation to industrial agglomeration, economists prefer to detect the effects of supplier sharing and local labor markets (Combes and Gobillon 2015), while geographers are interested in understanding the social structure of knowledge sharing (Li et al. 2012).

In relation to management, the engagement of geography was given huge impetus through the work of Michael Porter who highlighted the significance of location and industrial clusters for business strategy (see, for example, Porter 2000). Unlike economists who discovered geography at the macroeconomic level, management scholars recognized the importance of location in the micro-processes of individual firms. Strategy and management studies build on observations of actions and choices made by firms using a diverse range of research methods. Compared to deductive reasoning of economists, this empiricism by management scholars is more in line with geographers' research practices. As a result, research collaboration and cross-fertilization between geography and management scientists have been relatively smooth. For example, in the past several years, several major journals have published special issues to bridge the two fields, including *Regional Studies* (Knight et al. 2020), *Journal of Economic Geography* (Bathelt et al. 2018), and *Journal of International Business Studies* (Mudambi et al. 2018). Overlapping themes range from innovation strategies and industrial clustering to international investment and global knowledge flows. The advantage of management studies lies in a deep understanding of what happens inside the firm, while geographer's work tends to emphasize how the firm's business environment shapes its competitive advantages or disadvantages. Geographers benefit from the research exchange with management studies since the firm-centered approach addresses the concern that many regional development theories lack a solid foundation of microeconomic processes (Li and Bathelt 2018).

Recent decades have witnessed increasingly prominent interactions between economic geography and economics and management. Looking back, economic geography has benefited greatly from this interdisciplinary knowledge exchange that has broadened its frontiers, tightened research rigor, and enriched the understanding of the spatial dimension of economies. Looking forward, it seems likely that this endeavor will continue to bear fruit.

13.4 Geography, Critical Theory and the Humanities

Within the field of human geography, there has been a strong social science emphasis, as discussed above. Yet, also during this time, human geography has adopted more critical approaches that have widened and enriched the field. These more critical approaches look beyond the scientific pursuit of collecting more data and knowledge about the world and instead critically reflect on how social, political, economic, and cultural factors shape knowledge production, society, and space in ways that earlier approaches tended to ignore. This happened first with the sub-field of cultural geography beginning in the 1920s, and much more forcefully with the related sub-field of critical geography in the 1970s. To this day, engagement with critical theory and the humanities continues to reinvigorate the field of geography well beyond its classical borders. The analytical rigor of these approaches is not in big data or modeling, but rather in challenging key assumptions about space and power that more conventional geographic analyses leave unquestioned and unexplored.

13.4.1 Cultural and Humanistic Geography

Cultural geography represents one of the first major shifts away from the positivism and determinism that dominated early twentieth-century geographic thought. Rather than emphasizing how landscapes and available resources influence or determine the development of societies, cultural geography instead acknowledges that people affect their environments as much as their environments affect them. Starting in the 1920s, cultural geography began exploring these interconnections between societies and their wider landscape. The field posited that the constitution of a society—its tools, shelters, cuisine, and other elements typically associated with culture—is not determined by environmental factors, but in many ways the reverse. Culture influences how the world is understood and how societies interact with their environments, resulting in 'cultural landscapes' created as much by human culture as environmental constraints (Sauer 1925). Culture, in short, is not passively determined but is an active agent of environmental change.

In the 1970s and 80s, developments in the humanities and critical theory began to influence the field of geography even more profoundly. Humanistic geography, 'new humanism,' and 'new cultural geography' took the wider discipline of geography even further from its historical roots in positivism, determinism, and quantitative analysis (Tuan 1976; Relph 1976; Buttimer and Seamon 1980; Ley 1981; Cosgrove and Jackson 1987; Ley 1985). The focus on the particularities of human intention, meaning, and culture in these analyses became even stronger, shifting further away from the determinism associated with environmental context.

13.4.2 Critical Geography

From the 1960s onward, alongside developments in cultural and humanistic geography, the sub-field of 'critical geography' emerged. Similar in its rejection of positivism, critical geography builds heavily on key themes in critical theory, in particular Marxism and post-structuralism. Critical geography moves beyond an emphasis on culture or the humanities, to consider questions of power more directly. As diverse as this subfield is, it is nonetheless united in its commitment to reflect critically on how power shapes space and society in ways previously overlooked.

Critical geography has its roots most firmly in Marxist critique, which was experiencing a revival in the 1970s. At this time, heightened awareness of the geographic impacts of capitalism in terms of 'unequal exchange' (Emmanuel 1972), 'underdevelopment' (Frank 1966), and 'regionalism' (Massey 1978) began to infiltrate the field. Building on these early critiques, Harvey (1981) and Smith (1984) advanced the concepts of 'spatial fix' and 'uneven development' that have since become seminal in the field. Harvey uses the concept of a spatial fix to show how capitalism's inherent push for continued expansion manifests spatially. Once the accumulation of surplus-value is saturated in a given terrain, capital must continue expansion beyond its borders. This geographic expansion provides a temporary 'fix' to crises of capital overaccumulation, forging new outlets for the accumulation of surplus-value. Through the concept of 'uneven development', Smith further extended this work, applying it to the realm of nature and the natural. These ideas remain highly influential at the beginning of the 21st Century, with a new generation of critical geographers advancing various 'fixes' beyond the spatial: 'regulatory', 'environmental', 'biophysical', hydro-social', socioecological', and 'cultural' fixes all open new sites of accumulation to capitalist dynamics (Swyngedouw 2013; Cohen and Bakker 2014; Castree and Christophers 2015; Ekers and Prudham 2017; Zhu 2020).

In addition to the Marxist critique, critical geography increasingly relies on other strains of critical theory, including postmodernism, constructivism, and post-humanism. While there is no clear term that unites these diverse approaches, they are sometimes grouped together under the broad label of 'post-structuralism', given their mutual aversion to analyzing society in terms of its underlying structures (capitalist vs. proletariat, sign vs. signified, and so forth). Post-structural approaches reject the 'grand narratives' associated with Marxist critiques of capitalism (e.g., the singular logic of capital accumulation that drives its exploitative relations) and further maintain that there is no vantage through which one can objectively analyze the world to reveal its deeper truths. Indeed, for the post-structuralist, there is no deeper truth to speak of, but rather many truth statements that only make sense as such within a given discourse. Key post-structural thinkers include Michel Foucault on discourse and power/knowledge (Foucault 1984), Gilles Deleuze and Felix Guattari on assemblage (Deleuze and Guattari 1987), Donna Haraway on situated knowledges (Haraway 1988), and Bruno Latour on actor-network-theory (Latour 2005). While these scholars are not themselves geographers, their work has inspired a new turn in critical geography toward the post-structural, postmodern, and posthuman (Soja 1989; Minca 2001; Murdoch 2006; Castree and Nash 2006; Soja 2011).

Despite differences between post-structural and Marxist geographies, most contemporary critical geographers engage with both approaches. More than choosing one side or another, strains of critical geography such as feminism, post-colonialism, and political ecology draw from both Marxist and post-structural thought (Rose 1993; Peet and Watts 1996; Ashcroft et al. 1998; Robbins 2012; McDowell and Sharp 2016). This has resulted in a focus not only on the political-economic and material production of nature and space but also on their production through more intangible, discursive means. Critical geographers contend with a range of power structures from the material to the discursive, with no clear line distinguishing the two. This scholarship is at the forefront of critical geography today and is certainly a key research theme within the IGU Commission on Political Geography. United in its firm rejection of positivism and determinism, the emergence of critical geography represents a profound shift in the spirit and commitment of the field of geography.

13.5 Geography and Human-Environment Relationships

13.5.1 Evolution of Ideas on Human-Environment Interactions

Human-environment relationships can be partly reflected by the multiple conceptualizations of nature in different languages and countries (Coscoeme et al. 2020). Traditionally China views humans as part of nature and thought (天人合一), which seeks a harmony between people and nature. The conceptualization has also affected human behaviors and designing landscape architecture. Research on human-environment interactions (HEI) is not only a reflection on the history of societal development but also a reflection on the evolving human epistemology regarding the relationship between people and nature. The nature of the relationship between humans and their environment has changed since behaviorally modern humans, *Homo sapiens* first emerged. A reliance on natural systems for food and other resources evolved, in particular through advances in technology, that ultimately enabled the establishment of our modern substantially industrialized, and urbanized society. Just as the nature of the relationship itself has changed, so has the *interpretation* of the nature of that relationship.

According to Moran (2005), three main themes can be identified in Western intellectual history to explain human interactions with nature: (1) environmental determinism, which emphasizes the dominance of nature, (2) possibilism, which emphasizes the dominance of human culture, and (3) adaptationism, which bridges the gap between these two and emphasizes the mutual interactions of people with nature. Harden (2012) identifies five categories of HEI, among which both environmental determinism and natural hazards approaches portray humans as essentially passive agents that lie at the mercy of environmental change (Fig. 13.1). The 'human impact' perspective on the other hand recognizes the significant role of people in changing their environment (Marsh 1864; Carson 1962). The sustainability framework considers that interactions between the human and natural systems may be mutually supportive.

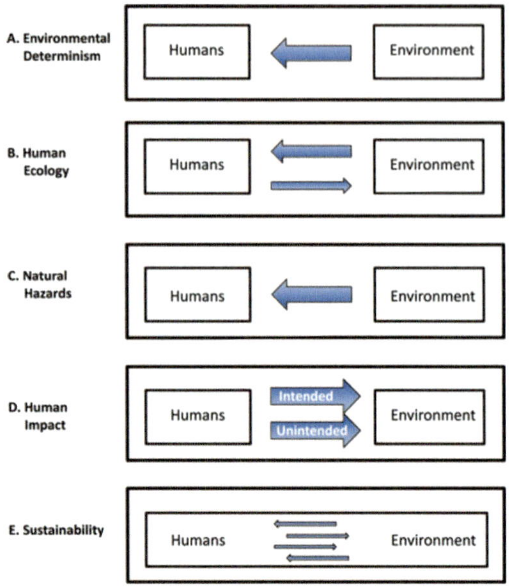

Fig. 13.1 Ways of framing human-environment interactions. The arrows represent the direction(s) of causation. *Source* After Harden 2012

13.5.2 Population, Resources, and Environment

Consideration of human-environment interactions initially focused largely on the relations between population and resources. Malthus (1798) envisaged that population growth would outstrip the ability of the environment to support it and that the lack of sufficient resources would result in population collapse. The Malthusian influence on theory was prominent for some time but in due course, counter-arguments prevailed, among others that improvements in agricultural productivity and efficiency through technology facilitate ongoing population growth. Von Thünen's model of agricultural land use also provided an important impetus for geographers in their attempts to understand resource-related spatial elements of the human-environment nexus

(O'Kelly and Bryan 1996; Van Wey et al. 2005). Public attention on resource constraints resurfaced in the 1960s and 70s with the publication of *The Population Bomb* (Ehrlich 1968) and *The Limits to Growth* (Meadows et al. 1972) prompted by emerging environmental problems and encapsulated in Rachel Carson's (1962) *Silent Spring*. Concerns around global environmental problems such as deforestation, pollution, biodiversity loss, greenhouse gas emissions, and land degradation were growing and from the 1980s onwards; these issues provided fertile ground for the boom in global environmental change approaches that geographers were quick to adopt (Mannion 1991) and in which the IGU community has made substantial contributions through the work of, for example, the Commissions on Population, Land Degradation, and Land Use and Land-Cover Change to name but three.

13.5.3 Integrated Research Initiatives

In the late 1980s, studies under the International Geosphere-Biosphere Program (IGBP) were led by largely natural science disciplines, not least geology, climatology, and biology. However, with growing evidence that socioeconomic uncertainties exacerbate physical environmental problems, the International Human Dimensions Program (IHDP) was created in parallel. Several influential reports, such as *Grand Challenges in Environmental Sciences* (National Research Council 2001), released in the following years greatly facilitated more integrated research in identified priorities such as energy, industrial metabolism, health, environmental security, and land-use/land-cover change) and to which geographers could make strong contributions (e.g., Moran 2005, 2010). The UN Sustainable Development Goals (UNSDGs have, of course, been driving the research agenda in many disciplines, including Geography (see below). Understanding cross-scale interactions is a rising challenge in sustainability science and transdisciplinary research in general, especially where solutions need to arise from the intersection of top-down and bottom-up activity (Smith et al. 2018). Physical geographers especially have made contributions to global environmental change and what is broadly termed earth system science, while—increasingly of late—they have turned to the Anthropocene (Meadows 2022) as a lens through which to view their subject, the concept itself spawned out of a realization of the sheer magnitude of human impact on the global environment.

Geography has long had a concern with the relationship between people and land and, using tools such as remote sensing and Geographic Information System (GIS), geographers combine both field observation and large-scale satellite observation to understand the human-environment interactions across scales. The advantage of these tools, coupled with increasing computational power, is that they enable analyses of a much wider range of factors to explain geographic processes and patterns. For example, it is possible to include household attributes, such as demographic information and decision-making, to develop spatially explicit maps (Evans et al. 2005; Moran and Brondizio 1998) and understand more completely the driving forces underlying land-use and the influence of socioeconomic and cultural dynamics.

In addition to understanding the patterns of changes in human-environment interactions, it is also critical to understand processes and to know more about how (and why) humans make decisions, and how they induce collective actions that may have deleterious consequences. The depletion of public goods/resources (such as fisheries, forests, and water resources) is a major global challenge and, as Hardin (1968) argues in *The Tragedy of the Commons*, individual users in a shared resource system acting independently according to their own self-interest may behave contrary to the common good and invoke resource depletion. To tackle this problem, Ostrom developed a framework for analyzing the sustainability of Social-Ecological Systems (SES) (Ostrom 2009, 1990). The basic structure of Ostrom's SES framework is organized into four main domains of analysis (resource systems, resource services and units, governance systems, and actors), each of which has a nested set of tiers of level-specific variables that affect the likelihood of self-organization in efforts to achieve a sustainable SES. In fact, multiple economic theories are also applied to human-environmental interaction research. For example, the management of the commons often builds on the assumption of 'economic man' which envisages humans as consistently rational, narrowly self-interested agents. Economics that examines production, consumption, and exchange, provides important theory and methodology (e.g., general equilibrium models) for analyzing international trade and its impacts on telecoupled human-environment systems (Andrew and Peters 2013; Hertel and Tsigas 2000; Lenzen et al. 2013).

In the meantime, with the development of human ecology, cultural ecology, political ecology, and ecosystem ecology at the interfaces of geography, anthropology, and ecology, research on human-environment interactions has evolved to be interdisciplinary and transdisciplinary. In 1995, the U.S. National Science Foundation announced a competition for national centers of excellence on the human dimensions of global environmental change. Carnegie Mellon and Indiana Universities received center-level awards. Of the two centers, Indiana University focused more on land-use/land-cover change and on forest ecosystems in particular, while the Carnegie Mellon center focused on integrated assessment issues. In 2007 a formal standing program in 'Dynamics of Coupled Natural and Human Systems' was created (National Science Foundation 2014) that supports research projects that advance basic scientific understanding about the complex interactions among natural physical and/or biological systems and human social and behavioral systems. Several other programs or initiatives on the studies of HEI have emerged globally (Table 13.1). These initiatives and funding support have greatly facilitated HEI research and a number of these have been taken up by IGU Commissions.

13.5.4 Frameworks for Studying Human-Environment Interactions

A framework approach is helpful in understanding complex research questions and can provide the basic vocabulary of concepts and terms for constructing causal explanations, thereby assisting in identifying the universal elements of a theory (McGinnis and Ostrom 2014). A number of such frameworks have been developed in relation to human-environment interactions (Binder et al. 2013; Liu et al. 2007a; Fishcher-Kowalski and Weisz 2016) which, despite the shared goal of understanding the complex relationships toward the goal of sustainability, differ in their disciplinary roots, applicability, and spatio-temporal scale. The 'human-earth' system was proposed in the 1970s by Chuanjun Wu, a former IGU Vice-President from China, which stresses that the human-earth relationship is core to geographical study in all developmental stages of the discipline (Lu and Guo 1998). The social-economic-natural complex ecosystem theory was also put forward by a Chinese ecologist and geographer in 1984 (Ma and Wang 1984), which represents an early attempt to understand the complex interaction between humans and environmental systems. Three of the most commonly applied frameworks are briefly reviewed here, viz. socioecological systems (SES), coupled human and natural systems (CHANS), and telecoupling/metacoupling.

SES, as noted above, is a widely applied general framework for analyzing sustainability (Ostrom 2009). Prominent applications of SES are those that focus on resilience, vulnerability, and adaptability (Folke 2006; Folke et al. 2005; Walker et al. 2004; Young et al. 2006). Specifically, this field investigates the nested cycles of adaptive change in SESs in which persistence and novelty are intertwined, leading finally to transformations. The CHANS framework (Turner et al. 2003b, 2003a) is similar in dealing with vulnerability (exposure, sensitivity, and resilience) to environmental hazards. Liu et al. (2007a, 2007b) show how CHANS characterizes the dynamical two-way interactions between

Table 13.1 Representative programs on the studies of coupled human and natural systems

Program name	Dynamics of coupled natural and human systems	Beijer Institute of Ecological Economics	Resilience alliance	Intergovernmental panel on climate change	Millennium ecosystem assessment	Intergovernmental science-policy platform on biodiversity and ecosystem services
Focus	Complex interactions among human and natural systems at diverse spatial, temporal and organizational scales	Ecological economics	Research on the dynamic of complex adaptive systems in order to discover foundations for sustainability	Assessment of scientific, technical, and socioeconomic information to understand climate change, its impacts, and choices for adaptation, and mitigation	An international program assessing conditions and consequences of ecosystem change for human well-being and options for responding to those changes	IPBES performs regular and timely assessments of knowledge on biodiversity and ecosystem services and their interlinkages
Major funding source	U.S. National Science Foundation	Kjell & Marta Beijers Foundation	Diverse grants from private foundations	World Meteorological Organization and United Nations Environment Program	Multiple sources	National governments, United Nations
Duration	2000–present	1991–present	1999–present	1988–present	2001–2005	2012–present
Web address	https://www.nsf.gov/pubs/2006/nsf06587/nsf06587.htm	http://www.beijer.kva.se/	http://www.resalliance.org	http://www.ipcc.ch/	https://www.millenniumassessment.org/en/index.html	https://www.ipbes.net

Updated after Liu et al. (2007b)

human systems (e.g., economic, social) and natural (e.g., hydrologic, atmospheric, biological, geological) systems. Of particular interest in studying these interactions is the understanding of feedbacks, surprises, nonlinearities, thresholds, time lags, legacy effects, path dependence, and emergence across multiple spatial, temporal and organizational scales (Liu et al. 2007a). Telecoupling (telecoupled human and natural systems) (Liu et al. 2015, 2013) builds on the key elements of CHANS but extends the spatial scope and scale of connections, while metacoupling is a more integrated framework that deals with socioeconomic and environmental interactions within a coupled human-natural system (intracoupling), between adjacent systems (pericoupling), and between distant systems (telecoupling) (Liu 2017). In all these frameworks, the selection of appropriate scale of analysis is important (Dietz 2017). The SES framework emphasizes individual scale, while many applications of CHANS emphasize the household scale (Liu et al. 2016). Compared to SES and CHANS, the metacoupling framework is more inclusive and has the capacity of analyzing HEI across different scales.

13.6 Geography and Sustainability Science

Sustainability science has emerged rapidly during the last four decades and focuses on the interactions between natural and social systems and how, without threatening the planet's life support systems, these systems impact on the dual challenge of meeting the needs of the world's population while substantially reducing poverty (Fig. 13.2; Kates 2011). Sustainability science focuses on the dynamic interactions between nature and society, which echoes the identity of geography: human-environment (nature) interaction (Turner 2002), or geography as human ecology (Barrows 1923). In the past several decades, geography and sustainability science have enriched each other and this has reshaped the dimensions and direction of geography.

The concept of sustainable development has experienced extraordinary success since its advent in the 1980s. Sustainability is often these days an integral part of the agenda of governments and corporations and has become central to the mission of research laboratories and universities worldwide (Bettencourt and Kaur 2011). Although the term sustainable development was promoted by the Brundtland Commission's report in 1987 (Brundtland 1987), the concept emerged much earlier. In the early 1970s, Paul Ehrlich raised the issue of a sustainable civilization in several of his influential works (for example, Ehrlich and Ehrlich 1970). The term sustainable development is enshrined in the Agenda 21 action plan that emerged from the United Nations Conference on Environment and Development in 1992 (Piel 1992) and soon became a major focus. The US National Research Council of the National Academies of Sciences, Engineering, Medicine issued a major report about transitioning to sustainability in 1999, which aimed at building a foundation for the development of science and technology for the new century (National Research Council 1999). Geographer Robert Kates' ideas on sustainability science (Kates et al. 2001) were influential and followed up by many. In 2004, the journal *Proceedings of the National Academy of Sciences* (PNAS) initiated a special section devoted to sustainability science, and this greatly promoted the field (https://www.pnas.org/portal/sustainability). Then, in 2015, the United Nations adopted the Sustainable Development Goals (SDGs) as a universal call to action to end poverty, protect the planet and ensure that all people enjoy peace and prosperity by 2030 (https://sdgs.un.org/goals) which has proved to be a milestone in promoting the development of sustainability science. All these efforts are aimed at transitioning from an economy-centered to an environment-centered world (see also Messerli et al. 2019) and the Sustainable Development Report (Independent Group of Scientists 2019).

Geographers have proved to be major players in the field of sustainability science. The 1994 and 1995 presidential addresses of the American Association of Geographers (AAG) were all

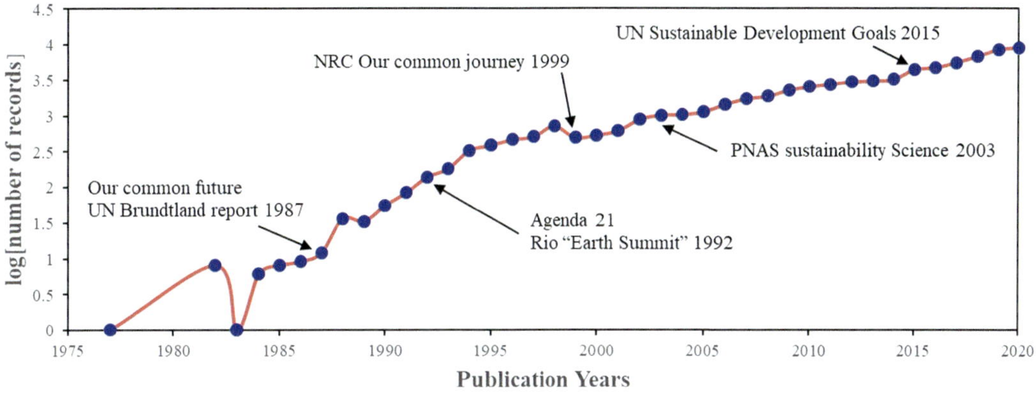

Fig. 13.2 Research publications with the terms 'sustainability' or 'sustainable development' in the title, 1977–2020. *Source* Web of Science: https://www.webofknowledge.com/

directly focused on sustainable development. Thomas Wilbanks spoke of 'sustainable development in geographic perspective' in 1994 (Wilbanks 1994), while Kates entitled his address 'Lab notes from the Jeremiah Experiment: hope for a sustainable transition' (Kates 1995). The International Geographical Union (IGU) has also promoted many kinds of activities related to sustainability and established a Study Group on Rural Sustainability as early as 1992, which was ultimately to be accepted as a Commission (Ehlers 1993). The IGU Commission on Geographical Education (CGE) published the Lucerne Declaration on Education for Sustainable Development in July 2007 (Reinfried 2009). The recently established Commission on Geography for Future Earth: Coupled Human-Earth Systems for Sustainability clearly addresses such issues head-on. As 'the science for sustainability', Geography has an increasingly important role to play in developing the knowledge and the skills to equip future generations with the tools to adapt to and mitigate potentially catastrophic global environmental change (Meadows 2020).

The challenge of sustainable development is achieving society's development goals within the planet's environmental limits over the long term. In seeking to help meet this sustainability challenge, geographers and researchers from other disciplines are focusing on the dynamic interactions between nature and society, with equal attention to how social change shapes the environment and how environmental change shapes society. These questions are problem-driven, with the goal of creating and applying knowledge in support of decision-making for sustainable development (Clark and Dickson 2003). To address the most important challenges we face today, geography and sustainability science must continue to support each other mutually with a view to achieving the SDGs and, perhaps, even the global 'prosperity' that Moore (2015) calls for.

13.7 Conclusions: Geography as a Bridge

In pondering the discipline of Geography in the context of the other sciences and social sciences, it becomes evident that, while geography indeed has enormous integrative potential, many academic departments have diverged, resulting in the separation of 'human' from 'physical' geography. This has been expressed by the observation that many university geography departments have changed their names or split altogether, as physical geographers joined their earth science partners and human geographers teamed up with, among others, anthropologists and sociologists (Meadows 2018). So, is the integrating capacity of geography a 'historical relic' and '...more rhetoric than reality' (Sharpe 2009)?

The world is changing more rapidly and more profoundly than ever before, all the more obviously so given the rampant spread of the COVID-19 pandemic. The unprecedented dispersal of the virus has led to disruption of global supply chains, the closing of national borders, and widely imposed lockdown restrictions with diverse implications for the economy, livelihoods and – all of which demonstrate the imperative of geography which deals with the interconnectedness of people and places among all the components of the earth system, and also among remote places in the telecoupled world systems. Issues ranging from the local to global scales that relate to, for example, poverty, environmental hazards, migration, refugees, border walls, trade wars, climate change, biodiversity loss – all have a geographic component. In the post-pandemic world, a green transition is already on the horizon, COP 15 of the UN Convention on Biological Diversity and COP 26 of UN Climate Change Conference have called for a sustainable transition. Carbon neutrality and sustainable development need to be embraced, not only by scholars and researchers from diverse disciplines but also the politicians, decision-makers, and the global population at large. Against this background, geography is likely to be regarded by other disciplines as increasingly indispensable.

As global actions are converging to integrative, interdisciplinary, and future-oriented research, this is a good time for geography. Future earth, the world's largest community of sustainability scientists—is transforming its governance and management structures in pursuit of deeper efficiency, inclusivity, and impact. This shift will broaden its reach and open new inroads for transformative sustainability science around the world. The National Science Foundation of the United States has supported convergence research which integrates knowledge, methods, and expertise from different disciplines, thus forming novel frameworks to catalyze scientific discovery and innovation. The IGU is moving into its second century and we find ourselves at a turning point for humanity. We have entered a decade of action to transform the world and achieve the SDGs outlined in Agenda 2030. IGU should build on its heritage, and its voices must be heard on the many critical global challenges; joining hands with other disciplines is an important step toward addressing the current crises.

The challenges posed by the range of human-environmental problems that characterize the Anthropocene should be a 'call to arms' for geographers, soliciting the rediscovery of the importance of the diverse and complex connections between nature and society and the recognition that explanations rooted in both the physical and the human (social, economic, political, cultural) domains are essential. However, this should entail more than simply collaborating on problems of common interest but rather a fundamental recognition of the concept that environments are 'as much the product of unequal power relations, histories of colonialism, and racial and gender disparities as they are of hydrology, ecology, and climate change' (Lave et al. 2014: 2). Geography continues to develop through its vibrant connections with other fields and geographers should continue to show interdisciplinary leadership by embracing different perspectives, by supporting institutional arrangements that foster interdisciplinary activity, and by seeking the knowledge and techniques that other fields can contribute to geographic perspectives, approaches, and insights to the collective effort. Indeed, we still have much to learn from (and to teach) each other.

References

Andrew RM, Peters GP (2013) A multi-region input–output table based on the global trade analysis project database (GTAP-MRIO). Econ Syst Res 25:99–121

Ashcroft B, Griffiths G, Tiffin H (1998) Key concepts in post-colonial studies. Routledge, London

Baerwald TJ (2010) Prospects for geography as an interdisciplinary discipline. Ann Assoc Am Geogr 100:493–501

Baker JNL (1931) A history of geographical discovery and exploration. Harrap & Co., London

Barrows HH (1923) Geography as human ecology. Ann Assoc Am Geogr 13:1–1

Bathelt H, Cantwell JA, Mudambi R (2018) Overcoming frictions in transnational knowledge flows: challenges of connecting, sense-making and integrating. J Econ Geogr 18(5):1001–1022

Bednarz S (2019) Geography's secret powers to change the world. Can Geogr 63(4):520–529
Bettencourt LMA, Kaur J (2011) Evolution and structure of sustainability science. Proc Natl Acad Sci 108:19540–19545
Binder CR et al (2013) Comparison of frameworks for analyzing social-ecological systems. Ecol Soc 18 (4):26
Brundtland GH (1987) Our common future. World Commission on Environment and Development, Brussels, Brussels
Buttimer A, Seamon D (1980) The human experience of space and place. Croom Helm, London
Carson R (1962) Silent spring. Houghton Mifflin, Boston
Castells M (1996) The network society: the information age—economy, society and culture, vol 1. Blackwell, Oxford
Castree N, Christophers B (2015) Banking spatially on the future: capital switching, infrastructure and the ecological fix. Ann Assoc Am Geogr 105(2):378–386
Castree N, Nash C (2006) Posthuman geographies. Soc Cult Geogr 7(4):501–504
Chorley RJ (1973) Geography as human ecology. In: Chorley RJ (ed) Directions in geography. Methuen, London, pp 155–169
Chorley RJ, Dunn AJ, Beckinsale RP (1964) The history of the study of landforms, volume 1: geomorphology before Davis. Methuen, London
Clark J (2018) Policy through Practice: local communities, self-organization, and policy. In: Clark G, Feldman M, Gertler M, Wojcik D (eds) The new oxford handbook of economic geography. Oxford University Press, Oxford, pp 810–825
Clark WC, Dickson NM (2003) Sustainability science: the emerging research program. Proc Natl Acad Sci 100:8059–8061
Cohen A, Bakker K (2014) The eco-scalar fix: rescaling environmental governance and the politics of ecological boundaries in Alberta, Canada. Environ Plann D Soc Space 32(1):128–146
Combes PP, Gobillon L (2015) The empirics of agglomeration economies. Handb Reg Urban Econ 5:47–348
Coscoeme L et al (2020) Multiple conceptualizations of nature are key to inclusivity and legitimacy in global environmental governance. Environ Sci Policy 104:36–42
Cosgrove D, Jackson P (1987) New directions in cultural geography. Area 19(2):95–101
Deleuze F, Guattari D (1987) A thousand plateaus: capitalism and schizophrenia. University of Minnesota Press, Minneapolis
Dietz T (2017) Drivers of human stress on the environment in the twenty-first century. Annu Rev Environ Resour 42:189–213
Ekers M, Prudham S (2017) The metabolism of socioecological fixes: capital switching, spatial fixes, and the production of nature. Ann Am Assoc Geogr 107 (6):1370–1388
Ehlers E (ed) (1993) International geographical union bulletin 43(1–2):87–88
Ehrlich PR (1968) The population bomb. Ballantine Books, New York
Ehrlich PR, Ehrlich AH (1970) Population, resources, environment: issues in human ecology. Freeman, San Francisco, W.H
Emmanuel A (1972) Unequal exchange: a study of the imperialism of trade. Monthly Review Press, New York
Evans TP, VanWey LK, Moran EF (2005) Human-environment research, spatially explicit data analysis, and geographic information systems. In: Moran EF, Ostrom E (eds) Seeing the forest and the trees: human-environment interactions in forest ecosystems. MIT Press, Cambridge, MA, pp 161–185
Fink LD (1979) The changing location of academic geographers in the United States. Professional Geographer 31:217–226
Fischer-Kowalski M, Weisz H (2016) The archipelago of social ecology and the island of the Vienna School. In: Haberl H et al (eds) Social ecology. Springer International, Dordrecht, pp 3–28
Folke C (2006) Resilience: the emergence of a perspective for social–ecological systems analyses. Glob Environ Chang 16:253–267
Folke C et al (2005) Adaptive governance of social-ecological systems. Ann Rev Environ Resour 30:441–473
Foucault M (1984) The Foucault reader. Pantheon, New York
Frank AG (1966) The development of underdevelopment. Mon Rev 18(4):17–31
Fujita M, Krugman P, Venables AJ (1999) The spatial economy: cities, regions and international trade. MIT Press, Cambridge MA
Giddens A (1971) Capitalism and modern social theory. Cambridge University Press, Cambridge
Gregory JK (2000) The changing nature of physical geography. Arnold, London
Hall T et al (2015) Where's the geography department? The changing administrative place of geography in UK higher education. Area 47(1):56–64
Haraway D (1988) Situated knowledges: the science question in feminism and the privilege of partial perspective. Fem Stud 14(3):575–599
Harvey D (1981) The spatial fix—Hegel, Von Thünen, and Marx. Antipode 13(3):1–12
Harden CP (2012) Framing and reframing questions of human–environment interactions. Ann Assoc Am Geogr 102:737–747
Hardin G (1968) The tragedy of the commons. Science 162:1243–1248
Hertel T, Tsigas M (2000) Structure of GTAP. In: Hertel TW (ed) Global trade analysis: modeling and applications. Cambridge University Press, Cambridge
Huxley TH (1877) Physiography: an introduction to the study of nature. Macmillan, London
Independent Group of Scientists (2019) Global sustainable development report 2019: the future is now—science for achieving sustainable development (United Nations, Independent Group of Scientists appointed

by the Secretary-General). United Nations, New York https://sustainabledevelopment.un.org/content/documents/24797GSDR_report_2019.pdf

Johnston RJ (2009) Geography and the social science tradition. In: Clifford NJ, Holloway SL, Rice S, Valentine G (eds) Key concepts in geography. Sage, London, pp 46–65

Johnston RJ, Sidaway JD (eds) (2004) Geography and geographers, Anglo-American human geography since 1945, London, Arnold, p 527

Kates RW (1995) Labnotes from the Jeremiah experiment: hope for a sustainable transition. Ann Assoc Am Geogr 85:623–640

Kates RW (2011) What kind of science is sustainability science? Proc Natl Acad Sci 108(49):19449–19450

Kates RW et al (2001) Sustainability science. Science 292:641–642

Knight E et al (2020) The competitive advantage of regions: economic geography and strategic management intersections. Reg Stud 54(5):591–595

Krugman P (1991) Geography and trade. MIT Press, Cambridge MA

Krugman P (1993) On the relationship between trade theory and location theory. Rev Int Econ 1(10):110–122

Latour B (2005) Reassembling the social: an introduction to actor-network-theory. Oxford University Press, Oxford

Lave R et al (2014) Intervention: critical physical geography. Can Geogr 58(1):1–10

Lenzen M et al (2013) Building Eora: a global multi-region input–output database at high country and sector resolution. Econ Syst Res 25:20–49

Ley D (1981) Cultural/humanistic geography. Prog Hum Geogr 5(2):249–257

Ley D (1985) Cultural/humanistic geography. Prog Human Geogr 9(3):415–423

Li P, Bathelt H, Wang J (2012) Network dynamics and cluster evolution: changing trajectories of the aluminum extrusion industry in Dali, China. J Econ Geogr 12(1):127–155

Li P, Bathelt H (2018) Location strategy in cluster networks. J Int Bus Stud 49:967–898

Liu J (2017) Integration across a metacoupled world. Ecol Soc 22(4):29

Liu J et al (2007a) Complexity of coupled human and natural systems. Science 317:1513–1516

Liu J et al (2007b) Coupled human and natural systems. Ambio 36:639–649

Liu J et al (2013) Framing sustainability in a telecoupled world. Ecol Soc 18(2):26

Liu J et al (2016) Pandas and people: coupling human and natural systems for sustainability. Oxford University Press, Oxford

Liu J et al (2015) Systems integration for global sustainability. Science 347:125–132

Lu D, Guo L (1998) Man-earth areal system: the core of geographical study—on the geographical thoughts and academic contributions of academician Chuanjun Wu. Acta Geogr Sin 53(2):97–105

Ma S, Wang R (1984) The social-economic-natural complex ecosystem. Acta Ecol Sin 4(1):1–9

Malthus TR (1798) An essay on the principle of population as it affects the future improvement of society. J. Johnson, London

Mannion AM (1991) Global environmental change: a natural and cultural environmental history. Longman, Harlow, UK

Martin R, Sunley P (1996) Paul Krugman's geographical economics and its implications for regional development theory: a critical assessment. Econ Geogr 72 (3):259–292

Massey D (1978) Regionalism: some current issues. Cap Class 6:106–125

Marsh GP (1864) Man and nature. Scribner, New York

McDowell L, Sharp J (eds) (2016) Space, gender, knowledge: feminist readings. Routledge, London

McGinnis M, Ostrom E (2014) Social-ecological system framework: initial changes and continuing challenges. Ecol Soc 19(2):30

Meadows DH et al (1972) The limits to growth: a report for the club of Rome's project on the predicament of mankind. Universe Books, New York

Meadows ME (2018) Geographical research and future earth. In Beer T, Li J, Alverson K (eds) Global change and future earth: the geoscience perspective (Special Publications of the International Union of Geodesy and Geophysics). Cambridge University Press, Cambridge, pp 114–120

Meadows ME (2020) Geography education for sustainable development. Geogr Sustain 1:88–92

Meadows ME (2022) Where is the anthropocene? In: Gonencgil B, Meadows ME (eds) Geography and the anthropocene. Istanbul University Press, Istanbul (in press)

Messerli P et al (2019) Expansion of sustainability science needed for the SDGs. Nature Sustainability 2:892–894

Minca C (2001) Postmodern geography: theory and Praxis. Blackwell, Oxford

Moore HL (2015) Global prosperity and sustainable development goals. J Int Dev 27:801–815

Moran EF (2005) Human–environment interactions in forest ecosystems: an introduction. In: Moran EF, Ostrom E (eds) Seeing the forest and the trees: human-environment interactions in forest ecosystems. MIT Press, Cambridge MA, pp 3–21

Moran EF (2010) Environmental social science: human-environment interactions and sustainability. Wiley, London

Moran EF, Brondizio E (1998) Land-use change after deforestation in Amazonia. In: Liverman D et al (eds) People and pixels: linking remote sensing and social science. National Academy of Science, Washington DC, pp 94–120

Mudambi R et al (2018) Zoom in, zoom out: geographic scale and multinational activity. J Int Bus Stud 49:929–941

Murdoch J (2006) Post-structuralist geography: a guide to relational space. Sage, London

National Research Council (1997) Rediscovering geography: new relevance for science and society. National Academy Press, Washington DC. https://doi.org/10.17226/4913

National Research Council (1999) Our common journey: a transition toward sustainability. The National Academies Press, Washington, DC

National Research Council (2001) Grand challenges in environmental sciences. National Academies Press, Washington DC

National Science Foundation (2014) Dynamics of coupled natural and human systems. https://www.nsf.gov/pubs/2014/nsf14601/nsf14601.htm (accessed 3 Apr 2020)

O'Kelly M, Bryan D (1996) Agricultural location theory: von Thunen's contribution to economic geography. Prog Hum Geogr 20:457–475

Ostrom E (2009) A general framework for analyzing sustainability of social-ecological systems. Science 325:419–422

Ostrom E (1990) Governing the commons: the evolution of institutions for collective action. Cambridge University Press, Cambridge

Overman HG (2004) Can we learn anything from economic geography proper? J Econ Geogr 4(5):501–516

Peet R, Watts M (1996) Liberation ecologies: environment, development, social movements. Routledge, London

Piel G (1992) Agenda 21: sustainable development. Sci Am 267:128–128

Porter ME (2000) Location, competition, and economic development: local clusters in a global economy. Econ Dev Q 14(1):15–34

Reinfried S (2009) Education for sustainable development and the Lucerne declaration. Int Res Geogr Environ Educ 18:229–232

Richards K (2008) Geography and the physical sciences tradition. In: Clifford NJ, Holloway SL, Rice S, Valentine G (eds) Key concepts in geography. Sage, London, pp 21–45

Relph E (1976) Place and placelessness. Pion, London

Robbins P (2012) Political ecology: a critical introduction. Wiley, London

Rose G (1993) Feminism and geography: the limits of geographical knowledge. University of Minnesota Press, Minneapolois

Sauer C (1925) The morphology of landscape. Univ Calif Publ Geogr 2(2):19–53

Sharpe C (2009) Past-president's address: is geography (the discipline) sustainable without geography (the subject). Canadian Geogr 53(2):123–138

Smith N (1984) Uneven development: nature, capital and the production of space. University of Georgia Press, Athens

Soja EW (1989) Postmodern geographies: the reassertion of space in critical social theory. Verso, London

Soja EW (2011) Beyond postmetropolis. Urban Geogr 32(4):451–469

Smith SM et al (2018) Advancing sustainability science for the SDGs. Sustain Sci 13(6):1483–1487

Stoddart DR (1966) Darwin's impact on geography. Ann Assoc Am Geogr 56:683–698

Stoddart DR (1986) On geography. Blackwell, Oxford

Storper (2013) Keys to the city: how economics, institutions, social interaction and politics shape development. Oxford, UK and Princeton, NJ, USA, Princeton University Press

Swyngedouw E (2013) Into the sea: desalination as hydro-social fix in Spain. Ann Assoc Am Geogr 103(2):261–270

Tuan Y-F (1976) Humanistic geography. Ann Assoc Am Geogr 66(2):266–276

Turner BL (2002) Contested identities: human-environment geography and disciplinary implications in a restructuring academy. Ann Assoc Am Geogr 92:52–74

Turner BL et al (2003a) A framework for vulnerability analysis in sustainability science. Proc Natl Acad Sci 100:8074–8079

Turner BL et al (2003b) Illustrating the coupled human–environment system for vulnerability analysis: three case studies. Proc Natl Acad Sci 100:8080–8085

Van Wey LK, Ostrom E, Meretsky V (2005) Theories underlying the study of human-environment interactions. In: Moran EF, Ostrom E (eds) Seeing the forest and the trees: human-environment interactions in forest ecosystems. MIT Press, Cambridge MA, pp 23–56

Walker B et al (2004) Resilience, adaptability and transformability in social–ecological systems. Ecol Soc 9(2):5

Wilbanks TJ (1994) Sustainable development in geographic perspective. Ann Assoc Am Geogr 84:541–556

Young OR et al (2006) Global environmental change, resilience, vulnerability, and adaptation: a cross-cutting theme of the international human dimensions programme. Glob Environ Chang 16:304–316

Zhu AL (2020) China's rosewood boom: a cultural fix to capital overaccumulation. Ann Am Assoc Geogr 110(1):277–296

Ruishan Chen is Professor of land system change at the Center of Land Use Planning and Rural Development in the School of Design at Shanghai Jiao Tong University. He received his PhD in physical geography from Peking University, China. He has been a visiting Ph.D. student and was recently research fellow at the Earth Institute at Columbia University. His research focuses on land system change, natural resources and climate change. Chen has worked with many international organizations and agencies including IPBES, ISSC, the Urban Climate Change Research Network (UCCRN), Network of Sustainable Hydropower Development in Mekong river countries (NSHD-M) and the World Bank.

Annah Zhu is an Assistant Professor of environmental globalization at Wageningen University in the Netherlands. She received her Master's in environmental management from Duke University and her Ph.D. in society and environment from the University of California, Berkeley. Her research focuses on the role of China in the processes of environmental globalization and China's overseas environmental impacts. She is a veteran of the United Nations' Environment Program in Geneva, and a former Peace Corps volunteer in Madagascar.

Yingjie Li is a Ph.D. Candidate in the Center for Systems Integration and Sustainability (CSIS) at Michigan State University. His research focuses on understanding the interactions of human and environmental systems to advance the Sustainable Development Goals (SDGs). His current research examines how telecoupling shapes global sustainability. He also has broad research interests in ecosystem service flows, international trade, the food-energy-water nexus, and environmental policy.

Pengfei Li is an Associate Professor of strategy and global management at the Haskayne School of Business, University of Calgary, Canada. He received his Ph.D. from Peking University and taught at HEC Montréal, University of Toronto, and East China Normal University in Shanghai. His research and publications focus on industrial clusters, business strategy, and industrial evolution, ad has been recognized in awards by the American Association of Geographers, Regional Studies Association (UK), and the Academy of International Business.

Chao Ye is Professor in human geography at the School of Geographic Sciences, East China Normal University. He obtained his Master's in economic geography from Northwest Normal University, Lanzhou, and his Ph.D. in urban–rural interaction from the Institute of Geographic Sciences and Natural Resources Research, Chinese Academy of Sciences, Beijing. He was visiting professor in human geography at the University of British Columbia, Canada. Ye's research focuses on the production of space, geographic thought, urban–rural governance, and sustainability science. He contributes regularly to blogs and articles for readers outside of geography, which illustrates his commitment to introducing geography to others.

Michael E. Meadows is Emeritus Professor and Visiting Research Scholar in the Department of Environmental & Geographical Science at the University of Cape Town, South Africa, and Professor in the School of Geographic and Ocean Sciences at Nanjing University. He was Visiting Professor in the School of Geographic Sciences at East China Normal University from 2016 to 2020. Meadows obtained his undergraduate degree from the University of Sussex and a Ph. D. from the University of Cambridge, United Kingdom. Research interests and publications lie broadly in the field of physical geography and more specifically concern Quaternary environmental change and the geomorphological and biogeographical impacts of natural and anthropogenic changes. Meadows was formerly Secretary-General and Treasurer of the International Geographical Union (IGU) and is currently IGU President 2020–2024).

Global Understanding: A New Geographical Paradigm for the Twenty-First Century?

14

Benno Werlen

Abstract

With the ongoing globalization processes, geographical living conditions are changing profoundly, for everyone, everywhere in the world. The ways we live locally, how we eat, work, travel or communicate, shape our world on a global scale. The International Year of Global Understanding—submitted in the name of IGU to and approved by UNESCO—focuses on making the connection between local action and global challenges understandable for everyone, and to fostering on that basis a new geographical worldview. The paper presents one of the most important projects in the history of the IGU, institutional framework, outcomes, as well as its theoretical background. The theoretical frame includes the practice-centered approach of everyday 'geography-making', proposed for research, education, and information. This program intends to make visible the ways people transform nature locally and globally. It addresses the why and how of peoples' distinct attitudes on sustainability dilemmas and relevant implications. Socio-cultural backgrounds and regional conditions, as well as lifestyle consequences, are considered the core dimensions of deep societal transformations. The project follows the leitmotif that you cannot change the world toward global sustainability if you do not understand what your everyday actions mean for the world as a whole. Global thinking demands global understanding.

Keywords

Geographical worldview · Globalization · Acceleration · Sustainability · Geography-making · Societal transformation

14.1 Introduction

Virtually every domain that shapes our everyday lives politically, economically, and socially has been altered over the last decades by the rapidly increasing interdependence of the world's economies, cultures, and populations. Processes of globalization have altered our perspectives of both space and time. Many everyday practices and spatial distances are now easily traversed by new technologies. Far-flung places and people are in ever-closer contact. New kinds of personal global communities and networks are emerging. Nevertheless, this altered space-time constellation does not diminish the importance of the local. In fact, in some respects the opposite is true.

B. Werlen (✉)
Department of Geography, UNESCO Chair on Global Understanding for Sustainability, Friedrich Schiller University Jena, Loebdergraben 32, 07743 Jena, Germany
e-mail: benno.werlen@gmail.com

V. Kolosov et al. (eds.), *A Geographical Century*,
https://doi.org/10.1007/978-3-031-05419-8_14

Mainly due to the Digital Revolution, local communities and regions have lost, in nearly all respects, their independence from broader events and interrelations. Places and regions are no longer 'container spaces' of cultural, social, or economic life (Werlen 1993a 1–4; 1995 169–174). Rather, they are becoming transit stations of globally connected processes. While the determination and identification of places and regions remain important for communication and practical activities, the content of what happens at and through these stations has become less significant. Indeed, an exclusive focus on spatial arrangements at the local or the regional scales misses important dimensions of today's geographical realities. Consequently, a recalibration of the geographical view of the world is needed: a turn from spatial research to a practice-centered global perspective. Such a broader perspective of local and regional living conditions opens the possibility of a global understanding of these processes. And this kind of understanding is urgently required if we are to find pathways to tackle global issues at the local and regional levels. With the ongoing Digital Revolution, and the corresponding and the associated process of globalization, we are experiencing a new expansion of the spatial scope of action, and a subsequent deep transformation of local living circumstances which for many is pulverizing an established 'ontological security' (Laing 2010, 39).

Why are so many people experiencing globalization processes as disruptive and challenging? First, if these viewpoints are not synchronous, finding one's place and reaching a satisfactory world orientation becomes difficult. ''I don't recognize this world anymore' may in these circumstances be an appropriate response. Second, we can also see the reasons why geographical literacy is fundamental to addressing the challenges that come with globalization. Given that the experiences of globalization may be disruptive of a traditional geographical worldview, based on the concept of hermetically contained spaces, the quest to re-establish traditional geographical living conditions can hardly be the solution. But such a 'solution' is exactly what nationalist discourses promise when promoting a backward-looking strategy based on an imagined past. Instead of taking the new geographical conditions into account and adjusting policies to them, these discourses call for a return to the 'pre-given' unity of nature, culture, society, economy, and political order as the 'true', or 'natural' spatial reality.

All in all, in face of the Digital Revolution there are only two viable options: re-establishing historical-geographical conditions as they once were, or developing an understanding of the new. This latter option requires a commitment to change on all levels, both within and outside geography, and even within the sciences. A well-established tool that can be used to overcome the barriers between scientific disciplines that impede globally interconnected understanding of contemporary problems, including those between science and the everyday world, can be found in the International Years UNESCO framework.

14.2 The Frame: The International Year

Today's most pressing existential problems can only be addressed through an inclusive, global perspective that encompasses understanding the conditions and consequences of everyday actions. Meeting the challenges of the twenty-first century not only requires new ways of knowledge production but even more so knowledge dissemination and implementation. Finding solutions to global challenges certainly cannot be met by a single scientific discipline. In fact, such challenges require new alliances and approaches to research which include innovative partnerships across a wide range of scientific disciplines, as well as collaboration with local communities, and all sectors of civil society. Taking these necessities together calls for transdisciplinary cooperation. Although these new approaches to scientific research have been studied and elaborated for several decades, scientific institutions, organizations, and funding institutions are still not well-prepared for this. Greater attention must therefore focus on establishing the requisite conditions to overcome the limitations of a traditional disciplinary division of scientific work.

Collaboration with local communities requires raising awareness of the global nature of the world's most pressing challenges, and of how the living conditions of the world's populations affect these problems. Building such awareness calls for the broadest possible approach. As education, in general, is structured around traditional scientific disciplines, and learning contents depend on disciplinary knowledge, serious time-lags in comprehensive knowledge production are inevitable. This is especially problematic when swift change is needed. Consequently, an approach that can bridge inclusiveness and independence is indispensable.

Such a platform for action exists in an International Year within the framework of the United Nations and its sub-organizations, such as the Economic and Social Forum (ECOSOC) and UNESCO. The International Year concept was established to draw attention to major issues affecting the global community that are relevant to most of the world population, and that are consistent with the purposes and principles of the United Nations, as stated in the Charter of the United Nations. In addition, its subject 'should be a priority concern in the political, social, economic, cultural, humanitarian, or human rights fields... and should be of concern to all or the majority of countries, regardless of their economic and social systems, and should contribute to the development of international cooperation in solving global problems' (United Nations 1980, 24). There are also some practical requirements that need to be fulfilled, of course, such as contributing to solving international problems and promoting universal peace. A carefully elaborated and financially supported program is needed during the declared year as well as follow-up activities at all levels, international, national, and local. In addition, it should lead to clearly identifiable and practical results.

The program of a proposed International Year needs broad support from topically relevant organizations and institutions to make the relevance of the subject visible and recognizable to the UN bodies. In the case of the International Year of Global Understanding, the topic and the first general program lines were designed within the framework of the International Geographical Union in a meeting at the Home of Geography by a group of internationally reputed geographers in June 2008. It was developed further by interdisciplinary working groups, representing more than 20 disciplines of the human, social, and natural sciences from all continents in meetings at Weimar and Jena in 2011. The necessary strong support—established over the years and based on an extended consultation process—was provided by the International Social Science Council (ISSC), the International Council for Philosophy and Human Sciences (CIPSH), and the International Council for Science (ICSU) and others. To proceed, the proposal then needed a member state (preferably supported by a voting group), to present the proposal to the UNESCO Executive Board, and, after a positive evaluation, to the General Conference for unanimous approval. In this case, the submitting member state was Rwanda and the supporting voting group was Africa. They submitted the project to the UNESCO Executive Board in June 2013 and the proposal was adopted at the General Conference the following December. When UNESCO Resolution 37 C/63 proclaimed 2016 the International Year of Global Understanding (IYGU), the way was clear to build awareness of a new geographical reality in the making. But what is the meaning of 'global understanding', and to what precisely does it refer?

14.3 The Claim: Global Understanding

In the context of the 2016 International Year, 'global understanding' refers specifically to raising awareness of the implications of the Digital Revolution with respect to the radical extension of the spatial reach of human actions and their consequences (Werlen et al. 2016). The widespread application of digital technologies and the exuberant transformation of natural conditions on the 'time altar' of the Anthropocene have fundamentally altered living conditions for most of the world's population (Crutzen

2002). Consequently, one of the important tasks for the IY project is to develop and promote an understanding of the new, globalized conditions of everyday life-worlds around the globe, and to develop an adapted geographical worldview. The theoretical background of the program has a practice-centered focus (Werlen 1987, 1993a, 1997; Giddens 1984). The theory is rooted in the increasing importance of global communication and information flows in real-time, together with the dissolution of distance for many aspects of everyday life. Given these new dynamics, a simple or flat spatial focus is less pertinent from socio-cultural respect. Instead of focusing on 'the' geography of the earth's surface, it focuses on how these geographies are produced and reproduced daily by everyone from different positions of power.

This approach shifts from an understanding of geography as the study of space, regions, and places to a focus on the locally and regionally embedded everyday makings of geography along with their global consequences. A global understanding aims to promote an appreciation of current social and cultural realities as the outcome of new forms of geography-making and proposes new ways of addressing human-induced global challenges.

Basing the program on real-world actors and their social and cultural as well as regional and local contexts makes it possible to develop a highly differentiated and culturally sensitive practice-centered perspective. Instead of presupposing space to be a 'container' in which life is conducted, the subject and its relations with the real world become the focus of attention. Putting the everyday activities at the center of interest allows the researcher to take the local consequences of global phenomena into account and focus on their 'relations with the world' (Werlen 1993b, 1997, 2021; Rosa 2013, 2019).

From this perspective, a geographic vision can be established that captures globalized living conditions and ways of living that are enabled by the digital revolution. Research and education can then unfold a new representation of the world —or more precisely, a new worldview—that reveals the changing geographical conditions of everyday practices, e.g., how local air pollution or the consumption of goods reflects and shapes global conditions. Based on this, a geographical awareness of one's own living conditions and ways of living can develop into an attempt to reveal and enhance understanding of the globalized 'geographical imagination' (Gregory 1994, 282) as 'the subject's whole creation of a world for itself' (Castoriadis 1997, 115).

World-binding capacity is ultimately expressed in the degree to which one succeeds or fails in integrating resources, goods, and other people into one's activities. It can be seen most directly in the contents of our shopping basket as consumers, in the number of goods that are consumed for our own way of life, and its implication in remote places. But it can also be seen in the strategies for the control of resources, goods, and people by producers, with their mostly global world relations and ties. Productive as well as consumptive spheres are each linked to global implications, which is particularly evident in the field of sustainability. In particular, we also share the negative consequences of our actions globally. And since escaping these is not in sight, there is no alternative to changing practices at the local level.

14.4 The Background: Societal Relations of Space

As noted above, one of the key features of the Digital Revolution is to enable actions across continents in real-time. Of course, this is only possible for actions that are digitally transferred, such as electronic communications, money transfers, and the like. But because communication is a fundamental form of social interaction, changes in the ways and means of communication will impact societal processes whether traditional, modern, or beyond. The pre-given potential of acting over distance—what I would call 'gesellschaftliche Raumverhältnisse' or 'societal-spatial relations' (Werlen 2010)—can even be seen as the central axis for the forms that social worlds can take. To understand the impacts of globalization on various societal realities, and the main access points that increase global understanding in the long run, it is important to approach this issue theoretically as well as practically.

If one agrees with Ferdinand Tönnies (2001 [1887]) that acting and interacting over distance is what distinguishes society from community, then it becomes clear that the possibilities and forms for overcoming spatiality contribute directly to the constitution of societal realities. In other words: the existence and shape of social worlds depend per se on the possibilities for acting and communicating over distance.

According to Tönnies (2001 [1887], §19), 'society' is to be understood as 'a circle of people who are essentially separated from each other', but who nevertheless cooperate. Of course, 'essentially separated' not only refers to spatial separation, but also to 'social separation' (Clausen et al. 1985; Merz-Benz 1995). Nevertheless, the available capabilities to act over spatial distance are central to determining possible social forms. We can even see them as constitutive. At the same time, they are important for understanding the impacts of the Digital Revolution on the formation of societal realities. To this and the related relevance of global understanding, the concept of 'societal-spatial relations' is central.

As already indicated, 'societal-spatial relations' can be described in the first instance as the spatial conditions of action within which societal realities are created and the natural pre-given 'Mitwelt', or 'co-world', is shaped for specific purposes (Werlen 2010, 327). Thus 'spatial conditions' refer exclusively to the means, media, and possibilities created socially over time to facilitate cooperation with others, despite the spatiality of the world–in short: to be able to act and communicate over distance, to master spatiality.

Historically, mastering spatiality did not proceed in a linear, evolutionary fashion but rather in revolutionary leaps. Broadly speaking, three radical transformations of societal relations of space can be noted in the Neolithic, the Industrial Revolution, and the current Digital Revolution of spatio-temporal world relations (Werlen 1997, 200ff.).

With each revolution, the *modus operandi* of reality production changed as new societal relations of space were established and a deep transformation in the formation and shape of social and cultural worlds took place. This allowed a shift in the spatial and temporal reach of people's actions and, by extension, all forms of the social. For example, the pre-Neolithic mode was dominated by the primacy of the present and orality, requiring almost all interactions to involve bodily co-presence. It did not allow a broader societal constellation to emerge as defined by Tönnies, just a community.

By contrast, modern social structures such as the nation-state draw on analog written communication and past arrangements; they operate in the diachronic/distanced mode. The Digital Age —based on numeric digitality—is rooted in the synchronic/distanced modus, changing radically many institutional and informal features of the well-established and the deeply rooted. The Digital Revolution is reshaping societal relations of space more radically and more quickly than its two predecessors. The vanishing of distance (Flusser 1992, 31) in many areas of life is taking place at the expense of several established routines and traditional undertakings. For instance, everyone can experience the breath-taking speed at which the spatial conditions of our actions and life contexts change. Changes that took centuries after the Neolithic, like the speed of human mobility and animal power, and decades after the Industrial Revolution, now take less than a few months such as the shift from mobile to smartphones and the rise of social media. Not only is it difficult to comprehend the acceleration of technological innovation, especially in the field of communication, it is even more problematic to fully grasp the social consequences of these changes.

What we can already observe is a loss of the expected connection between proximity and familiarity, as well as between distance and unfamiliarity. In addition, we also witness a colonization of private spheres by means of digital technologies. These changes are possibly just the first signs of the newly established societal-spatial relations and their consequences for modes of construction of geographical, cultural, social, economic, and political realities. And most of all, digital technologies enable the detachment of the symbolic from the material vehicles of data and information.

The profound implications of the detachment of the symbolic from material carriers become particularly apparent with regard to the basis of the modern state, the principle of territorialization, and territorialized core practices of the nation-state such as border controls and taxation. The tight binding of society and space in the form of state and territory is largely based on fixed relations between material vehicles and symbol or symbolic meaning. This is as much the case with national currencies as with writing and written communication in the form of newspapers, or the shift from print media to electronic communication, from coins to digits, and so on.

The connection becomes even clearer when tariffs are imposed in relation to the symbolic monetary value of exchanged goods. The most visible manifestation of the relationship of meaning and matter (Gren 1994) is perhaps the customs gate. The barrier at the border marks the spatial location of the state's scope and at the same time stands for all forms of territorial control. With the replacement of material carriers by digital transactions, the basis of previously known forms of territorial control is significantly weakened.

The ongoing weakening of territoriality as one of the basic principles of modern states is an indication of the depth of the transformation of the spatial conditions, and therefore of geographical everyday practices. Perhaps it is no exaggeration to say that 'society' as we know it is at stake. But what form the new 'societal' could take is not apparent.

Societal-spatial relations always imply specific 'space-time-relations'. Thus, it is no coincidence that 'globalization' and the 'acceleration' of life are observable simultaneously (Giddens 1990; Rosa 2013). They are an expression of a highly relevant mutual dependency. But since societal relations of time ultimately reflect established societal relations of space, there is good reason to give the latter priority when analyzing globalization processes. In any case, together both point to the consequences of the emerging reconstitution of societal-spatial relations and corresponding space-time relations.

14.5 The Program: The International Year of Global Understanding

The design of the program of the International Year of Global Understanding began with the insight that humanity's grand challenges are self-made and affect the entire planet. The program was consequently developed on the premise that global sustainability can be reached best if sustainable everyday actions are established locally. Everyday practice is where the local and global become one, and where scientific insights are applied. Therefore, a widespread awareness of how everyday actions create global challenges that impact humanity is necessary. This includes the ability to understand connections between actions that may seem disconnected across time and space. Grasping the global condition of local and regional living contexts of one's own life in this way constitutes Global Understanding. It means making clear the global impact of local and regional ways of living. At the same time, fostering an understanding of everyday practices in the context of 'world relations' with all its implications in terms of reach, bindings, and connectivity, is a prerequisite for reaching a geographical worldview appropriate to the new societal relations of space. This is the aim of the program of Global Understanding (Werlen 2012; Werlen et al. 2016).

Global Understanding is a comprehensive science-, education, and information-oriented approach. The education program aims to raise awareness of the global embeddedness of everyday life around the globe and increase understanding of the inextricable links between local action and global phenomena. Teachers, students, and pupils as well as citizens are invited to take responsibility for their actions and to make daily contributions to achieving global sustainability. In scientific and academic terms,

Global Understanding aims to strengthen integrative transdisciplinary thinking. It is geared to overcoming the widely established barriers between the natural sciences, social sciences, and humanities. Joint work should, whenever possible, be carried out in partnership with engaged citizens promptly and directly. Research findings need to be translated not only for use in classrooms at all levels, but also into broader public awareness campaigns through easy-to-understand publications, computer games, movies, documentaries, TV shows, and the like (Fig. 14.1).

In a more specific sense, the program of Global Understanding addresses three key interfaces. The first refers to the local-global nexus of human activities, the second to the relationship between culture, society, and nature, and the third to everyday life and science (Fig. 14.2).

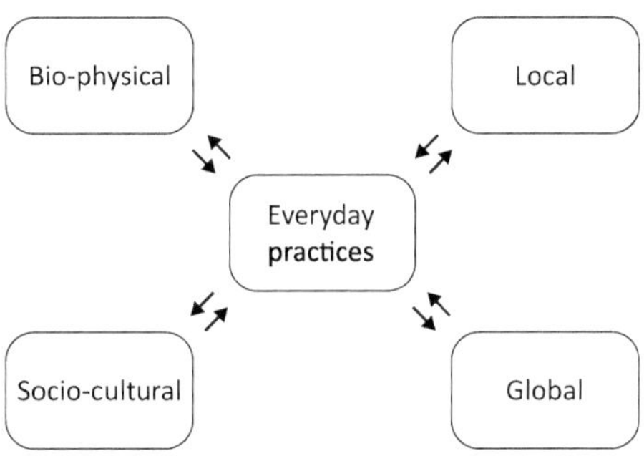

Fig. 14.1 Key dimensions of geographical practices (Werlen et al. 2016: 607)

14.6 The Local and the Global

Because people are engaged under globalized conditions to cope with their everyday lives and to show the ways in which they bring the world into their lives, the central tasks are, first, to establish awareness of the global implications of everyday living and, second, to identify the culturally, socially, and economically unequal circumstances of people and their relationship to spatio-temporal structures and flows. Productive and consumptive forms of world-binding, for instance, can be approached through 'natural resource stories' (Doolittle 2010; Galafassi et al. 2018) or the 'follow the thing' approach (Cook 2004; Ermann 2012). These approaches are geared toward identifying global consequences of world-binding processes and, in the process, promoting awareness of the implications of personally defined lifestyles for the structuration of global flows of products and natural resources.

The same local-global perspective can be applied with respect to political, legal, informative, and symbolic appropriations as forms of world relations. In the case of the political and legal dimensions, especially important are the aforementioned weakening of the territorial principle and resulting legal gaps in the protection of global commons. Despite the fact that a fully binding decision-making body with global jurisdiction is extremely unlikely to be established in the time frame available to address humanity's most pressing problems, global

Local || Global
Global impact of local action
Social-cultural || Natural
Culturally adapted, ecologically and socially sustainable lifestyles
Everyday || Science
Sustainable patterns of action and knowledge for local use

Fig. 14.2 Interfaces of global understanding

necessities require swift, yet deliberate action from the bottom up. One possibility is to promote global justice while integrating that dimension into the implementation of local interests and concerns. Here again, Global Understanding is an indispensable precondition for targeted action. Another case of this synoptic local-global view pertains to regionalist and nationalistic movements, often fed by the unsettling implications of globalization processes. Of course, by emphasizing the 'we', nationalistic as well as regionalist discourses have a strong identity-forming capacity. However obvious or 'logical' such backward-oriented compensatory solutions may appear, the exclusion and marginalization of others can undermine security. Feelings of belonging and security need to be achieved in new, less contradictory, and more cosmopolitan ways.

Finally, a third important dimension is the interrelation of global information and knowledge flows, local meaningful symbolic appropriations, and communication. The informational world bindings arise from global telecommunications (Castells 1996) and global information networks. If the actors' stock of knowledge and information constitutes the basis of symbolic appropriation, then it is helpful to know how the new geographies of information affect meaningful forms of world-binding. It is also important to make a clear distinction between a Global Understanding of local conditions and a devaluation of the local against the global.

14.7 The Socio-cultural and the Natural

Local everyday activities are not just embedded in global processes; they are simultaneously intertwined with the socio-cultural and bio-physical realms. Practice-centered conceptualizations of sustainability include, first, embracing the shift in focus from 'sustainability' to 'living sustainably', and, second, treating the natural sphere as embedded in the socio-cultural world, not as external to it, as 'environment'.

In this perspective, currently dominant ecologically oriented approaches toward sustainability need to be expanded. Sustainability is not 'only' an ecological issue, but a much broader social and cultural one. The integrated analysis of everyday activities as a starting point reflects this turn in perspective. Sustainability research and policies should always begin with the social-cultural side and then look for the bio-physical consequences. In keeping with this approach, objects like 'habitats' or 'ecotopes' are no longer the primary starting point in the development of sustainability policies.

Global Understanding seeks to explore the ways in which people with specific socio-cultural and historical backgrounds transform nature. These backgrounds and associated lifestyles are at the core of unsustainable practices. A widespread awareness of how everyday actions create the challenges that impact humanity is a prerequisite to finding solutions in all realms (Fig. 14.3).

All in all, the synoptic view of the global and the local is as much a part of Global Understanding as that of the integration of the cultural and the natural. To this end, the program proposes to focus on essential everyday activities such as eating and drinking, housing, working, travel, and communicating. The first step in that direction has been undertaken by Jackson et al. (2016) on the topic of eating and drinking, Robertson (2016) on communicating, and Robinson et al. (2016) on global urbanization processes. They highlight the ways in which the Global Understanding program can be implemented.

14.8 Everyday World and Science

The development of an integrative understanding of the connections between the local and the global as well as the cultural and the natural and its implementation also requires a reshaping of scientific work. Mobilizing citizens for a profound transformation of lifestyles toward sustainability in a global perspective requires the generation of appropriate knowledge. Making such knowledge available is a task that goes

Fig. 14.3 Everyday practices of globalizing geography-making (after Werlen 2015: 23)

	Human actions as	
Socio-Cultural	**Everyday Practices**	**Natural**
Cultural Values	**Eating/Drinking => Surviving**	Soil Degradation
Division of Labor	**Working/Housing => Urbanizing**	Climate Change
Cultural Diversity	**Moving/Staying => Belonging**	Biodiversity
Schemes of Interpretation	**Wasting/Recycling => Preserving**	Air Pollution
Glocalized Lifestyles	**Communicating/Networking => Interacting**	Sea Level
Leisure	**Sports/Entertaining => Recreating**	Hazards

beyond a simple optimization of established scientific strategies, like the refinement of interdisciplinary cooperation. So far, the dominant strategy to improve the quality of scientific explanations and problem solutions has been to push disciplinary specialization, with negative consequences for interdisciplinary collaboration and communication with the everyday world.

Considering the socio-cultural embeddedness of nature, the high variety of socio-cultural and regional conditions for local actions, and the crucial role of local actions for global constellations, no 'one-size-fits-all' problem solutions are available. Global Understanding offers an alternative, one that recognizes connections across traditionally fragmented domains of knowledge and avoids disciplinary tunnel vision.

Transdisciplinary cooperation offers a way out. This path opens up a variety of propositions, for research as well as teaching. On the teaching front, the preparation of students for transdisciplinary work is the first step. As for research, transdisciplinary cooperation needs to be promoted, along with curriculum reform aimed at fostering collaborations with engaged citizens.

New research and teaching formats such as living labs can promote experimental exchange between actors from different sectors. Learners and teachers as well as researchers can cooperate in specific local or regional contexts to find sustainable solutions for local problems from a global perspective. Such approaches can be expected not only to yield new scientific insights and learning experiences but also to generate a stronger social commitment on the part of science and the larger academic world.

14.9 The Role of the IGU

The IYGU was the first major IGU project aimed at broad international cooperation in its hundred-year history. Previously, the IGU often played a supporting role in projects such as the International Year of Mountains in 2002, Planet Earth in 2008, and the UNESCO Man & Biosphere Program in the 1980s. Lessons learned from these projects and ties with the International Council for Philosophy and the Human Sciences (CIPSH) furthered the IYGU initiative.

The International Sciences Councils and the International Scientific Unions are facilitating cooperation and dissemination. Together with their participation at major events of the

International Councils in Nagoya (2010), Paris (2011), and Durban (2015), recent IGU meetings allowed for program coordination, mobilization, and presentation of the aims and claims of the IYGU to the international community of scientists, academics, and especially geographers. Through its organizational structure of 43 IGU commissions and more than 70 National Committees (with their national conferences), they helped mobilize the international community. The IGU established the Commission on Global Understanding in 2016 to pursue essential follow-up activities for the International Year.

14.10 Outcomes and Conclusions

With the UNESCO Proclamation of 2016 as the International Year of Global Understanding (IYGU) (UNESCO 2013a, b, c), the concept of Global Understanding found its first practical application (www.global-understanding.info/en). With the overarching claim that 'to think globally calls for global understanding', the initiative provided a framework and foundation for a new geographical worldview in an age of Digital Revolution and generated the energy needed to propel transformation.

Perhaps the most important achievement of the IYGU was to connect persons from diverse fields around the globe. The resulting interactions, together with the IGU's membership in the International Council of Philosophy and the Human Sciences, are creating platforms for transdisciplinary research and education. A global network of more than 40 Regional Action Centers has been founded, bringing together concerned partners and activists from a wide range of fields, all working toward a global understanding of sustainability. The participants include NGOs, universities, academies, schools, libraries, youth organizations, international sports organizations, intercontinental state organizations, embassies, national chambers of industry and commerce, and the Erasmus Mundus program.

Following the opening ceremony in February 2016 at the Volkshaus in Jena, more than 1000 events have been organized around the globe to promote the concept of 'Global Understanding', reaching millions of people on all continents. A major IYGU program event took place in the framework of the Glastonbury Festival in the UK, one of the world's largest music and performing arts events attended by c. 200,000 people. Another cooperation with the arts was established in association with Most-UNESCO, Mémoire de l'Avenir and CIPSH (UNESCO 2018). The IYGU also played an important role in the World Humanities Conference in Liège 2017, leading to the creation of seven UNESCO Chairs fostering Global Understanding around the world. The designation of UNESCO's World Science Day in 2017 as 'Science for Global Understanding' reflects the institutional impacts of the program. Public participation was reflected in a series of public outreach campaigns through initiatives such as a 'story map competition' in cooperation with ESRI and the Federal State of Thuringia, one of the principal sponsors of IYGU.

All these achievements represent the beginning of a global mobilization, spurred on by the Jena Declaration, to enable citizens around the world in all walks of life to understand global interconnectedness and to act for global sustainability (www.thejenadeclaration.org). They represent vital contributions the IGU and other international scientific organizations make in achieving a more sustainable future.

References

Castells M (1996) The rise of the network society. Blackwell, Oxford

Castoriadis C (1997) The imaginary institution of society. MIT Press, Cambridge

Clausen L, von Borries W, Dombrowsky R, Prahl HW (eds) (1985) Tönnies heute. Zur Aktualität von Ferdinand Tönnies. Mühlau, Kiel

Cook I (2004) Follow the thing: papaya. Antipode 36:642–664

Crutzen PJ (2002) Geology of mankind. Nature 415(3):23

Doolittle AA (2010) Stories and maps, images and archives: multimethod approach to the political ecology of native property rights and natural resource management in Sabah, Malaysia. Environ Manage 45:67–81

Ermann U (2012) Follow the Thing! Ein Überblick über einige Geographien der Warenwelt. GeoGraz 51:6–11
Flusser V (1992) Das Verschwinden Der Ferne. Archplus 111:31–32
Galafassi D et al (2018) Stories in social-ecological knowledge co-creation. Ecol Soc 23(1):22–51
Giddens A (1984) The constitution of society: outline of the theory of structuration. Polity Press, Cambridge
Giddens A (1990) The consequences of modernity. Polity Press, Cambridge
Gregory D (1994) Geographical imaginations. Blackwell, Oxford
Gren M (1994) Earth writing: exploring representation and social geography between meaning/matter. Kulturgeografsk serie Nr 85. Department of Geography, Gothenburg
Jackson P, Spiess W, Sultana F (eds) (2016) Eating, drinking: surviving. The international year of global understanding (IYGU). SpringerBriefs in Global Understanding. Springer, Heidelberg
Laing RD (2010) The divided self. Penguin, London
Merz-Benz P-U (1995) Tiefsinn und Scharfsinn. Ferdinand Tönnies' begriffliche Konstitution der Sozialwelt. Suhrkamp, Frankfurt a. M.
Robertson M (2016) Communicating, networking, interacting: the international year of global understanding (IYGU). SpringerBriefs in Global Understanding. Springer, Heidelberg
Robinson J, Scott AJ, Taylor PJ (2016) Working, housing, urbanizing: the international year of global understanding (IYGU). SpringerBriefs in Global Understanding. Springer, Heidelberg
Rosa H (2013) Social acceleration: a new theory of modernity. Columbia University Press, New York
Rosa H (2019) Resonance: a sociology of our relationship to the world. Polity Press, Cambridge
Tönnies F (2001 [1887]) Community and civil society. Cambridge University Press, Cambridge
UNESCO (2013a) Proclamation of 2016 as international year of global understanding (IYGU). Executive board. Hundred and ninety-second session. Item 39. 192 EX/39, Paris, 14th August 2013a. UNESCO, Paris
UNESCO (2013b) Proclamation of 2016 as international year of global understanding. General conference 37th session, Item 5.24. 37 C/63, Paris, 5th November 2013b. UNESCO, Paris
UNESCO (2013c) Proclamation of 2016 as international year of global understanding (IYGU). Records of the general conference. 37th session. Resolutions. UNESCO, Paris
UNESCO (2018) Would humanities conference: outcome document—a new humanities agenda for the 21st Century. UNESCO, Paris
United Nations (1980) Economic and social council. Resolutions 1980/67: guidelines for international years and anniversaries. United Nations, New York
Werlen B (1987) Gesellschaft, Handlung und Raum. Grundlagen handlungstheoretischer Sozialgeographie. Franz Steiner Verlag, Stuttgart
Werlen B (1993a) Society, action and space: an alternative human geography. Routledge, London
Werlen B (1993b) Gibt es eine Geographie ohne Raum? Zum Verhältnis von traditioneller Geographie und zeitgenössischen Gesellschaften. Erdkunde 47(4):241–255
Werlen B (1995) Sozialgeographie alltäglicher Regionalisierungen. Bd. 1: Zur Ontologie von Gesellschaft und Raum. Franz Steiner Verlag, Stuttgart
Werlen B (1997) Sozialgeographie alltäglicher Regionalisierungen. Bd. 2: Globalisierung, Region und Regionalisierung. Franz Steiner Verlag, Stuttgart
Werlen B (2010) Epilog. In: Werlen B (ed) Gesellschaftliche Räumlichkeit. Vol. 2: Konstruktion gesellschaftlicher Wirklichkeiten. Franz Steiner Verlag, Stuttgart, pp 321–338
Werlen B (2012) True global understanding and pertinent sustainability policies. In: Scheunemann I, Oosterbeek L (eds) A new paradigm of sustainability. IBIO, Rio de Janeiro, pp 163–172
Werlen B (2015) From local to global sustainability: transdiciplinary integrated research in the digital age. In Werlen B (ed) Global sustainability. cultural perspectives and challenges for transdisciplinary integrated research. Springer, Dordrecht, pp 3–16
Werlen B, Oosterbeek L, Henriques MH (2016) International year of global understanding: building bridges between global thinking and local action. Episodes 39 (4):604–611
Werlen B, Blumenthal H (2020) Global understanding, education and sustainability. In: UNESCO (ed) Humanistic futures of learning. Perspectives from UNESCO Chairs and UNITWIN Networks. UNESCO, Paris, pp 87–90
Werlen B, Kauffman E, Gäbler K (2020) Future knowledge mobilization for deep societal transformations. In Canadian Commission for UNESCO (ed) Imagining the future of knowledge mobilization: perspectives from UNESCO Chairs, pp 113–126
Werlen B (2021) World relations and the production of urban realities: on space and action, city and urbanity. In Rüpke J, Rau S (eds) Religion and urbanity: reciprocal formations. De Gruyter, Berlin

Benno Werlen currently holds the UNESCO-Chair on Global Understanding for Sustainability at the Friedrich Schiller University, Jena (Germany). From 1998 to 2018, he held the chair of social geography at Jena University. He has served as a panel member for the European Research Council from 2008 to 15, and for the French National Research Agency (ANR) from 2008 to 12. From 2004 to 2016, he chaired the IGU commission "Cultural approach in geography", and since 2016, the commission "Global understanding". He was the coordinator of the "UNESCO 2016 International Year of Global Understanding" and in 2016, became the IGU Lauréat d'Honneur for his lifetime achievements.

Geography and Environmental Issues 15

Jorge Olcina Cantos

Abstract

Territorial sustainability is one of the major concerns of Geography in recent decades. The concern for environmental issues and their impact on land use has acquired a very important role in the geographic discipline in recent decades. Basically, it is a reaffirmation of one of the epistemological principles of geography: the study of the relationship between the physical environment and the human being (Murphy, Geography: Why it matters? Polity Press, Cambridge, 2018). The current ecological footprint exceeds the natural regeneration capacity of natural resources. This is combined with the problem of global warming, a consequence of the massive use of fossil fuels as the main source of energy supply throughout the world. Geography has opened up new analysis fields and has promoted critical visions in the face of environmental problems and risks. The use of new tools (automated mapping, remote sensing, big data) has allowed improving the results, betting on the modeling and creation of future scenarios necessary for the allocation of new land uses. Geography must keep on studying the natural environment from the precision of its analysis and its proposals. The twenty-first century is the century of the environment. And geography is the ideal scientific discipline to meet this challenge. This is not about wishful thinking, a dream, it must be a reality.

Keyword

Geography · Twenty-first century · Sustainable development goals

Highlighting and further promoting the work of IGU Commissions and Task Forces in relation to key global initiatives, including inter alia, the UN Sustainable Development Goals (SDGs), the UN Decade of Ocean Science for Sustainable Development, Disaster Risk Reduction, Paris Climate Action, and Habitat-III, and ISC Future Earth.
IGU Strategy 2020-2024

15.1 Geography in the Twenty-First Century: A Science for Territorial Sustainability

The discipline of Geography, which required so many centuries to establish its conceptual and methodological bases as a true science, is essential in the twenty-first century. Geography is defined as the "where" within the basic

J. Olcina Cantos (✉)
Department of Regional and Physical Geography, University of Alicante, Campus de San Vicente del Raspeig s/n., 03690 Alicante, Spain
e-mail: jorge.olcina@ua.es

V. Kolosov et al. (eds.), *A Geographical Century*,
https://doi.org/10.1007/978-3-031-05419-8_15

reference system, of the fundamental questions that structure the knowledge of human beings, that give meaning to their stay on the earth's surface and to their social relations (Murphy 2018). Civilizations always develop in a territory. This has been the case throughout human history. The territory is the one that ultimately determines, sometimes conditioning, the ways of thinking and acting by human beings who participate in a civilized action.

Today, geography is the science of global change, environmental sustainability, and territorial information. There are three areas of work that include the main branches of geography specialization (physical geography, human geography, regional geography, geographic information technologies) and are closely related to each other (Fig. 15.1).

The study and analysis of geographic space have become increasingly specialized since the middle of the last century. Paradigms, concepts, methods, and tools have been incorporated that have diversified and intensified geographic research. The environmental approach, which places the natural environment at the heart of the analysis, has been taking prominence until it became a main subject of study for geographic science. The 1980s was decisive in this advance with the concern for the deterioration of the natural environment. The consequences of the oil crises of the 1970s, the Bhopal chemical accident (1983), the Chernobyl nuclear disaster (1986), the "discovery" of the ozone hole, and the introduction of the hypothesis of climate change due to the greenhouse effect, highlighted the mismatches of economic growth throughout the world and the need to rethink the importance of the environment (Sloterdijk 2018).

During these years, different schools of thought regarding the relationship between human beings and nature (liberal, cultural, eco-socialist) emerged, which were to be strongly presented at the Rio de Janeiro Summit in 1992. This United Nations world summit on the environment ended by supporting the West's vision of nature protection, which had in Gro Harlem Brundtland's essay "Our Common Future" the conceptual principles of sustainable development. This report belongs to the International Commission on Environment and Development of the United Nations.

Taking into account what the principle of sustainable development action has meant for the global state of the environment, there are some highlights and challenges. In many countries, regulations and plans have been activated that consider the natural environment as a fundamental element in land-use planning processes.

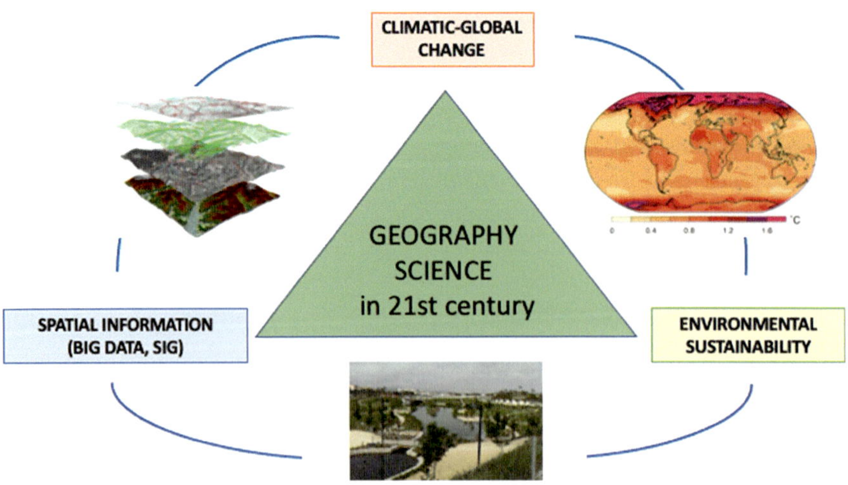

Fig. 15.1 Main fields of action of geography in twenty-first century. Own elaboration. Images reproduced with permission

These laws contain mechanisms for environmental evaluation, risk assessment, and citizen participation. Social awareness in favor of environmental protection has been generated, in which the work of environmental defense organizations has played an important role. International organizations (UN, UNESCO, World Bank) have made a firm commitment to sustainability as a guiding principle for actions and policies, establishing it as a medium-term objective (17 SDGs). The Davos Forum, on the other hand, has been including environmental issues (climate change, atmospheric extremes, water management) as major "hotspots" of the world economy for years to come. In short, "sustainable development", understood as social progress in accordance with the conditions and limits imposed by the natural environment, has been assumed (socially, economically, and politically) as a final objective in the processes involving the transformation of geographic space.

However, the major environmental issues affecting our planet continue to increase in severity. Global warming has been relentless over the last two decades, related to the increase in greenhouse gas emissions, despite international agreements (Kyoto, Paris). Air pollution in some regions and cities of the world is intolerable and causes more than eight million deaths worldwide according to the WHO. Plastics have become a problem for the world's oceans, acting as an obstacle to marine life and even entering the fish food chain. Deforestation of tropical forests has continued at an unstoppable rate in recent decades, particularly worrisome in rainforests. Commercial agricultural clearing is alarmingly reducing forest area, threatening the dynamics of unique ecosystems fundamentally important to the biogeographic and climatic operation of the planet (Stern 2015). Such examples question the effectiveness of the sustainable development concept as an objective of environmental actions, especially large-scale management of global problems affecting the natural environment.

Geography, as a territorial science, has made a commitment to sustainability as a principle of action in the different thematic areas that compose it, physical, agricultural, industrial, economic, urban, and so on. Contents concerning sustainability are included in geography teaching, at different educational levels; research is developed with sustainability as the main or secondary axis; professional geography, on the other hand, participates in territorial processes that are marked by the fulfillment of sustainability criteria.

15.2 Geography and Environment: The Important Contribution of the IGU

The International Geographical Union (IGU) has dedicated outstanding attention to environmental issues since its foundation in 1922, as shown by the topics discussed in different International Congresses (IGC), especially since the second half of the last century (Kish 1992). Indeed, by the middle of the twentieth-century basic questions of the physical environment (geomorphology, climatology, biogeography, hydrography) were discussed (Martin 2016, 2017). It was from 1950, after the Washington IGC (1952), that commissions and task forces directly related to environmental issues were created and specifically tasked in IGU International Congresses or in publications to develop these issues (Schelhaas 2021).

Starting in 1950, commissions were developed on Erosion Surfaces Around the Atlantic, Applied Geomorphology, Evolution of Slopes, and Study of Human Impact on Mountain Ecosystems. In 1968, the Commission on Man and Environment used, for the first time, the concept of environment. And in 1976, in the framework of the celebration of the Congress of the IGU in Moscow, a group on Environmental Problems was created and an Environmental Atlas was promoted in collaboration with the ICA (International Cartographic Association). Subsequently, in 1984, a Study Group was created on Climate Change, while a commission on Coastal Environment was created which followed the tradition of other working groups that already existed in the IGU like Coastal Geomorphology and Coastal Development.

For its part, the IGU-Climatology Commission which had been active since the 1950s began to worry about the problem of climate change in the late 1980s (Yoshino et al 2013). The terms "climate change" or "global change" were incorporated into the denomination of some thematic groups created since 1990, including Critical Environmental Zones and Global Change (1988–92), Gender and Geography, and Regional Hydrological Response to Global Warming (1990–1992), Climate Changes and Periglacial Environments, and Geomorphological Response to Environmental Change (both in 1996) and have also been added to the name of some ongoing working commissions not directly related to climatologies, such as Commission on Geography of Tourism, Leisure, and Global Change or Commission on Global Change and Human Mobility. In recent years the Commission on Climatology has proposed the use of a very interesting concept "anthroposphere" for the study of alterations caused by humans on the Earth's gaseous cover (Commission on Climatology report 2019–20).

1992 proved decisive for the development of environmental issues in geography. This was the year of the Rio de Janeiro Earth Summit and the Washington DC IGC, which dedicated several sessions to environmental problems under the title Our Shared Global Future (Liverman 1999). This year the Commissions on Natural Hazards Studies, Mountain Geoecology and Sustainable Development, History Monitoring of Environmental Change, and Environmental and Development were created. Likewise, the study groups on Environmental Changes and Conservation in Karst Areas and Critical Environment Situations and Regions were instituted. A year later, a study group on Sustainability of Rural Systems was created (1993–98), which expanded its theme in 1996 as Sustainability of Rural Systems, Population, and Environment. An environmental issue of global importance, which included land degradation and desertification, deserved the creation of the latter commission in 1996. Step by step, "sustainability" was consolidated in the IGU and across the geographic community in general, both within the committees and in the IGCs organized from 1992 to the present.

In the task of developing environmental issues in geography, there has been an outstanding range of work by national committees of the IGU in developed countries and, especially, less developed countries in Asia, Africa, and South America. The latter countries observed and were exposed to very radical changes which occurred in their natural environment, often caused by decisions from outside their countries. It is worth highlighting the work carried out by some members of the IGU who have been very active in their Commissions or in the Executive Council. In this regard, the role played by Professor R. G. Singh of the University of Delhi was significant, a great defender of the geographical treatment of the environment from his positions of responsibility in the UGI (Biogeography Commission, Executive Committee) and in other organizations of research (ISC) and promoter of the book series Advances in Geographical and Environmental Sciences (Singh 2016). The recent IGU Presidents (Professors Abler, Kolosov, Himiyama, and Meadows) have given a definitive boost to environmental issues, making it the main objective of the UGI in its relations with society and other international organizations (Bellezza 2000).

The teaching of environmental issues in geography also obtained a scientific journal *International Research in Geography and Environmental Education* (IRGEE), edited by the Geographical Education Commission and indexed in the international impact indexes (SJR and ESCI index). For its part, the series of monographs published by Springer which had been developed by various IGU Commissions, contains numerous titles on environmental issues, effects of climate change, or application from geography of the United Nations Sustainable Development Goals (Kumar 2020).

Since the mid-twentieth-century, the IGU has developed relationships with scientific organizations and associations directly related to issues of environmental sustainability (International Association of Cartography) and, specifically, climate change (International Science Council or World Meteorological Organization and International Association on Meteorology) among others. Some of its prominent members have participated in international activities and events on environmental issues organized by the United Nations, UNESCO (e.g., International Year of the Earth 2008), and IPCC (e.g., participating in climate change reports) (IPCC 2021).

15.3 Geography's Focus on Environmental and Territorial Sustainability: Climate Change as a Key Issue

Over the last three decades, which have been critical for the planet in environmental matters, geography has been conceptually and methodologically incorporating and developing new ideas that have emerged in geographic science itself or in other disciplines of knowledge. Moreover, a series of world events linked to the environment have served as an ideal context for the expansion of environmental concern in geography (Fig. 15.2).

In the last hundred years, the natural environment has gone from being the framework for human action to becoming the most important problem to be solved in order to guarantee the survival of human life on Earth. This is precisely due to the intensive use of natural resources and transformation of natural landscapes, implemented in the developed world and exported with intensity in its results to less advanced countries (Patel and Moore 2018). Fortunately, in recent decades there have been demonstrations of defense and sustainable management of the environment with the participation of geography as a discipline that defends, from its origins, the harmonious relationship between the environment and human life.

It is possible to highlight five steps since the 1970s in which geography has actively shaped academic consideration of the environment, incorporating new concepts and methods of analysis (Fig. 15.2). First, through the diffusion of the concept "environment" after the first summit of the United Nations on the Environment in Stockholm in 1972; second, through the development of risk analysis after the publication of Ulrich Beck's work in 1985 (Beck 1999) which recovered the geographical tradition of work on natural hazards which emerged in the 1940s (Olcina 2008); third, through the publication of the first IPCC Report (1990) that established a "before and after" in the prioritization of environmental issues in the world and created the

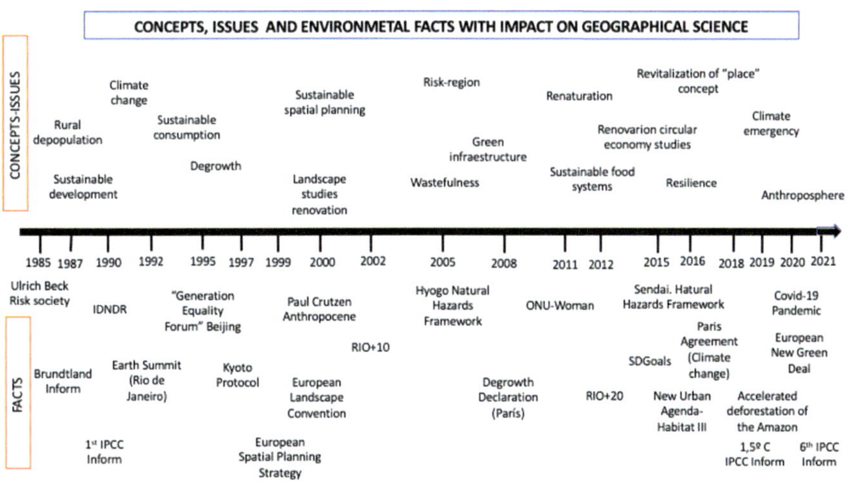

Fig. 15.2 Concepts, issues, and environmental facts with impact on geography in the last decades. Own elaboration. Without permission for its reproduction. In the elaboration of this diagram-synthesis, it has been important to consult the following books: Cook et al. (2015), Crutzen and Stoermer (2000), Crutzen (2002), Harvey (2009), Jameson (2015), Mannion (2014), Svatek Ziegler (2019). These are refered in the bibliographic list

idea of current "climate emergency"; fourth, through the diffusion of the concept "sustainable development" which first appeared in 1987 and was popularized after the Rio de Janeiro Environment Summit (1992); and fifth, through the assessment of the global effects of environmental transformation and climate change and the appearance or recovery of concepts in relation to the development of socioeconomic and territorial sustainability practices (renaturation, global-local, circular economy, resilience) (Murray 2005).

Climate change and its associated effects (atmospheric hazard) (IPCC 2019, 2021; Cramer et al. 2019) will occupy, from different geographic approaches, a great part of research and geographic practice in the next decades. As noted, several concepts that have emerged in recent decades that geography has incorporated into its research corpus—resilience, decrease, renaturalization—are closely related to the objective that geography should pursue in environmental and territorial matters (sustainable territorial development) and with the guiding principle of policies on a global scale (mitigation and adaptation to climate change) (Malm 2018; Kahn et al. 2019; Hayes 2020). In this way, geography is the most suitable scientific discipline for understanding the territorial diversity of climate change and for the development of proposals to reduce its effects. Geography, all over the world, is developing research that rigorously acts in the fight against climate change, with a vision of integration and collaboration with other disciplines.

One of the areas of study and professional practice where geography can demonstrate its commitment to sustainability and adaptation to climate change is territorial planning. Thermal warming and the expected effects on temperatures and precipitation in different regions of the world will condition territorial planning. The loss of thermal comfort expected in the coming decades may be mitigated in the design of buildings (bioclimatic architecture) or green areas in cities, reducing its impact. On the other hand, the increase in intense rainfall and flooding processes, already evident in several regions of the world, must be mitigated with territorial planning proposals based on the use of risk mapping and the accurate delimitation of areas where high-risk intensive land use should not be considered. Additionally, the effects that global warming is having on coastal areas due to rising seawater levels will determine territorial actions in coastal areas in the coming decades (Feyen et al. 2020).

Spatial planning has been improving its procedures and working methods in recent decades, with the appearance of standards that have improved the consideration of the natural and cultural elements of the environment as important factors in the planning of new uses. In many western countries, there has been a transition from economic conceptions of spatial planning, which considered land as a space of possibilities and of direct assignment of new uses, to positions that mainly value natural resources and the historical-artistic heritage of territories which, through delimitation and protection, can be designed with new uses that are compatible with those.

The trust in supposed capacities of resistance and control of nature by means of structural actions (dams, canalizations, etc.), motivated, on the one hand, the promotion of irrigated agriculture in climatically unsuitable territories and, on the other hand, the forced integration of the final sections of river courses and the encroachment of their flood beds in urban areas, with the subsequently added problems that this caused (Pérez et al. 2018) like increased vulnerability and exposure to extreme natural events, especially floods. This has been especially intense and marked in coastal areas across the world, where a large part of global economic activity and main urban settlements are concentrated.

Two new elements have been added to the spatial planning of European countries in recent decades. The landscape, on the one hand, has become an operational instrument for establishing new uses in the territory. The principles contained in the European Landscape Convention (2000) have been integrated into the urban and land use regulations of the countries and regions in the European Union. The landscape units have become a primary object of work in environmental studies that are incorporated into spatial planning processes (Jones and Stenseke 2011). On the other hand, territorial green infrastructure is the basic and initial instrument of any planning process. It is a conception that arose in North American landscape architecture in the first decades of the twentieth-century (Law Olmstead), which brought together environmentalist ideas from the eighteenth and nineteenth centuries (Benedict and McMahon 2002), and which had been incorporated into spatial planning at mainly regional and local scales over the last three decades (Lafortezza et al. 2013). As a conceptual and applied connection between both elements, green infrastructure is defined as an interconnected network of landscapes of significant environmental, cultural and visual value. Consequently, green infrastructure integrates, as a planning object, the set of landscapes defined in a territory, but also designs the connection between them based on existing or proposed natural or artificial connectors (Comisión Europea 2014).

The spread of the green infrastructure concept has been considerably rapid internationally and particularly in Europe where, in addition to a long history of environmental policies, there is significant concern about recent transformations in land use. Studies conducted by the European Environment Agency reveal that European territory is suffering a progressive process of biodiversity loss and soil artificialization which, in addition to urbanized surface area growing at a higher rate than the urban population itself (EEA 2011), is a clear reflection of an increasingly dispersed urbanization model with direct impact on landscape fragmentation. In this context, the EU is undertaking different projects aimed at monitoring and protecting Europe's biodiversity and landscapes as well as managing green infrastructure. But the incorporation of the green infrastructure tool into spatial planning is a process not exclusive to European territory. There are excellent examples in the United States, Japan, Australia, New Zealand, Singapore, South Korea, among others (Breuste et al. 2015).

Green infrastructure is used in two complementary ways: as a tool for sustainable spatial planning, as a set of urban infrastructures designed under sustainability criteria for rainwater drainage; and as a cartographic tool for sustainable spatial planning (Elorrieta and Olcina 2021). Several cities around the world have designed sustainable stormwater circulation systems (SDUs) through the construction of large-capacity storm sewers, installation of storm tanks, or design of flood parks in order to reduce the risk of urban flooding.

In turn, elements for risk and climate change management are beginning to be incorporated into spatial planning processes based on the use of green infrastructure. The aspects of green infrastructure which should be specifically incorporated into urban and spatial planning in order to reduce the effects of global warming are the following: (a) rising temperatures and loss of thermal comfort, whose effects can be mitigated through urban design measures such as an increase in public parks and green spaces in housing (terraces and green facades): (b) rising sea levels in coastal areas, whose effects should be reduced with structural actions in some cases, and with spatial planning (regulation of uses along the coastline, removal of the first coastlines). And (c) changes in rainfall, with an increase in its intensity and irregularity, which makes it necessary to design large-capacity water drainage areas, as well as water storage reservoirs of greater capacity than existing ones to guarantee the supply for urban demands.

All of this should be based on the development of precise models of climatic or environmental (sea level) behavior that allows for the periodic updating of projections for their application in spatial planning (Fig. 15.3).

A fundamental element for incorporating climate change and associated risks into spatial planning processes is the map. In fact, green infrastructure as a territorial planning tool must include cartography, using detailed scales to integrate all the elements that compose it. Risk is probably the most important component due to its impact on human life (Global Commission on Adaptation 2019). This is why in many countries the map has become the legal accreditation document of the risk condition of a territory

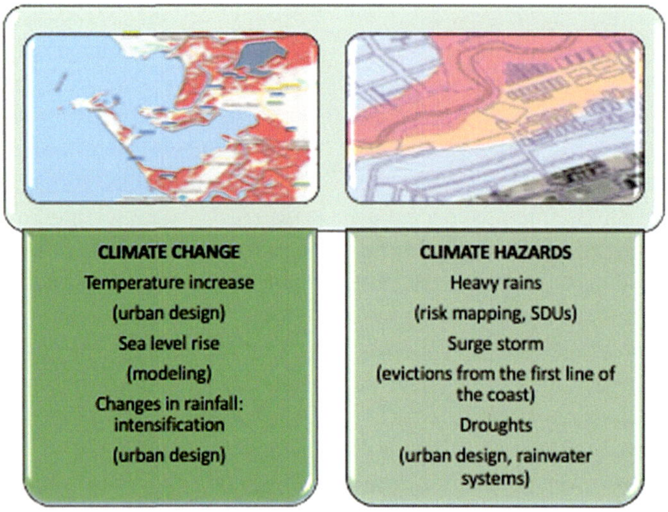

Fig. 15.3 Aspects of climate change and climate risks that can be incorporated into green infrastructure and territorial planning. *Source* Prepared by the author. Images reproduced with permission. Maps taken from: New York 2 °C warning and sea-level rise. Available at: https://ss6m.climatecentral.org/#11/40.6643/-73.9385; and Territorial Action Plan against the Risk of Floods in the Valencian Community (PATRICOVA)

subject to planning. In general, the map made to be included in a territorial information system for green infrastructure that must be consulted in spatial planning should be a rigorous document. This is an important challenge for academic and applied geography. In fact, sustainable spatial planning is a good example for geography of the integration of academic content and applied action.

But the task of geography in this matter goes further. Along with the map, spatial planning carried out from geography incorporates concepts and working methods characteristic of its scientific corpus that is especially suitable in this task, such as changes of scale, fieldwork, personal interviews, and integration of physical and human aspects of the territory. Finally, it should be noted that territorial planning is not the exclusive competence of geography. Other disciplines participate in these processes (engineering, architecture, geology, environment, sociology, law, economics). Geography has established collaborative ties with all of them and shares concepts and methods in the development of spatial planning projects. This work has been carried out since the 1940s in countries such as France, the United Kingdom, Germany, and Russia. In others, it has been a more recent collaborative process (1970–80s). Since the end of the twentieth-century, the incorporation of computer tools for the study of environmental, social, and territorial processes has made it possible to establish collaborative relationships with the mathematical and computer sciences. Process behavior models were, and continue to be, developed, and in the specific context of climate change, climate projections are incorporated into land-use planning processes, which make it possible to adjust land-use decisions at different scales. Nevertheless, the capacity to integrate facts, observe on the ground, elaborate realistic proposals, and make geography the most suitable discipline for the integration of principles of sustainability and adaptation to climate change in spatial planning.

15.4 Geography and Sustainable Development Goals: An Ethical Responsibility

Geography has an ethical commitment in its research and in its action toward the fulfillment of the Sustainable Development Goals (SDGs), approved in 2015 by the United Nations. It is also a science that is particularly well adapted to research and applied proposals that favor SDGs. The SDGs address issues related to the physical environment, living beings, economic activities, scientific research, culture, and education. These factors are closely related to the objectives and goals of geography in the twenty-first century: the integral development of the territory under the principles of environmental sustainability and adaptation to climate change.

Different fields of geographic research (physical geography, human and regional geography, mapping and data processing) address the goals and targets included in each of the 17 SDGs. Sustainable development cannot be understood without the territory; and the territory includes the physical environment and human action, the two working objects of geography. Four Sustainable Development Goals (Goals 6, 11, 13, and 15) are directly related to the concepts and methods developed by environmental geography in the last three decades (Fig. 15.4).

In addition to research, geography is used as part of many countries' social communication of the SDGs, particularly in education. At the same time, geographic entities in each country have adopted the sustainable development objectives as principles of action for geographic work in the current decade. The IGU, as a member of the International Science Council (ISC), strongly supports SDGs. The 2020–2024 strategy specifically states that the IGU will further promote the work of the commissions and working groups in relation to key global initiatives including, among others, the UN SDGs, the UN Decade of Ocean Sciences for Sustainable Development, Disaster Risk Reduction, the Paris Agreement, and the Habitat-III conference. It has launched the Future Earth Geography initiative whose mission is to promote geography for people to thrive in a sustainable and fair world.

This commission seeks to build and link knowledge to increase the impact of research, explore new directions for development, and contribute to the achievement of global sustainable development goals by seeking new forms of sustainable development. This commission is initially focused on human-land systems and their interfaces with the coastal ocean and atmosphere, promoting broader analysis and innovative thinking about the Earth's global

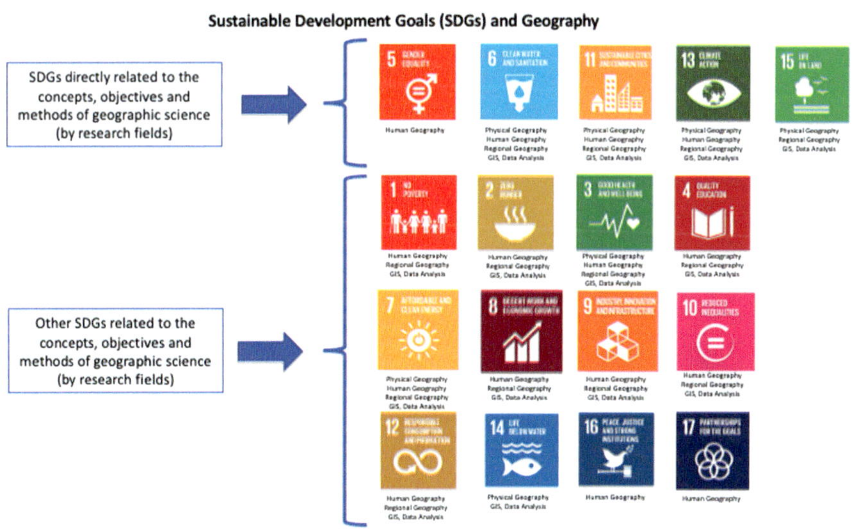

Fig. 15.4 Sustainable Development Goals (SDGs) and Geography. Own elaboration. Source of images (https://sdgs.un.org/es/goals); reproduced with permission

sustainability through the bridge and synthesis of physical geography, human geography, ecology, hydrology, atmosphere, climate, and social sciences. To this end, six thematic areas have been identified for development in the coming years:

(1) Conceptual foundations for future Earth Geography, and especially new theories and hypotheses on coupled human-land systems for sustainability.
(2) Open and inclusive platforms for geospatial Big Data and observations of coupled human-earth systems.
(3) Integrated earth system models to deepen our understanding of complex earth systems and human dynamics at different scales.
(4) Relationships and dynamic analysis between ecological processes, services, and human well-being.
(5) Human contributions and responses to global climate/environmental changes and sustainability.
(6) Linkages and dynamic analysis between ecological processes, services, and human well-being.

In any case, if the governments of each country do not promote, as they should, the goals and purposes of the UN's SDGs, the science of geography must set itself up as the "guardian" discipline for their fulfillment, promoting actions or, if necessary, protest at lack of institutional commitment.

It is certain that not all of the goals and targets of the Sustainable Development Goals will have been met by 2030. These are challenging and ambitious goals, which require cultural changes in many societies around the world and would require economic budgets and medium-term planning to develop them. Regardless of the future evolution of the international commitment to sustainable development, institutional geography (IGU, entities, and associations of each country) must promote the maintenance of territorial sustainability objectives throughout the twenty-first century as axes of political action at all scales of work.

15.5 Geography: The Realizable Dream of Territorial Sustainability and an Agenda for the Future

The social role of geography as a science of sustainable territory is set to play an increasingly important role in the following decades. The twenty-first century is the century of the environment and climate change mitigation and adaptation (Romero and Olcina 2021). Geography is the most suitable scientific discipline for the research, planning, and applied management of the elements behind these two major global issues.

As mentioned previously, new concepts have been incorporated from different disciplines across the natural and social sciences and from engineering and adapted to geographical inquiry, including the idea of the circular economy, food justice, and many others.

Geography faces the third decade of the twenty-first century with a great scientific and social potential. It is a solid academic discipline with growing public recognition, without denying the fact that it has an important path to consolidate its social position at a global level. The IGU and national geographic associations and entities have played an important role in showing the suitability of geography as a territorial science that has taken on the principles of sustainability and led the fight against climate change.

The upcoming decades will be the ones of global management of climate change, based on measures adopted by governments across the world to reduce the effects of global warming. Geography must be there, promoting research, educating and communicating the expected consequences of global warming, and proposing solutions. This is the main challenge for the future of geography, for which the discipline must provide a serious and solvent response (Table 15.1).

The IGU centennial in 2022 is an excellent opportunity to further promote the discipline of geography on a planetary scale. It is necessary

Table 15.1 Key themes for geography in the twenty-first century

Physical geography	• Climate change • Deforestation • Soil degradation • Green infrastructure • Ice loss • Natural risks
Human and regional geography	• Decrease • Climate migrations • Aging • Economic impact of climate change • Sustainable cities • Food justice • Technological and health risks
Geographic information and cartography	• Big data analysis • New satellites for environmental purposes • Modeling • Territorial intelligence

for geography to have a more prominent presence in international organizations related to the management of the earth's territories. And it is necessary for a greater presence in the media and social networks. The IGU must improve information on its website and on its social networks. Regarding environmental issues, they must be made available to all those institutional documents or academic works prepared by the IGU that have contributed to the promotion of environmental and sustainable development. The IGU Newsletter's online archive, from the "Home of Geography", is a good example of what can be achieved (Bellezza 2000) and must be supported by all the national committees. For their part, the national committees must improve the open dissemination of all contributions on environmental issues presented at IGU International Con0resses or its thematic meetings. Geography, in the IGU centennial year, must become the scientific discipline that best studies environmental issues because of its global understanding of the facts, due to its working methods, and due to the elaboration of cartography, and the reasonable and rational proposals for improvement it puts forward.

For the communication of advances and proposals in geographic science, simplicity has a greater social effect than overly complicated language. Scientific rigor can be communicated and can be taught with clear ideas and using a comprehensible language for the public. Scientific dissemination of environmental issues should become one more activity in the teaching and applied work of geography. The precise use of social networks can contribute toward a greater social knowledge of geography and its working themes. The indisputable principles of action for geographical research, professional activity, and teaching must be equality, justice, professional ethics, honesty, and peace.

The next years and decades will be marked by the development of actions promoting climate change mitigation and adaptation. This is the great challenge facing humanity this century. Geography is the science particularly appropriate for tackling this issue, proposing solutions and managing policies aimed at fighting global warming and its multiple environmental, social, economic, and cultural effects.

Environmental sustainability and climate change, with all their related aspects, will be the main thematic axes of geography in this century. As the motto of the 1992 IGC in Washington DC noted, "Geography is Discovery", a discovery that has become more complex since the origins of the discipline, incorporating reflection, epistemological approaches, and new processes. In the nineteenth century, after the scientific and academic consolidation of geography, the discovery of territories continued to be, as in ancient times, the "dream" of geography. In the twentieth-century, with its socioeconomic ups and downs, with its geopolitical highlights, and challenges, the "dream" of geography was the discovery of new epistemological approaches to understand the complexity of human processes on the earth's surface (Tuan 1998). In the current century, the dream of geography must be the discovery of solutions to environmental deterioration and climate change.

In 1992, Tim Unwin stated that geography was responsible for studying some of the most important issues for humanity (environmental deterioration, pollution control, environmental conservation, climate change, and resource management), where global knowledge of physical processes and social practices was necessary (Unwin 1992). All these questions continue to occupy a major place among geography's research and in its professional practice three decades later. A few more have since been added. Then in the early nineties, a concept that was going to be a major turning point in the epistemological evolution of geography was gaining importance: sustainability. Territories and places, at any scale, must be "livable" and geography must provide knowledge and practices to achieve this.

In 2022, as we celebrate the IGU centennial, the health of our planet, in general terms, has worsened. Territorial sustainability is less achievable. And the effects of global warming are already being felt. There is no time to lose. From universities, in schools, and in professional practice, geography must be the science that helps societies across the world make our planet livable, with its research, its citizen actions, its scientific complaints of bad practices and always from the strictness of its work. Only then will we be able to maintain the professional credibility of geographic science, the social utility of our discipline, the dream of discovering a world that must necessarily be better.

References

Beck U (1999) World risk society. Polity Press, Cambridge

Bellezza G (ed. since 2000) IGU E-newsletter. Available at: http://www.homeofgeography.org/

Benedict MA, McMahon ET (2002) Green infrastructure: smart conservation for the 21st century. Renew Resour J 20:12–17

Breuste J et al (2015) Introduction (special issue on Green Infrastructure for Urban Sustainability). J Urban Plann Dev 141(3)

Comisión Europea (2014) Construir una infraestructura verde para Europa. Unión Europea, Bruselas, Bélgica

Cook B, Rickards L, Rutherfurd ID (2015) Geographies of the anthropocene. Geogr Res 53(3):1–13

Cramer W et al (2019) Risks associated to climate and environmental changes in the Mediterranean region. MedEEC Network. Available at: https://ufmsecretariat.org/wp-content/uploads/2019/10/MedECC-Booklet_EN_WEB.pdf

Crutzen PJ, Stoermer EF (2000) The 'Anthropocene.' Glob Change Newslett 41:17–18

Crutzen P (2002) Geology of mankind. Nature 415(23) https://doi.org/10.1038/415023a

EEA (European Environment Agency) (2011) Green infrastructure and territorial cohesion. The concept of green infrastructure and its integration into policies using monitoring systems. Unión Europea, Luxemburg

Elorrieta B, Olcina J (2021) Infraestructura verde y Ordenación del Territorio en España. Ciudad Y Territorio Estudios Territoriales (CyTET) 53(207):23–46

Feyen L et al (eds) (2020) Climate change impacts and adaptation in Europe. JRC PESETA IV final report. Publications Office of the European Union, Luxembourg

Global Commission on Adaptation (2019) Adapt now: a global call for leadership on climate resilience. Global Center Adaptation and World Resources Institute, Sept. 2019. Available on: https://cdn.gca.org/assets/2019-09/GlobalCommission_Report_FINAL.pdf

Harvey D (2009) Cosmopolitanism and the geographies of freedom. Columbia University Press, New York

Hayes DJ (2020) Flatten the (climate) curve. State Energy and Environmental Impact Center, New York School of Law, 31 Mar 2020

Intergovernmental Panel on Climate Change (2019) IPCC special report on the ocean and cryosphere in a changing climate (SROCC), IPCC, 2019: summary for policymakers. In: Pörtner HO et al (eds) IPCC special report on the ocean and cryosphere in a changing climate. Geneva

IPCC (2021) Climate Change 2021: the physical science basis: contribution of working group I to the sixth assessment report of the intergovernmental panel on climate change. In: Masson-Delmotte V et al (eds) Cambridge University Press, Cambridge (In Press)

Jameson F (2015) The aesthetics of singularity. New left review. https://newleftreview.org/issues/II92/articles/fredric-jameson-the-aesthetics-of-singularity
Jones M, Stenseke M (eds) (2011) The European landscape convention: challenges of participation. Springer, Dordrecht, NL
Kahn M et al (2019) Long-term macroeconomic effects of climate change: a cross-country analysis. IMF Working Paper, WP, 19/215
Kish G (1992) International geographical union: a brief history. GeoJournal 26(2):224–228 http://www.jstor.org/stable/41145364
Kumar P (2020) Publication. IGU-Springer. https://igu-online.org/publication-invitation-for-publication-through-book-series/ Accessed: August 2021
Lafortezza R et al (2013) Green infrastructure as a tool to support spatial planning in European urban regions. IForest: Biogeosciences Forestry 6:102–108
Liverman D (1999) Geography and the global environment. Ann Assoc Am Geogr 89(1):107–120
Malm A (2018) The progress of this storm: nature and society in a warming world. Verso, London
Mannion A (2014) Global environmental change: a natural and cultural environmental history. Routledge, London
Martin Martín T (2016–2017) Un pasado que reivindicar: España en los Congresos Internacionales de Geografía, I y II. Boletín de la Real Sociedad Geográfica, CLI y CLII, 149–180 y 283–346
Murphy AB (2018) Geography: why it matters? Polity Press, Cambridge
Murray WE (2005) Geographies of globalization. Routledge contemporary human geography series. Routledge, London
Olcina Cantos J (2008) Cambios en la consideración territorial, conceptual y de método de los riesgos naturales. Scripta Nova. Revista Electrónica de Geografía y Ciencias Sociales. Universidad de Barcelona, Barcelona, vol XII, núm 270 (24). http://www.ub.es/geocrit/sn/sn-270/sn-270-24.htm
Patel R, Moore JW (2018) A history of the world in seven cheap things. A guide to capitalism, nature, and the future of the planet. Verso, London
Pérez Morales A, Gil-Guirado S, Olcina J (2018) Housing bubbles and the increase of flood exposure. Failures in flood risk management on the Spanish south-eastern coast (1975–2013). J Flood Risk Manag 11:302–313
Romero J, Olcina J (eds) (2021) Cambio climático en el Mediterráneo. Procesos, riesgos y políticas. Ed. Tirant Lo Blanc, Valencia
Schelhaas B (2021) Archive for geography, finding aid (preliminary). International geographical union. Available on: https://igu-online.org/igu-archives/ Retrieved: August 2021
Singh RB (ed) (2016–) Advances in geographical and environmental sciences, IGU-springer series. All titles are available on: https://www.springer.com/series/13113. Accessed: Aug 2021
Sloterdijk P (2018) ¿Qué sucedió en el siglo XX? Ed. Siruela, Madrid
Stern N (2015) Why are we waiting? The logic, urgency and promise of tackling climate change. MIT Press, Cambridge, Mass
Svatek Ziegler S (2019) The anthropocene in geography. Geogr Rev 109:271–280. https://doi.org/10.1111/gere.12343
Tuan YF (1998) Escapism. Johns Hopkins University Press, Baltimore
Unwin T (1992) The place of geography. Routledge, London
Yoshino M et al (2013) Climates and societies—a climatological perspective: a contribution on global change and related problems prepared by the commission on climatology of the international geographical union 33: GeoJournal Library. Springer

Jorge Olcina Cantos (Alicante, Spain 1966) is full professor of Regional Geographical Analysis at Alicante University, where he teaches Spatial Planning, Climatology and Natural Hazards. He was visiting professor at universities in Spain, Europe and Latin America. Member of the Editorial Board of various scientific journals on geographical and environmental issues. Spanish key speaker at the International Year of Planet Earth (UNESCO, 2008). President of Spanish Geographical Association (2017–2021).

Geography and the Information Society

16

Michael F. Goodchild

Abstract

Geographic information in the form of maps and text and increasingly of digital data has always played a fundamental role in the discipline of geography. The chapter provides a brief outline of the history of GIS, including the role played by a commission of the IGU. Significant events in its development are discussed, including the social critique that began in the late 1980s and the Internet that emerged in the early 1990s. Spatial data infrastructure and Digital Earth are comparatively recent reformulations of the vision of GIS. The chapter ends with a new and comprehensive vision of geospatial infrastructure and with a suggested new role for the IGU.

Keywords

Geographic information system (GIS) · Geographic information · Social critique · Geospatial infrastructure

16.1 Introduction

Information has always played a central role in the discipline of geography. Early humans relied on spoken narratives to share information about food sources, hazards, and enemies and used song and gesture to add effectiveness (Chatwin 1988). The idea of a map, a simple and approximate rendering of nearby geographic features, would have emerged very early in human society, perhaps as lines scratched on dirt surfaces or preserved on rock walls. By the Age of Discovery which began in Europe in the late fifteenth century, the science of map-making had advanced to the point where maps could be *planimetric*, scaling the surface of the Earth to a model globe or a sheet of paper, such that distances on the map or globe were approximately proportional to distances on the Earth. Such maps had become a very efficient and valuable means of compiling, storing, and sharing geographic information, which we can define simply as information about *what* is *where* (and sometimes *when*). Mercator (Crane 2003) made some of the first globes and invented a way of creating a flat representation of the curved surface of the Earth such that a ship sailing on a constant bearing would follow a straight line (a *rhumb* line or *loxodrome*) on the map. Gutenberg's invention of printing ensured that maps could be reproduced and distributed in quantity. Books also became important repositories of geographic

M. F. Goodchild (✉)
University of California, Santa Barbara, USA
e-mail: good@geog.ucsb.edu

V. Kolosov et al. (eds.), *A Geographical Century*,
https://doi.org/10.1007/978-3-031-05419-8_16

information, in the form of narrative descriptions of places: descriptions that could be reproduced, stored, bought, and sold with little concern for international boundaries.

This chapter traces the relationship between geographic information and the discipline of geography, from those early beginnings to today, and speculates on where all of this is leading us. First, I describe geographic information as we think of it today, following what we might call the digital revolution and the advent of the information society. Second, I describe its role in everyday human activities, and how the technical world of geographic information technologies is attempting to engage with the human world of verbal description. Third, I discuss various ethical and societal issues that arise as a result of the massive investments that have been made in geographic information and associated technologies in recent decades. Finally, I discuss how visions of the role of geographic information have changed over the past half-century, and speculate on what the future might hold.

References have been included to provide further explanation where appropriate. One further point should also be made at the outset: while I have attempted as far as possible to explore the international dimensions of the topic, this perspective inevitably reflects my experience as an academic geographer based in Canada and the US.

16.2 Geographic Information Today

16.2.1 The Roots of GIS

By the mid-1960s, computers were becoming ubiquitous in universities and large organizations, and ideas for using them to process geographic information began to take shape. Perhaps the best-known of these was the Canada Geographic Information System, a component of a federal-provincial project known as the Canada Land Inventory. Roger Tomlinson, a British–Canadian geographer, conceived of using computers to measure the quantities of land that could be available for specific uses, and led a contract that the Government of Canada signed with IBM. Measurement of area was the only purpose and statistics were the only product; at the time there was no available means to produce output in map form (Tomlinson and Toomey 1999).

Tomlinson began to promote the idea of a geographic information system (GIS), an integrated computer application that would acquire, store, and process many types of geographic information for a range of purposes. Those purposes became clearer as Tomlinson built a worldwide network of researchers with interests in computer applications of geographic information, and convened international conferences in 1970 and 1972 (Tomlinson 1971, 1972) under the auspices of the IGU's Commission on Geographical Data Sensing and Processing. Digital maps could be created and edited during the map compilation process, just as word processors are used to compose text. Images from satellite-based or aerial remote sensing could be digitized, and computers could be used to classify the raw images and to search for specific features. Digital maps could be used for planning, by combining layers representing variables such as groundwater, surficial hydrology and geology, transportation networks, human settlements, and soil characteristics, following the ideas then being advanced by the landscape architect McHarg (1969). Geographic information about transportation could be converted to digital form and processed in the development of transportation plans. All of these projects and more began to fuel the development of GIS (for histories of the early development of GIS see Coppock and Rhind 1991; Foresman 1998), and the IGU played a significant role as a host for international discussions and exchange (see, for example, Mounsey 1988). By the end of the 1970s, several companies were marketing GIS software, forming the beginning of what is now a global GIS industry with annual sales of software, data, and expertise in the hundreds of billions of US dollars.

16.2.2 New Types of Data

The geographic information that drove the development of CGIS was of a single type which today we would describe as an *area-class* map: lines on a map that partitioned the area into irregularly shaped zones of specific classes. Area-class maps include themes of soil, forest types, current land use, and land cover, and constitute a very significant percentage of all maps. But there are many other types of geographic information, and many of the developments of the 1970s and 1980s were directed at advancing GIS to accommodate them, thus opening an array of new applications and allowing geographers to ask new kinds of questions.

Many other developments followed. While the map remains the dominant metaphor for the contents of a GIS, there are many types of geographic information—information about what is where (and perhaps when)—that are not easily expressed in map form using standard map-making tools (Goodchild 1988). They include the third spatial dimension, since maps are of necessity two-dimensional; dynamics since maps are inherently static; and gradients, which are difficult to portray with a pen. All of these were addressed in the 1990s and early 2000s, driven in part by some well-publicized critiques of GIS in the early 1990s, and enabling a range of new applications that used types of data that had never been mapped in the past. Today, GIS has advanced to the point where it can accommodate virtually any type of geographic information, but there remain important exceptions. There may be subjective feelings about places that are better expressed in other media, such as text or song, and there may even be esthetic aspects of maps that are hard to capture and reproduce in the algorithmic environment of a GIS.

16.2.3 Sources of Geographic Information

One of the factors driving interest in geographic information in the past three decades has been the increasing ease with which it can now be produced. The advent of the Global Positioning System (GPS) in the late 1980s had dramatic influence, especially after 2010 with the inclusion of GPS chips in smartphones and vehicles. With advances in software, it became possible for the small organization and even the individual to acquire the means of map-making, and the newly available Internet was clearly capable of supporting the widespread sharing of digital information of all kinds. The vision of a spatial data infrastructure emerged in response to these changes, first in the US (NRC 1993) and subsequently in many other parts of the world. Its central argument was that the federal government would adopt a new role, as a setter of standards and coordinator of activities, rather than as a dominant producer. It would sponsor a national clearinghouse of digital geographic data (Goodchild et al. 2007), a national standard for metadata (methods for describing the essential features of data sets), and national standards for the quality of geographic information.

With GPS, drones with aerial cameras that could capture fine detail, easy access to existing digital data via the Internet, and inexpensive mapping software, it was clear that the citizen could become both a consumer and producer of geographic information. Projects such as OpenStreetMap (openstreetmap.org) recruited volunteers to create digital maps, by traveling around their own neighborhoods, and by interpreting fine-resolution imagery of other parts of the world. It soon became apparent that any distinction between the trained professional and the amateur was disappearing, as citizens began to acquire skills that had previously been limited to experts. The term *neogeography* (Turner 2006) provided a suitable way of describing this development, and the term *volunteered geographic information* (Goodchild 2007; Sui et al. 2012) was coined to describe geography's own subset of crowdsourcing and citizen science.

Other ways of determining what is where began to emerge following further technical developments. Vast amounts of geographic information were being generated by the GPS receivers being carried by individuals and vehicles. Largely unbeknownst to their users, their locations were generating "pings" at frequent intervals that were being captured by vendors, aggregated, and sold on to a rapidly growing industry where they would be used to build models of the owner's spatial behavior. In principle, the user has the ability to turn off these functions, but few are aware of their implications and sufficiently skilled to do so. While we are assured that the pings have been anonymized by stripping them of any reference to an individual, it is nevertheless easy to string together the pings generated by an individual device and to make accurate inferences about the identity of the owner, his or her home and work locations, the identity of his or her doctor, and many other attributes that people would not normally be willing to share (Valentino-DeVries et al. 2018).

GPS is a powerful way of determining location, but many other tools now exist with a similar purpose. RFID (radio frequency identification) is the technology underlying many smartcards and can be used to determine when and where these cards are used: when taking money out of a bank machine, or when boarding public transit, for example. QR codes are increasingly popular and have a similar effect. Bluetooth also allows the determination of position and can be used to determine the exact location of a shopper's phone inside a store. Wifi is another emerging technology for determining position indoors (Chen and Chen 2021).

With GPS, it has also become easy to tag many kinds of records with the geographic location of the user. Social media services such as Facebook and Twitter allow the user to do this, and geographers have been quick to explore the implications of knowing where and when a particular message was posted. It is possible now to trace the spread of a disease as people post messages or to provide early detection of disasters (Issa et al. 2017). Other projects have shown how it is possible to attach additional sensors to smartphones or to Internet-connected vehicles; for example, suitable sensors can be used to create detailed maps of air pollution or urban heat islands as they follow their users around a city (Schneider et al. 2017).

Finally, a large group of technologies form what is known as the Internet of Things (IoT), and contribute to the concept of the "smart city." Many new vehicles are now fitted with sensing devices and access to the Internet, allowing others such as truck fleet managers or insurance agencies to know much about the driver's locations and driving habits. Home security devices may be connected to the Internet, allowing homeowners to monitor the home while away. Sensor networks are widely deployed in major cities, to monitor and control traffic, urban air pollution, or noise. Closed-circuit television (CCTV) cameras image cities and highways and provide information that is widely used in fighting crime. By combining CCTV with facial recognition, cities are now able to identify and track a specific individual and to signal alerts when any unusual activity occurs. At time of writing, there were an estimated 630,000 CCTV cameras in London and 1.15 million in Beijing. Perhaps most frightening is the imbalance between the credibility of CCTV video on the one hand, and the ease with which it can now be fabricated on the other. In short, it is now possible to imagine a world in which we know the location of everything, at all times, and in which any form of personal privacy has become impossible. Alarms have been sounded for many years about the implications of all of this, from the Panopticon (Schofield 2009) to 1984 (Orwell 1949). I discuss these implications in detail below.

16.3 Consumerization

As geographic information and its uses have become more ubiquitous it has been necessary to pay more attention to how humans learn and think about their surroundings. One effect has been a growing importance for concepts of *place* alongside the expert's concept of *space*. Another has been a remarkable shift in mapping practice, as geographers and others have tried to escape the limitations of past practice by acquiring geographic information about an entirely new range of phenomena. These two topics form the subsections of this discussion of consumerization.

16.3.1 Space and Place

In cartography, surveying, and geodesy the question of "where" is addressed with measurements, to an accuracy determined by the measuring instrument. Latitude and longitude are perhaps the most universal measurements, using the Equator and the Prime Meridian, and expressing distances from these reference lines as angles; but there are many other coordinate systems in use. These coordinates are the basis for representing all data types in GIS, and in geographic information technologies in general.

Yet all of this has little to no direct relevance to the day-to-day life of humans. Almost no one can recall the latitude and longitude of his or her home when asked, though they almost certainly use mapping and wayfinding technologies every day. Instead, humans learn about and discuss the world through a hierarchy of named places: room, house, street, neighborhood, city, county, state, country, or continent. When humans interact with geographic information technologies, they do so in the language of place names and rely on the technology to translate these into coordinates. Thus, in using a wayfinding app the user will first specify a destination as a place name or by pointing on a map and perhaps work through a number of steps designed to autocorrect and disambiguate in order to confirm the exact desired destination.

By the 1990s it had become essential to be able to convert a street address to coordinates, a process known as geocoding, in support of census taking, mail delivery, and the compilation of health statistics. GIS applications for vehicle routing and scheduling also became popular in the 1990s, leading to the development of "navigable" databases that would know not only where the roads and streets were, but how they were connected, what the speed limits and turn restrictions might be, and all of the other information needed to successfully route oneself from an origin to a destination.

As these tools became more available to the public, and especially following the introduction of the smartphone, more and more reliance was placed on what became known as point-of-interest (POI) databases. These link named places with coordinates, and include all businesses, all housing units, all churches, all named landscape features such as mountains and parks, and anywhere else that a person might specify as a trip origin or destination. Today the POI database for a major city will have an order of magnitude more records than the city's population. In short, the consumerization of GIS has led to a radical rebalancing of space and place, or coordinates and place names. The essential vagueness of human discourse is now encountering the precision of digital technologies, leading in turn to a host of interesting research questions.

16.3.2 What Should Be Mapped?

The traditional answer to this question reflected many realities: the difficulty and cost of obtaining detailed information about potentially remote parts of the geographic world; the limited accuracy of measuring instruments; and the high cost of compiling information in map form. Producers of maps responded by ensuring that these costs could be mitigated by the widest possible range of benefits from applications, for as long as possible. Accordingly, maps should represent only those aspects of the Earth's surface that were essentially static: physical features, settlements, the transportation infrastructure but not the vehicles that use it, and residential populations expressed as densities, but not migrants or journeys to work. Mapping practices largely originated in Europe and North America and were often imposed on other parts of the world with little respect for local cultures, concepts of land ownership and use, or the features of the local landscape that were important to its inhabitants. Mapping became a tool used by the colonial powers to impose their rule (Keay 2000), often by replacing traditional place names with references to explorers, colonial masters, or the map-makers themselves.

Underlying these practices is the belief that mapping is an objective scientific process; that the process is replicable, such that two people asked to produce the same map independently would arrive at the same result (within the bounds of measurement error). This may be an acceptable position with respect to political boundaries and property ownership, and even for many well-defined physical features. But it begins to fall apart in the case of soil maps or maps of land cover; indeed, for most of the maps that were included in the Canada Geographic Information System that was discussed above. In the 1990s a number of researchers launched a broader challenge (Harley 2001; Wood 1992), arguing that maps were social constructions that could be interrogated to reveal aspects of the agenda of their makers. From this perspective, there could be no single, true map, but only a series of maps reflecting the different perspectives of individuals, cultures, or social groups.

16.4 Ethical and Societal Issues

16.4.1 The Critique of GIS

Geographers everywhere were quick to explore the opportunities offered by digital geographic information and its associated technologies. Computers appeared to offer endless opportunities for analyzing data using the rapidly expanding set of methods commonly known as spatial analysis (Berry and Marble 1968); and it seemed that GIS might serve to integrate all of these methods in a computational infrastructure for the discipline. Yet the 1970s and 1980s also saw a swelling movement against this apparently wholesale adoption of a positivist methodology, which appeared to be reducing humans to predictable automata, and by 1990 much of this critique had come to focus on GIS and its roots in geographic information.

Another line of critique challenged the degree to which digital geographic information could capture useful representations of the geographic world. The area-class map which dominated CGIS and the early commercial software products such as Esri's ARC/INFO forced the world into a very simple model in which sharp changes occurred at boundaries, and the area within each boundary was treated as homogeneous. This "Boolean" view of the world was clearly a vast oversimplification, yet widely applied in the practices of many agencies. In effect, the world was being simplified to fit the needs of GIS.

Yet another line of critique concerned the cost of GIS. Although the "cost of entry" was declining rapidly, in the early 1990s GIS was essentially limited to governments, corporations, and universities. This fed a sense that it was being used to strengthen the power of the already powerful, and to further marginalize those individuals and groups who were least able to afford the cost. Moreover, the perspectives of those groups, over such issues as planning proposals, were likely to involve the kinds of information—feelings, attitudes, and subjective judgments—that GIS was least able to represent. All of these critiques were assembled in a ground-breaking book edited by Pickles (1995) and titled *Ground Truth: The Social Implications of Geographic Information Systems*. The critiques led to a string of new research communities—Alt-GIS, GIS/2, Critical GIS, etc.—and to a significant degree of collaboration between the traditional GIS community, with its developers, educators and advocates on the one hand, and leading critical geographers on the other (Nyerges et al. 2011).

16.4.2 GIScience

For Tomlinson and others, the central issue of geographic information technologies was what they termed *spatial data handling*: the challenges in adapting computers, their software, and their processes of input and output to the special characteristics of geographical data. The term suggested that building and operating GIS would be a problem in engineering, although it was reasonable to expect that some of its functions could lead to new scientific knowledge. By the mid-1980s, interest was growing among scientifically oriented GIS users in the implications of some of the assumptions that had been made in building the technology. Most notable among these was the assumption that the map was the truth.

Research into these issues began in earnest in the 1980s (Burrough 1986; Goodchild and Gopal

1989), with the objectives of describing the errors in GIS input, representing the errors in GIS databases, and propagating the errors into GIS output. By the late 1980s, it was clear that the term "error" was inadequate, implying as it does the existence of a true value. Many properties being input to GIS lacked precise definitions; in the case of soil maps or land-cover maps, for example, the definitions of classes are commonly vague, using terms such as "mostly". The term "error" was replaced with "uncertainty", and researchers began to explore the potential of fuzzy sets (Frank and Mark 1991).

Interest in uncertainty led quickly to the realization that GIS was more than a simple application of computing: that it raised many issues of an intellectual nature, and contained the possibility of its own body of theory. In 1992 I published a discussion (Goodchild 1992) with the title "Geographical information science", which is now commonly abbreviated to GIScience. There are several ways of defining GIScience: as the scientific knowledge that enables GIS technology; as what the intelligent and experienced user of GIS thinks about when employing the software; as a set of principles that are generally known to be true of geographic information; and as the use of GIS to acquire scientific knowledge.

Underlying all of these is the essential truth that GIS deals not with reality, but with representations of reality that differ from reality in important ways. Korzybski (1933) expressed this succinctly as "the map is not the territory". GIS representations will always approximate, generalize, and abstract, sometimes omitting detail that turns out in hindsight to have significant impact on the results and the decisions that they support. Thus the intelligent and experienced user of GIS knows that all results must be evaluated carefully, and appreciates the value of ground truth, general geographic knowledge, and fieldwork, in their ability to reveal potentially important differences in the representation.

Two empirical principles stand out as particularly relevant to GIS (Anselin 1989). The principle of spatial dependence holds, in the words of Tobler (1970; and see also the forum edited by Sui 2004), that "nearby things are more similar than distant things". The effects of this very simple principle are profound: it enables the making of intelligent guesses about unobserved properties, by assuming that they are similar to known properties that are nearby (the function known as spatial interpolation); it enables the plotting of contours on maps of elevation; and it enables representations such as area-class maps that aggregate areas into polygons based on their degree of internal similarity. At the same time, it raises issues for any application of the methods of inferential statistics, since geographers must often deal with samples that have not been drawn independently from a parent distribution. The principle of spatial heterogeneity describes the essentially variable nature of the Earth's surface, and implies that it will always be difficult to generalize from one study about other studies in other locations, and potentially at other times.

16.4.3 Artificial Intelligence

With the growth of computer applications in geographic research, it was perhaps inevitable that the idea of automating geographic research would surface. In the 1980s Dobson (1983) advanced the notion of automated geography, and Openshaw advocated what he called a geographic analysis machine, that would take over the task of searching for suitable models based on their degree of fit to data (Openshaw et al. 1987; Openshaw and Openshaw 1997). In the case of spatial interaction, for example, where tables are compiled of the number of travelers, migrants, telephone calls, or commuting trips between origin areas and destinations, he suggested that a machine could take over the process of model formulation and analysis. The researcher would identify the likely explanatory variables—distance, travel cost, travel time, population of origin area, attractiveness of destination—and would explore every possible mathematical combination of these variables, ranking the combinations by their goodness of fit to the data.

Today these early ideas of automated research as a form of artificial intelligence have evolved into the field of machine learning, which has scored many successes in classification and prediction. Geographers have begun using machine learning (or *GeoAI*) to classify images, and to search for features in vast collections of images (Janowicz et al. 2020; Li 2020). But many questions have been raised about the use of machine learning in science, and its implication that the human actor is playing less and less of a role in research. The kinds of scientific knowledge, explanation, and understanding that have traditionally formed the goal of science are hardly satisfied by classification and prediction, and Pearl and Mackenzie (2018) have described machine learning as an elaborate form of curve fitting. The detailed outputs are hard to interpret, consisting of large collections of weights, and the concept of replicability, which lies at the heart of the scientific tradition, may be beyond the capabilities of machine learning: how can one compare the results of machine learning in order to determine if a result has indeed been replicated?

16.4.4 The Role of Geographers

Geographers played a key role in the development of GIS, as discussed above, and many methods of spatial analysis have been invented by geographers (see, for example, Anselin 1995; Fotheringham et al. 2002, 2017). But by the advent of the twenty-first century, many other disciplines had staked out their claim to geographic information technologies, leaving geographers to wonder what their long-term role might be. Computer scientists had contributed many of the original ideas to projects such as CGIS, and have long seen such topics as spatial (and spatiotemporal) databases, computational geometry, spatial reasoning, and human-computer interface design as important topics within their discipline. As applications of GIS spread across many university campuses, courses and programs were established in many departments, including geology, archaeology, and even religious studies. Many universities responded by establishing interdisciplinary centers to provide essential support to what had become a truly cross-disciplinary activity.

If ever there was a case for geography as the owner of GIS, that case has long been lost. One response has been to argue that geographers have the greatest expertise in the technology and its use, and can and should play essential roles in the kinds of team-based research that are now increasingly common. But this is to argue that GIS is a service, just as the library is a service, which should therefore be the responsibility of units that are budgeted and managed as services, rather than by an academic department. On the other hand, the kinds of issues that were addressed above are well suited to the interests of geographers, and not only geographers of a technical bent. So also are the issues addressed above: the principles of GIScience, and the degree to which they enable the technology.

Geography attracts a wide range of students into its majors and graduate programs, from those with a strong technical interest to those whose preferred approach is more humanistic. This has always had a strong beneficial effect on GIS, as students come to combine the mindset reflected in the technology with one that responds to the issues that have been addressed in this section. Moreover, it implies a willingness to reach out beyond the limits of the discipline, into engineering, the social and environmental sciences, and even the arts and humanities. Geography is in many ways an ideal home for the integrating technology of GIS, ideally suited to guiding the development of the technology so that it reflects all dimensions of human activity and concern, and ideally suited to taking a critical perspective on its societal impacts.

16.5 Evolving Visions

The early vision of GIS was grounded in maps, and much of the early content of GIS was derived from maps. By the late 1980s, however, advances in computer graphics and display devices had opened the possibility of displaying, rotating, and zooming into a globe in real-time. The term *Digital Earth* (Guo et al. 2020) was mentioned by then-Vice-President Al Gore in his book *Earth in the Balance* (Gore 1992) and fleshed out in a speech prepared for his delivery in 1998. By 2000 the average personal computer was capable of wrapping an image over a simulated sphere, and in 2005 Google released Google Earth, in effect the first consumer-oriented implementation of the Digital Earth vision. The early vision of GIS was enhanced again when the Internet became popular in the early 1990s, allowing the development of a national spatial data infrastructure. Now geographic information was seen as something to be shared, enabled by the Internet as the channel of communication.

In the future, the prevailing vision of GIS and the relationship between GIS, the discipline of geography, and the information society might best be captured in the concept of an international geospatial infrastructure (Dangermond and Goodchild 2019). Designed to support a vast range of human activities, from courses and programs in universities to wayfinding by the citizen and decision-making by planners, geospatial infrastructure will encompass the data, the software, the users, and the devices that they use to interface with the technology. It will be supported by standards: standards of accuracy and metadata, communications protocols, credentialing of users and developers, and standard data formats. Geographers will play an essential role in this infrastructure, in reflecting critically on its societal impacts, researching standards, acquiring scientific knowledge in the domain of GIScience, and applying GIS to the solution of problems and the support of decisions.

The vision of GIS has always been international, based in part on the early efforts of Tomlinson to build a global network of committed GIS developers, and in part by the global nature of GIS subject matter. Although the IGU played an essential role in those early efforts, the momentum they established has long sustained itself. But there continues to be a need for active participation by the IGU, especially in overcoming international differences in access to software and data, in addressing issues of language in a field that continues to be dominated by English, and in ensuring worldwide access to GIS education.

References

Anselin L (1989) What is special about spatial data? Alternative perspectives on spatial data analysis. Technical Paper 89-4. National Center for Geographic Information and Analysis, Santa Barbara, CA

Anselin L (1995) Local indicators of spatial association—LISA. Geogr Anal 27(2):93–115

Berry BJL, Marble DF (eds) (1968) Spatial analysis: a reader in statistical geography. Prentice Hall, Englewood Cliffs, NJ

Burrough PA (1986) Principles of geographical information systems for land resources assessment. Clarendon, Oxford

Chatwin B (1988) The songlines. Picador, London

Chen F, Chen L (2021) Smartphone-based indoor positioning technologies. In: Shi WZ et al (eds) Urban informatics. Springer, Beijing

Coppock JT, Rhind DW (1991) The history of GIS. In: Maguire DJ, Goodchild MF, Rhind DW (eds) Geographical information systems: principles and applications. Longman Scientific and Technical, Harlow, UK, pp 21–43

Crane N (2003) Mercator: the man who mapped the planet. Henry Holt, New York

Dangermond J, Goodchild MF (2019) Building geospatial infrastructure. Geo-Spatial Inf Sci 23(1):1–9. https://doi.org/10.1080/10095020.2019.1698274

Dobson JE (1983) Automated geography. Prof Geogr 35(2):135–143

Foresman TW (ed) (1998) The history of GIS: perspectives of the pioneers. Prentice Hall, Englewood Cliffs, NJ

Fotheringham AS, Brunsdon C, Charlton M (2002) Geographically weighted regression: the analysis of spatially varying relationships. Wiley, Hoboken, NJ

Fotheringham AS, Yang W, Kang W (2017) Multiscale geographically weighted regression (MGWR). Ann Am Assoc Geogr 107(6):1247–1265

Frank AU, Mark DM (1991) Language issues for GIS. In: Maguire DJ, Goodchild MF, Rhind DW (eds) Geographical information systems: principles and applications. Longman Scientific and Technical, Harlow, UK, pp 147–163

Goodchild MF (1988) Stepping over the line: technological constraints and the new cartography. Am Cartogr 15:311–319

Goodchild MF (1992) Geographical information science. Int J Geogr Inf Syst 6(1):31–45

Goodchild MF (2007) Citizens as sensors: the world of volunteered geography. GeoJournal 69(4):211–221

Goodchild MF, Gopal S (eds) (1989) Accuracy of spatial databases. Taylor and Francis, Basingstoke

Goodchild MF, Fu P, Rich P (2007) Sharing geographic information: an assessment of the Geospatial One-Stop. Ann Assoc Am Geogr 97(2):249–265

Gore A (1992) Earth in the balance: ecology and the human spirit. Houghton Mifflin, Boston, MA

Guo H, Goodchild MF, Annoni A (eds) (2020) Manual of digital earth. Springer, Singapore. https://doi.org/10.1007/978-981-32-9915-3
Harley JB (2001) In: Laxton P (ed) The new nature of maps: essays in the history of cartography. The Johns Hopkins University Press, Baltimore and London
Issa E et al (2017) Understanding spatio-temporal characteristics of Twitter data with geotagged and non-geotagged content: two case studies with topic of flu and ted-movie. Ann GIS 23(3):219–235. https://doi.org/10.1080/19475683.2017.1343257
Janowicz K et al (2020) GeoAI: spatially explicit artificial intelligence techniques for geographic knowledge discovery and beyond. Int J Geogr Inf Sci 34(4):625–636
Keay J (2000) The great arc: the dramatic tale of how India was mapped and Everest was named. Harper Collins, London
Korzybski A (1933) Science and sanity: an introduction to non-Aristotelian systems and general semantics. International Non-Aristotelian Publishing Co, Lakeville, CT
Li W (2020) GeoAI: where machine learning and big data converge in GIScience. J Spat Inf Sci 20:71–77
McHarg IL (1969) Design with nature. American Museum of Natural History, New York
Mounsey H (ed) (1988) Building databases for global science. Taylor and Francis, London
National Research Council (1993) Toward a coordinated spatial data infrastructure for the nation. National Academy Press, Washington, DC
Nyerges TL, Couclelis H, McMaster R (eds) (2011) The SAGE handbook of GIS and society. Sage, Los Angeles
Openshaw S et al (1987) A Mark I Geographical Analysis Machine for the automated analysis of point data sets. Int J Geogr Inf Syst 1:359–377
Openshaw S, Openshaw C (1997) artificial intelligence in geography. Wiley, Chichester UK
Orwell G (1949) 1984: a novel. New American Library, New York
Pearl J, Mackenzie D (2018) The book of why: the new science of cause and effect. Basic Books, New York
Pickles J (ed) (1995) Ground truth: the social implications of geographic information systems. Guilford, New York
Schneider P et al (2017) Mapping urban air quality in near real-time using observations from low-cost sensors and model information. Environ Int 106:234–247
Schofield P (2009) Bentham: a guide for the perplexed. Continuum, London
Sui DZ (2004) Tobler's first law of geography: a big idea for a small world? Ann Assoc Am Geogr 94(2):269–277
Sui DZ, Elwood S, Goodchild MF (eds) (2012) Crowdsourcing geographic knowledge: volunteered geographic information (VGI) in theory and practice. Springer, New York
Tobler WR (1970) A computer movie simulating urban growth in the Detroit region. Econ Geogr 46:234–240
Tomlinson RF (1971) International Geographical Union. Commission on Geographical Data Sensing and Processing, Ottawa
Tomlinson RF (1972) Geographical data handling. International Geographical Union, Commission on Geographical Data Sensing and Processing, Ottawa
Tomlinson RF, Toomey MAG (1999) GIS and LIS in Canada. In: McGrath G, Sebert L (eds) Mapping a northern land: the survey of Canada 1947–1994. McGill–Queen's University Press, Montreal (chapter 15)
Turner A (2006) Introduction to neogeography. O'Reilly, Sebastopol, CA
Valentino-DeVries J et al (2018) Your apps know where you were last night, and they are not keeping it secret. New York Times, 10 Dec. https://www.nytimes.com/interactive/2018/12/10/business/location-data-privacy-apps.html?searchResultPosition=1. Accessed 21 Feb 2021
Wood D (with Fels J) (1992) The power of maps. Guilford, New York

Michael F. Goodchild is Emeritus Professor of Geography at the University of California, Santa Barbara. He received his BA degree from Cambridge University in Physics in 1965 and his Ph.D. in geography from McMaster University in 1969, and has received five honorary doctorates. He was elected member of the National Academy of Sciences and Foreign Member of the Royal Society of Canada in 2002, member of the American Academy of Arts and Sciences in 2006, and Foreign Member of the Royal Society and Corresponding Fellow of the British Academy in 2010; and in 2007 he received the Prix Vautrin Lud.

Geography and Social Issues

17

Alexander B. Murphy and Virginie Mamadouh

Abstract

Geography's way of looking at the world and the kinds of international collaborations promoted by the IGU will become ever more significant in the social arena in the years ahead because the broad trends affecting life on Earth are international in scope and resonate with geography's core perspectival and analytical approaches. Key demographic, socio-economic, socio-cultural, and geopolitical trends point to the value of studies focused on spatial variability, human-environment interactions, space-time linkages, and interdependence across space and scale. If geography is to make important contributions to social well-being in the twenty-first century, its concern with these matters should be pursued in ways that serve to advance sustainability—an overarching goal that cannot be realized in the absence of increased international awareness and the sharing of geographically informed information and ideas. Nonetheless, geography's influence will remain marginal unless progress is made in combating widespread geographical ignorance on the part of the general public and influential political actors. As such, addressing educational and outreach challenges will be of critical importance to the future of the discipline.

Keywords

Future geographies · Social trends · Sustainability · Geographical perspectives · Geography education

17.1 Introduction

The pace and extent of social change during the first fifth of the twenty-first century portend a social world that will look very different in 2100 than it does today. In just the first twenty years of the century, Google has gone from a new start-up to a global behemoth, the percentage of the East African population with access to a cell phone has grown from 10% to upwards of 75%, China's nominal GDP has exploded, talk of a turn toward democracy in parts of the world has all but vanished in the wake of the ascendance of right-wing populist regimes, pandemics have moved from afterthoughts to dominant foci of attention, internet platforms and social media have displaced newspapers and magazines as primary sources of information and news, and in many

A. B. Murphy (✉)
Department of Geography, University of Oregon, Eugene, USA
e-mail: abmurphy@uoregon.edu

V. Mamadouh
Department of Geography, Planning and International Development Studies, University of Amsterdam, The Netherlands

V. Kolosov et al. (eds.), *A Geographical Century*,
https://doi.org/10.1007/978-3-031-05419-8_17

countries, the share of wealth controlled by the top five percent of the population has risen sharply.

Given the massive changes of the past two-plus decades—and the fact that more than three-quarters of the twenty-first century still lies ahead—making any general claims about geography's potential to address this century's social issues is, on its face, foolhardy. The immensity of the problem comes into focus when one considers how few people understood what lay ahead some 80 years ago (a period roughly corresponding to how much remains of the twenty-first century). In the early 1940s, few people anticipated the wave of decolonization that was to come, the social upheavals of the 1960s, the ubiquitous spread of personal computers, the development of the internet, the intensification and then end of the Cold War, the invention and diffusion of smartphones, China's tremendous economic growth, the acceleration in international migration, the globalized concern with human-induced climate change, and much more.

The IGU itself was hardly a beacon of progressive, forward-looking thinking at the time. At the 16th International Geographical Congress in Amsterdam in July 1938 geographers linked to the military were prominent among the participants and, as the British geographer Arden-Close (1938: 64) reported, "the most lively interest was shown in the discussions on The Possibilities of Colonization by the White Race in the Tropics. All political references were excluded, and there was no intention of expressing an opinion for, or against, white colonization." Fortunately, the IGU has changed beyond recognition since then (see Robic and Rössler 1996 regarding the role of women, Huang et al. 2017 on gender and geography, or Schelhaas et al. 2020 on efforts to decolonize the discipline). Indeed, the IGU has played an important role in promoting international collaboration and geographical research—efforts that have been critical to advancing the contributions geography has made to understanding some of the major issues the world has faced in recent decades.

Geography's recent contributions have taken many forms. Mapping and spatial analysis have shed light on some of the most important environmental, health, and welfare challenges society has faced. The writings of radical/critical geographers have offered profound insights into the root causes of uneven development and the ways in which interconnections across space and scale affect social outcomes and reinforce socio-economic inequalities. Humanistically oriented geographers have heightened appreciation for the difference that place makes in human affairs—in the process drawing attention to the limitations of approaches and models that treat place-based differences as little more than noise. Geographic studies from a feminist perspective have drawn attention to the importance of marginalized peoples and issues while demonstrating the limitations of top-down approaches to the analysis of social issues. And geographically grounded studies in cultural and political ecology have focused attention on the inextricable interconnections that exist between the physical and human world.

Despite the progress that has been made, the importance of geographical inquiry remains underappreciated in many circles, and geography's influence on policymaking and civic activism is limited. Yet there is much to suggest that geography's way of looking at social issues and the international collaborations advanced by the IGU will become ever more significant—even if the precise contours of the social developments that lie ahead cannot be anticipated. That is because the broad trends affecting social life are international in scope and resonate with geography's perspectival and analytical approach. To be sure, sweeping statements about geography's importance are complicated by the fact that the discipline is intellectually wide-ranging and is practiced in very different ways depending on the cultural and political context. Yet when geography's analytical insights are considered alongside the extensive spatial reach of many geographic processes and concerns, the importance of both the discipline itself and the IGU's focus on international collaboration becomes clear.

The challenge the IGU, and indeed all geographers face, is to ensure that geography's eclecticism is treated as a source of strength, not of weakness. That, in turn, requires appreciating the creative and interpretive advantages that stem from geography's diversity while not losing sight of its core analytical-cum-perspectival approaches. Susan Hanson drew attention to the latter in an article setting forth a set of "geographic advantages" associated with the discipline's central heuristic orientations: geography's capacity to advance understanding of "relationships between people and the environment, the importance of spatial variability (the place-dependence of processes), processes operating at multiple and interlocking geographic scales and the integration of spatial and temporal analysis" (Hanson 2004, 720). To these, we would add interdependence across space and scale.

Whatever precise words are used to characterize geography's analytical approach and perspective, it is clear that the importance of the IGU's mission will loom ever larger in a social world characterized by increasing interconnectivity, crowding, volatility, and complexity. Indeed, the integrative, context-sensitive perspective geography brings to bear is arguably of particular relevance given the increasingly obvious limitations of looking at the world solely through the lens of the topically grounded disciplinary silos that emerged from nineteenth-century efforts to modernize universities.

The critical importance of bringing a geographical perspective to bear on social issues in the twenty-first century can best be appreciated by looking at key trends in the social arena and considering why geography can be a notably powerful platform for addressing those trends. Prior chapters in this section address trends in two realms with profound social implications: the environmental and technological. As such, this chapter focuses on developments in other realms: demographic, socio-economic, socio-cultural, and geopolitical. These categories fall far short of covering all aspects of social relevance, and there is more than a little overlap among and between them, but they serve as useful platforms for illustrating geography's interpretive value. That value is ultimately rooted in the contributions geography can make in the service of sustainability. Sustainability might well be seen as a social realm deserving of treatment on its own. But it is also an overarching theme and challenge that cuts across all of the realms we examine here and that cannot possibly be addressed in the absence of increased international awareness and the sharing of geographically informed information and ideas. A subtext of the discussion that follows, then, is that work at the intersection of the geographical and the international—the type of work the IGU seeks to promote—is critical to the struggle to create a more sustainable future.

No matter how powerful work at that intersection might be, its value will be limited if the perspective geography brings to bear on social matters or its analytical capacity is little understood or appreciated outside the academy. To put it another way, achieving "dream discipline" status for geography cannot be realized unless progress is made in combating widespread geographical ignorance on the part of the general public and influential public figures and social actors alike. This chapter ends, then, with a consideration of the educational and outreach challenges that will fundamentally influence geography's role as an interpreter of, and contributor to, the social world of the twenty-first century.

17.2 The Demographic Realm

The accuracy of long-term demographic projections is notoriously bad. Few people foresaw the extent of the global population explosion that occurred in the decades immediately following World War II. Then just as commentaries began to appear suggesting that Earth was headed toward demographic catastrophe, global population growth started to slow—in some places quite dramatically. Any claims about what the demographic future will look like, then, must be made with great caution. Nonetheless, certain developments have unfolded over a long enough period of time and under sufficiently diverse conditions to be at least suggestive of some of the more important realms of demographic adjustment the world will likely face in the decades to come. These include great variability in population growth across space, a notable increase in human migration, and significant shifts in where people live in more localized settings (rural vs. urban, urban vs. suburban, and the like).

All of these arenas of demographic change have fundamental geographical dimensions that societies ignore at their peril. Each is rooted in differences across space, suggesting the importance of spatial analysis to make sense of them. Yet geographical insights beyond the spatial-analytic are needed as well because these arenas will be driven by, and will in turn influence, a mix of mappable and unmappable circumstances present in individual places, a combination of tangible and intangible interdependencies that exist across space and scale, and an array of complex socio-spatial dynamics that shape human-environment relationships (Bailey 2014).

Understanding the complex circumstances that influence, and are influenced by, demographic shifts will require consideration of multiple tangible and intangible differences across space in socio-economic and environmental circumstances, approaches to reproductive rights, lifestyle preferences, and health challenges. Against this backdrop, the geographic concept of place looms large—a concept that foregrounds the conjunction of forces in particular locales that shape social outcomes in ways that are often overlooked in studies focused on generalized trends. Consider the case of migration. Many migration models provide limited insight into human mobility because the variables that influence migration decisions are multiple and differ greatly from place to place. In some places, ecological factors may be particularly important, in others socio-economic factors loom large, in yet others gender or ethnic relations play a significant role—and each of these sits alongside other factors that influence migration.

What is needed, then, are place-sensitive assessments of relevant dynamics that take seriously the geographer's concern with what Webb and Brown (2017) call "the difference that place makes" in human affairs. These types of approaches can also be of critical value to policymakers grappling with the social challenges associated with aging populations, urban sprawl, counter-urbanization, threats to human health, the lifestyle changes that go along with demographic shifts or the impacts of different population policies, to cite just a few examples. Moreover, since the nature and significance of place-based influences vary widely across the planet, international collaborations of the sort championed by the IGU are essential to any serious effort to confront the place dimensions of migration.

Geography's concern with interdependencies across space and scale will also be of critical importance in a world experiencing significant demographic changes, no matter what precise direction those changes take (see Sheppard and McMaster 2008). The linkages among and between places work together with a suite of global, regional, and local processes to influence the size and direction of migration streams, just as the power relations that exist among places and the access different communities have to technology and infrastructure produced in distant places can affect where people settle, rural-to-urban migration, and much more. It follows that understanding twenty-first-century demographic change requires consideration of spatial and scalar interdependencies, whether in the form of critical assessments of the large-scale political-economic structural circumstances shaping demographic dynamics, spatial analyses of circuits of capital and resources, or humanistic accounts of the ways in which people's geographical imaginations reflect and shape demographically consequential life decisions.

Finally, material circumstances influence demographic change, whether natural or human-made. Settlement, migration, and procreation all reflect and remake the landscape. In the process, cities expand and contract, cultivated areas shift location, natural areas are remade, and so on. Moreover, pandemics and more localized health challenges raise fundamental sustainability challenges that lie squarely at the intersection between the environmental and the social. Against this backdrop, the importance of geography's conceptual concern with nature-society relations—its proclivity to look across the human-physical divide—becomes clear. What is needed are collaborations among geographers in different countries and world regions using political-ecology frameworks to explore how power relations reflect and shape environmental circumstances, geospatial visualizations that facilitate assessments of the social and environmental impacts of demographic change, and analyses that focus attention on the mix of human and physical forces that shape the evolution of landscapes over time.

17.3 The Socio-economic Realm

The past 80 years have been marked by overall economic growth, along with a complex geography of inequality exhibiting greater or lesser disparities depending on the scale and the area being considered. Periods of rising welfare and upward social mobility have alternated with periods of crisis and turbulence—often exacerbating uneven development. Recent decades have also witnessed dramatic changes in the structure of the global economy, thanks in part to productivity booms in certain sectors and places and to innovations such as the standardized multimodal container, which has greatly decreased the cost of shipping (along with ever-larger cargo ships and trains, lower fuel taxes, and infrastructure improvements).

We now live in an era of extensive economic globalization made possible by complex, wide-ranging value chains. Geographical perspectives are critical to understanding the contemporary picture and the changes to come. Geographers have shown how studying global production networks can shed light on transnational interactions (Coe et al. 2008). Their work on the delocalization of industrial production in low-wage countries has shed light on how corporations are able to externalize the social and environmental costs of production—resulting, for example, in rapid industrialization in specific locations in developing countries (often in designated economic zones) under poor working conditions and wrenching deindustrialization in older industrial areas (Dicken 2014).

Will the trend toward uneven but ever stronger interdependence continue in the coming decades? Recent developments point in different directions. The early twenty-first century saw the limits of just-in-time production as state borders closed after 9–11 2001 and again during the COVID-19 pandemic. Some countries and localities turned away from global value chains in an effort to insulate themselves from political disruption and reduce the ecological footprint of economic exchange. On the other hand, new information and communication technologies are increasingly enabling delocalization in the service sector. Thanks to them, digital nomads and freelancers can work online from anywhere—in less crowded rural areas rather than in cities; from tourist havens where workers can enjoy sea, sun, and beach in their free time; or from call centers halfway around the planet from the customers they are serving.

Further changes can be expected as production shifts in the face of digitalization and automatization, and as the energy sector moves away from fossil fuels. Even greater challenges lie ahead as the exhaustion or degradation of raw materials, water, soils, seas, and biodiversity works together with population growth and climate change to push planet Earth toward the *Limits to growth* presaged in 1972 by the Club of Rome. Moreover, rising demand for new raw materials will create its own stresses—for example, a potential 18-fold rise in demand for lithium by 2030—just one of 30 Critical Raw Materials (CRMs) identified by the European Commission in 2020 as being critically needed for energy transition and the digital technology future (European Commission 2020).

Since the fundamentally geographical character of all these developments is inescapable, geographical inquiry has an important role to play in shedding light on the economic challenges to come, including the possibilities and challenges of transitioning to more sustainable approaches to development and societies based on renewable energy. Traditional economic geography can provide detailed knowledge of regional specificities. Spatial analysis can offer insights into the geographical distribution of (new) raw materials and the conditions under which they can be developed, as well as the uneven geographies of rising energy demand (electric cars, servers, internet of things) and shifting access to transport, energy, and ICTs. Geographical approaches inspired by political-economy perspectives (e.g., world-system analysis, the Regulation School) can question the contingencies of specific "spatial fixes" and the role of the national and local state in economic policies. Last but not least, economic approaches that are sensitive to cultural and gender differences can shed light on externalities, environmental costs, and aspects of well-being that are too often pushed to the side in conventional analyses.

17.4 The Socio-cultural Realm

The past 80 years have seen continuous but shifting tensions between homogenizing and differentiating influences. Modernization theory long treated global cultural homogenization as an inevitable outcome of modernity. Yet there are enormous differences in how that process has played out across space, and modernity itself has inspired active resistance to homogenization. The limits of modernization's homogenizing impact became evident as early as the 1960s when speakers of regional languages and other cultural minorities began aggressively championing their cultural specificity in opposition to state efforts to impose national norms. The subsequent emergence of many other visible cultural fault lines makes it clear that socio cultural diversity is far from disappearing.

To be sure, the post-World War II period saw significant homogenizing tendencies, including the diffusion of American and, to a lesser extent, Soviet culture—soft power offshoots of superpower rivalry. Moreover, the global dissemination of fast-food outlets, television, movies, shopping malls, and other norms associated with American lifestyles produced a degree of homogenization, along with an attendant commodification of culture. Yet such developments also fostered resistance and differentiation.

The intensification of international migration flows has further complicated the socio-cultural picture with, for example, the outmigration of significant numbers of people from former colonies. Migration inevitably brings with it cultural adjustments for migrants and migrant-destination societies—producing geographically variable patterns of acculturation, hybridization, and resistance. Importantly, the tensions associated with migration and other forms of cross-cultural interaction are fueled by new media landscapes. These have developed independent of state control and reflect both political decisions (the desire to liberalize and commodify the media) and technological innovations (cable and satellite television, the internet, smartphones, and the like). As a result, people living in the same place are no longer embedded in the same communication ecosphere; instead, they choose what media to follow according to individual linguistic, religion, political, and lifestyle preferences. In the process, the cultural, social, and informational common denominators that long-held local communities together are being undermined. At the same time, the so-called digital divide has fostered social fragmentation that may well become worse in the decades to come.

Taken as a whole, these different aspects of, and reactions to, globalization are transforming traditions and modes of social reproduction (see generally Kloosterman et al. 2018). Since these forces play out very differently from place to place, a sensitivity to local specificities is and will continue to be, a prerequisite to any serious effort to understand the nature and uneven impact of global homogenizing processes. That makes the geographical perspective, with its emphasis on differentiation across space, of signal importance in the socio-cultural realm. Geographical inquiry can also shed light on the diffusion of cultural phenomena, as well as the development and reach of identity politics (e.g., through poststructuralist approaches focused on geographical imaginations and geopolitical representations).

In a somewhat different vein, geographical inquiry is critical to an appreciation of the connections that exist between culture and place —its artifacts, landscapes, and iconographic representations (Zhou 2011). Identification with localities or territories can serve to mobilize individuals and groups to assist compatriots and fight outsiders. Powerful forces such as nationalism and national identity can only be understood by taking seriously the ways in which social spatialization and spatial socialization (Paasi 1996) serve to co-constitute the identity of a place and the identity of its dwellers.

Geographer's concern with the concept of scale matters as well. Cultural configurations and relations are manifest at different scales. They can be hierarchical (the local being subsumed into the national and the supranational), but they can also be rhizomatic (networked) in ways that link individual bodies to regional and global processes. Moreover, cultural networks that exist in individual places can have significant ties to larger-scale networks thanks to new information and communication technologies,

Geography is also important to the study of othering processes that arise out of differences in language, ethnicity, religion, nationality, race, age, ability, gender, and sexual orientation. The discrimination that often accompanies such othering frequently plays out against the backdrop of geographical tactics and strategies that foster segregation and produce zones of inclusion and exclusion (bordering, ordering, and othering; see Van Houtum and Van Naerssen 2002). Feminist, queer, and black geographies have much to contribute in this regard, thanks to their focus on the situatedness of knowledge, the attention they give to the perspectives and circumstances of those in marginalized positions, and their attempt to develop analyses that disrupt binaries and engage with differences in emancipatory and empowering ways (Datta et al. 2020).

17.5 The Geopolitical Realm

The modern territorial state remains a fundamental building block of international society despite the erosion of state sovereignty in the face of globalization, transnational interactions, and myriad internal challenges. To be sure, the world political map has undergone significant change over the past 80 years. Decolonization brought most European colonial empires to an end, the Soviet Union and other countries broke up, and successful independence movements within several countries led to the creation of new states. At the same time, more powerful superstate institutions emerged. At the global scale, the United Nations spawned a broad array of agreements, institutions, and mechanisms regulating diverse aspects of international relations, including the World Bank and the International Monetary Fund.

Other post-World War II geopolitical developments included the rise of a bipolar Cold War order, the emergence of a large Non-Aligned Movement, and then, post-1990, the emergence of a shifting panoply of new geopolitical alignments with no clear overarching architecture, but with some ties to regional supranational organizations that had sprung up in the second half of the twentieth century to promote economic and social agendas (e.g., the EU and its predecessors, ASEAN), advance strategic objectives (e.g., NATO, Warsaw Pact, ANZAC), and foster cultural connections (e.g., the Commonwealth of Nations, Francophonie). These developments went hand in hand with dramatic changes in relations between state and society, with state-led economies taking root in the Communist bloc, welfare states developing in the Western bloc, and state-led modernizing and nation-building projects taking root in newly independent states. The situation began to change in the late twentieth century, with a turn toward neoliberalism (the US under Ronald Reagan and the UK under Margaret Thatcher took the lead), a shock transition to capitalism in formerly communist states, and the need to abide by conditions set by the World Bank and the IMF in states in the global economic periphery that had amassed significant debts. With a World Trade Organization promoting free trade and limiting the steps states could take to protect their national economies, some commentators concluded that the territorial state was giving way to an increasingly homogenized "flat world" (Friedman 2006).

Few geographers, however, embraced this view. Their focus on spatial arrangements and developments on the ground led them to underscore the ongoing role of states in regulating the economy and the enduring importance of state borders. Thirty years later that line of thinking has proven to be prescient, with many politicians championing state nationalism, closing borders, and backing away from interdependencies in an attempt to protect domestic interests and limit immigration. Remarkably, this is happening at a time when many of the most urgent challenges (climate change, the pandemic, the impact of the internet) are global in nature and cannot be tackled by individual countries.

The geopolitical fluidity of the post-Cold War period makes it all but impossible to predict what lies ahead. To what extent will US power be diminished? What role will China come to play? In the long run, will transnational corporations, especially Big Tech, dominate global information and communication, or will states follow the approach being taken by some authoritarian regimes today and deploy ICT to surveil and control citizens? Whatever the answers to these questions might be, geography offers concepts and tools that can help us make sense of them, offer insights into the motives and strategies of geopolitical actors, shed light on factors connecting different places, and draw attention to the uneven geographical distribution of opportunities and constraints associated with different territorial arrangements (see, e.g., Storey 2020).

The geographic advantage on these fronts stems in part from geographers' attentiveness to the geopolitical features of foreign policy—including possible contradictions among and between what Flint (2021) calls the geostructural, the geostrategic, and the geophysical. Far from the environmental determinism of early-twentieth-century geopolitics, geographical inquiry today focuses on the mutual construction of political space and society (see Agnew et al. 2015). Geography's "critical geopolitics" approach, for example, is well suited to the task of exposing the underpinnings of different geographical imaginations and geopolitical representations—ones that are often simply taken for granted to justify particular policy choices (Dodds et al. 2013). In a related, but somewhat different, vein, feminist geographical approaches can direct attention to the impacts of changing geopolitical arrangements on people's everyday lives (Smith 2001) and the ways in which resistance to state influence can be sustained. As the twenty-first century continues to unfold, careful consideration of the material, spatial, and ideological dimensions of geopolitics will be critical to the struggle to forge a more stable, peaceful future.

17.6 The Critical Importance of Education and Outreach

Geography's potential contributions, no matter how great, will have limited impact if the discipline continues to be associated with little more than place-name memorization, if geographical perspectives and insights remain marginal to decision making, or if the problem of widespread geographical illiteracy is not seriously addressed. Apart from organizing scholarly meetings and field trips, the IGU has played an important role in advancing geographical understanding—fostering joint research projects, publications, and educational activities. On the latter front, the IGU has convened an annual International Geography Olympiad (iGeo). In 2016, an International Charter on Geographical Education drafted by the IGU Commission on Geographical Education was adopted by the IGU General Assembly Stoltman et al. (2017).

Continuing efforts to promote geographic education and outreach will be fundamental to any serious pursuit of a "dream discipline" goal for geography. Several interrelated challenges are of critical importance in this regard: (1) expanding research on geographic learning, including the comparative advantages of the pedagogical approaches being adopted in different local and national contexts, (2) broadening and deepening the teaching of geography at all levels, (3) expanding public awareness of geography's nature and value, and (4) increasing the participation of individuals who think geographically and are familiar with geographical concepts and techniques in formal institutions and in the movements seeking to advance social and environmental justice.

Turning to the first of these challenges, despite the growth of the geography education literature in recent decades, much remains to be learned about which approaches and skills are most important to foreground in geography education, which topics and concepts are deserving of particular attention, and what types of assessment mechanisms are best suited to capture geographic learning (Bednarz et al. 2013). Given the rapid pace of social, technological, and environmental change in the twenty-first century, the importance of research on such matters will only increase in the decades ahead. As geospatial technologies based on detailed geocoded information continue to seep into more and more realms of human practice, what do we need to be teaching the students of the future about how these technologies function, their usefulness, and the security and privacy threats they pose? More generally, how can we best nurture students who are better observers of the world around them, have the capacity to think carefully and critically about shifting geographic patterns and processes, and can grapple with the complex relationships that exist between the physical and social worlds? Given the enormity of the geographical changes facing the world of today and the increasingly ubiquitous use of geographical technologies, comparative international research of the sort championed by the IGU will be needed to address such questions.

Understanding how geography can best be taught is of limited value if there are not robust opportunities for students to be exposed to geography. The attention paid to geography in schools and universities varies widely from country to country. Despite a longstanding official endorsement of geography as a core academic subject by the U.S. government, the discipline occupies only a marginal position in the curricula of many American primary and secondary schools, and a significant number of universities lack geography programs. Yet no student graduates from high school in France or Japan without significant geography coursework, and the discipline is a fixture at most French and Japanese universities. Addressing such disparities is important, but the greater challenge it is to broaden and deepen geographic education everywhere. Paving the way for a more peaceful, hopeful future will require an expansion in people's attentiveness to the diversity of our planet, their understanding of what the world looks like elsewhere, their appreciation of the interrelationships that exist both across space and within places, and their awareness of the possibilities and challenges associated with geospatial technologies (Murphy 2018: 110–132). These are the hallmarks of good geographic education; moving them to the forefront of the IGU's agenda is vital in the face of the opportunities and challenges the twenty-first-century presents.

Schools may be the place where formal geography education takes place, but geography's power and potential will remain significantly limited if efforts are not made beyond the arena of formal education to address the simplified, caricatured image of geography that is shared by many people who are no longer in school. Such an effort will require geographers and those sympathetic to the cause to focus more attention on outreach—to look for opportunities to address general audiences, to translate the insights of academic work for a general audience, and to devote time and energy to demonstrate what is missing if geographical ways of thinking and techniques are not brought to bear on the questions facing society. Given the importance of an informed citizenry to a well-functioning civil society, humanity can ill afford a twenty-first-century future with a general public lacking the knowledge and capacity to think critically about, for example, geopolitical representations, the claims made on behalf of geospatial technologies, or assertions about the socio-spatial implications of welfare reform or environmental regulation. In the increasingly eclectic, unregulated, siloed media environment of the twenty-first century, these critical analytical abilities are likely to loom ever larger.

In a related vein, geography's "dream discipline" capacity will be seriously curtailed if those in a position to shape policy and institutional agendas, and those calling for more radical social change, do not bring geographical perspectives and insights to bear in their efforts. Here too the IGU has an important role to play since in many places geographers or individuals with geographic training are not well represented. To be sure, the situation is beginning to change in formal institutional settings, with geographically oriented thinkers taking on more prominent roles in various national research councils, IPCC working groups, government commissions, and the like. But more efforts are needed to ensure that geographical ideas and approaches are given due consideration in the public arena. That, in turn, will require more geographers and their conceptual allies to embrace opportunities for involvement in policymaking, as well as meaningful participation in the institutions and social movements that are, or hope to be, at the forefront of social change.

Many of the twists and turns the twenty-first century will take are unknowable, but it seems highly likely that the century will continue to play host to a fraught relationship between humanity and the natural environment, significant shifts in where and how people live, geopolitical tensions within and across world regions, changing patterns and scales of identity, and ongoing sustainability challenges. Against this backdrop, geographic education and outreach initiatives should not be seen as efforts to bolster the standing of a venerable discipline. They should be seen as foundational to the preservation and perpetuation of healthy societies and ecosystems. That is the dream to which geography should aspire. Geographic education and outreach are integral to that dream because its realization is dependent on the success of a global effort to nurture societies that are equipped to apply geographical modes of thinking, techniques, and analytical approaches in the service of understanding past and present geographic conditions and contemplating future possibilities. With its international remit and geographic focus, the IGU is positioned to play a seminal role in such an effort. Embracing that role in creative and effective ways would build on the momentum the IGU has established over the past century and would help to advance a cause of profound and far-reaching importance for the century to come.

References

Agnew JA et al (eds) (2015) The Wiley Blackwell companion to political geography. Wiley, Chichester

Arden-Close CF (1938) The International Geographical Congress at Amsterdam, July 1938. Notes Rec R Soc Lond 1(2):62–64

Bailey A (2014) Making population geography. Routledge, London

Bednarz SW, Heffron S, Huynh NT (eds) (2013) A road map for 21st century geography education: geography education research. Association of American Geographers, Washington, DC

Coe NM, Dicken P, Hess M (2008) Global production networks: realizing the potential. J Econ Geogr 8 (3):271–295

Datta A et al (eds) (2020) Handbook of gender and feminist geographies. Routledge, London

Dicken P (2014) Global shift: mapping the changing contours of the world economy. Sage, London

Dodds K, Kuus M, Sharp JP (eds) (2013) The Ashgate research companion to critical geopolitics. Ashgate, Farham

European Commission (2020) Critical raw materials resilience: charting a path towards greater security and sustainability. COM(2020)474 Brussels. https://eur-lex.europa.eu/legal-content/EN/TXT/?uri=CELEX:52020DC0474

Flint C (2021) Putting the "geo" into geopolitics: a heuristic framework and the example of Australian foreign policy. GeoJournal. https://doi.org/10.1007/s10708-021-10387-5

Friedman TL (2006) The world is flat: a brief history of the |twenty-first century. Macmillan, New York

Hanson S (2004) Who are "we"? An important question for geography's future. Ann Assoc Am Geogr 94 (4):715–722

Huang S et al (2017) A continuing agenda for gender: the role of the IGU Commission on gender and geography. Gend Place Cult 24(7):919–938

Kloosterman RC, Mamadouh V, Terhorst P (eds) (2018) Handbook on the geographies of globalisation. Edward Elgar, Cheltenham

Murphy AB (2018) Geography: why it matters. Polity Press, Cambridge

Paasi A (1996) Territories, boundaries and consciousness: the changing geographies of the Finnish-Russian border. Wiley, Chichester

Robic MC, Rössler M (1996) Sirens within the IGU—an analysis of the role of women at International Geographical Congresses (1871–1996). Cybergeo: Eur J Geogr. https://doi.org/10.4000/cybergeo.5257

Schelhaas B et al (eds) (2020) Decolonising and internationalising geography: essays in the history of contested science. Springer, Cham

Sheppard E, McMaster RB (eds) (2008) Scale and geographic inquiry: nature, society, and method. Wiley, New York

Smith F (2001) Refiguring the geopolitical landscape: nation, 'transition' and gendered subjects in post-cold war Germany. Space Polity 5(3):213–235

Stoltman JP, Lidstone J, Kidman G (2017) The 2016 International Charter on Geographical Education. Int Res Geogr Environ Educ 26(1):1–2

Storey D (ed) (2020) A research agenda for territory and territoriality. Edward Elgar, Cheltenham

van Houtum H, van Naerssen T (2002) Bordering, ordering and othering. Tijdschr Econ Soc Geogr 93 (2):125–136

Webb MD, Brown LA (2017) Foreclosure, inequality, and the difference that place makes. Rural Sociol 82 (4):688–712

Zhou S (2011) On the research framework and methodologies of cultural geography. Bull Chin Acad Sci 4:174–183

Alexander B. Murphy is a still-active Professor Emeritus in the Department of Geography at the University of Oregon (USA), where he long held the Rippey Chair in Liberal Arts and Sciences. A former President of the American Association of Geographers and co-chair of the IGU's Commission on Political Geography, he specializes in political and cultural geography, with regional emphases on Europe and the Middle East. In the early 2000s he chaired the U.S. National Academy of Sciences—National Research Council Committee charged with identifying "Strategic Directions for the Geographical Sciences."

Virginie Mamadouh is an Associate Professor of Political and Cultural Geography at the University of Amsterdam (the Netherlands). She is a co-chair of the IGU's Commission on Political Geography, an editor *of Territory, Politics, Governance* and a former editor-in-chief of *Geopolitics*. She specializes in political geography and geopolitics, with a regional emphasis on Europe and its neighborhood (especially the Arab World). Key themes include geopolitical representations, linguistic diversity, migration, social movements and urban geopolitics. She teaches political geography, geopolitics, and geographical thought.

Part IV

Concluding Essay

Looking Back to Look Ahead

18

Ronald F. Abler

Abstract

The authors and editors of A Geographical Century have appropriately summarized a century of the IGU's accomplishments. The volume confirms the centrality of geographers' traditional natural environment/human stewardship view of the earth at a time when humanity itself threatens to destroy its home and in which pestilence threatens.

Measured against the criteria used to assess the effectiveness of disciplinary organizations, the IGU has done well by its constituents during its first century of existence and is well prepared to address the problems and capitalize the opportunities the future will bring.

Keywords

IGU in Global Science · International Science Council (ISC) · Congresses · Finances · Advocacy · Validation · Coordination · Outlook

The authors and editors of *A Geographical Century* have woven a timely and informative tapestry of the evolution and current state of the International Geographical Union (IGU), one of the world's venerable global scholarly societies, and the first and only society devoted to primarily to the discipline of geography. Whether in individual human or in institutional terms, achievement of a century of operations is indeed a time to take stock, especially when—as now—the IGU's centenary has arrived amid two concurrent global crises: human-induced climatic change and what portends to be a series of deadly pandemic waves. Either will wreak wrenching changes in humankind's day-to-day lives. In combination, they raise existential questions about the future of humanity.

The reality of both crises is denied by many inhabitants and politicians in many nations despite indisputable evidence to the contrary, which lends urgency to discerning what can be learned from the successes the IGU has enjoyed during its first century, and to avoid repeating any shortcomings that arose while grappling with global and international problems in its past.

IGU President Michael Meadows reminds us in his Preface that the IGU has survived—some fits and starts notwithstanding—a century of tumult following its birth in the aftermath of the First World War. Through World War Two, the Cold War, and the more recent horrors humankind has inflicted upon itself the IGU has continued to advance geographical research and education in the service of humankind. Going forward, the IGU's leaders and members will need to muster all the strategic and operational wisdom they can, and this volume's chapters are

R. F. Abler (✉)
Department of Geography, Pennsylvania State University, University Park, PA, United States
e-mail: rabler@aag.org

V. Kolosov et al. (eds.), *A Geographical Century*,
https://doi.org/10.1007/978-3-031-05419-8_18

well seasoned with useful and applicable insights. Even more than in the past century of primary socio-economic-political catastrophes and clashes, the ecological and bio-behavioral challenges humanity now faces demand the application of geography's traditional perspectives if they are to be met and overcome.

18.1 Substance: *plus ça change, plus c'est la même chose*

It is gratifying to note the degree to which geography's persistent quest to understand the natural science and human components of places and regions is referenced in so many chapters (especially Chap. 13) in *A Geographical Century*, whether the focus is past, present, or future. Geography *is* distinctive within the family of sciences in bridging natural science and social science perspectives, a distinction it shares to some degree with anthropology and psychology. Geography has long been almost *unique* among the sciences in the scales at which it analyzes and seeks to explain the interplay of natural and human components of places and regions. Geography has become somewhat less nearly unique in that respect since the concept of coupled systems was applied in ecology and related specialties in the 1990s. Nevertheless, human use of the earth in its places and regions remains geography's fundamental and widely applicable perspective and analytic framework, a core component of the discipline. Its key role as a bridge cited implicitly or explicitly in so many of the chapters in *A Geographical Century* has and will continue to serve the IGU and the world's geographers well in the years and decades to come.

An exogenous recent trend catalyzing the IGU's increasing centrality in the global scientific community is the realization by natural science specialists that they cannot devise effective solutions to many of the problems they address without partnering with social scientists, as well as with behavioral scientists, engineers, humanities experts, and members of the design professions. Increasing openness to more collaborative approaches to solving ecological dilemmas is expanding opportunities for geographers to act as full partners in big Science. The 2018 amalgamation of the International Council for Science (ICSU) and the International Social Science Council (ISSC) into the International Science Council (ISC) portends a future in which multi- and trans-disciplinary research and action programs are populated by teams of scholars. The IGU (a longstanding member of both ICSU and ISSC) played a key role in promoting and achieving the ICS.

Bridging the natural and human sciences is geography's birthright, a legacy often cited throughout *A Geographical Century*. It is time to redouble the discipline's exercise of that birthright, as is suggested several times in the volume. Because most of the world's problems are to a greater or lesser degree rooted in human behavior, the IGU should seek and cement linkages with organizations that serve the behavioral sciences and the humanities. Furthermore, collaboration with engineering societies and those for design professionals (architects and planners, for example) could be productive. In that regard, the IGU's recent linkages with the International Council for Philosophy and Human Sciences CIPSH noted in Chapt. 14 represent progress in that direction.

18.2 Spirit

What *has not* been written in *A Geographical Century* is notable and commendable. Much has changed in the IGU since my first participation in the 23rd Congress in Moscow in 1976. In ethos and operations, the IGU through the 1980s was a nineteenth- and (in a few respects) even an eighteenth-century organization. This volume's authors and editors have unfailingly exhibited respect for the IGU's past without the nostalgia and ancestor worship that too often permeates milestone publications. The IGU's past is indeed prologue for its next century, but it is taken here as a platform for continuous improvement rather than a constraint on current and future practice. If that spirit is a response to the directions of volume planners and editors, many kudos to them. If that spirit emerged spontaneously in the crafting of the chapters, even better, and many more kudos to everyone.

18.3 Assessment

How well, in terms applicable to scholarly societies, has the IGU performed during its first century? John H. D'Arms (1934–2002) often spoke of evaluating the effectiveness of scholarly organizations in terms of the services they provide for members of their disciplines, that is, how well they do as Convenors, Funders, Advocates, and Validators.[1]

Convening. The IGU has excelled at convening the world's geographers, owing largely to its origins in the history and tradition of international geographical congresses that preceded its formal establishment as an ongoing enterprise. IGU Congresses (and its Regional Conferences) are usually executed with considerable grace and panache. But as its author noted in Chap. 6, participation in meetings is heavily influenced by congress location, and IGU meetings provide international benefits to hosts as well as their guests. Beyond location, the persistent income disparities among potential participants and the language barriers inherent in international/global meetings mean that in practice, actual participation falls woefully short of what might have been achieved absent those barriers. Furthermore, the inability of some countries to muster even the very nominal dues of the lowest membership category results in their status as observers—rather than voting members—even when a country does manage to send a delegate to an IGU Congress and its General Assembly. The advent of postal/email balloting in virtual general assemblies has broadened participation in IGU governance, but difficulties in paying dues persist for too many countries. Schemes to subsidize dues for low-income countries have been discussed but only a few have been implemented, with mixed success. A persistent difficulty has been that any subsidy by one member country of another, risks the appearance, if not a reality, of a patron-client relationship.

Whether the adoption of virtual or hybrid virtual-traditional congresses and regional conferences in the future will increase participation in the IGU's meetings remains to be seen. The success of the postponed Istanbul 2020 would seem to augur well for the use of digital technologies to augment and replace face-to-face interaction, and digital meeting offers significant ecological benefits. Given the ways innovative ideas are transmitted at IGU meetings (Chap. 5), it is difficult to envision virtual replacements for the after-hours informal exchanges where new ideas often arise.

Human beings are herd animals by nature. Establishing the trust that is prerequisite to collaboration is much more quickly done face-to-face than via digital media. Moreover, impediments rooted in language differences (Chaps. 7 and 9) will persist and may even be aggravated by the poor sound fidelity characteristic of virtual media software. Software and sound will improve with time, but the complications of conducting meetings among participants scattered across Earth's 24 time zones will remain even should digital simultaneous translation and high-fidelity sound and video become available sooner than expected. Commanding interest and attention are much more difficult when participants are in ten or twelve time zones rather than only one or two.

Funding. IGU finances are only peripherally addressed in *A Geographical Century*, primarily regarding the difficulties prospective delegates have in obtaining funds to attend IGU meetings, and that many countries have in paying IGU dues. Although not a focus here, the topic cannot be ignored, as the IGU Executive Committee's members and IGU Commission and Steering

[1] D'Arms was a seasoned scholar and academic administrator. After a distinguished career in Classics at the University of Michigan, he was appointed President of the American Council of Learned Societies (ACLS) in 1997 and served in that capacity until his untimely death in 2002. The ACLS is an umbrella organization of humanities and social science societies. As of 2021 it was comprised of 78 member societies. The American Association of Geographers was and remains a member of ACLS.

Committee members know only too well. The IGU would appear to have done as well as it could reasonably have been expected to in recent years. Its modest funding for its Commissions and Task Forces has been somewhat augmented. Competitions for limited grants to attend congresses and conferences are now available for individuals, Major grants have been won from external sources to support worthwhile IGU projects, and a permanent endowment has been established.

As welcome as these improvements are, the IGU remains hampered by the limited funds it commands, which come almost exclusively from the country dues it receives. Wealthier international scientific unions sometimes enjoy strong financial support from cognate industries, as, for example, the petroleum industry and petroleum exporting countries that underwrite the operations of the International Union of Geological Sciences (IUGS).

The IGU badly needs a "sugar industry" a role that was sometimes fulfilled by national mapping agencies in the past. The travel industry would seem to be a logical financial partner, but it is highly decentralized, and seeking donations amid sometimes draconian travel restrictions is unlikely to be rewarding. The hazard insurance industries and especially major reinsurance providers might be targets for the IGU to consider in future funding quests, given their exposure to losses caused by environmental disasters.

Advocating. The IGU's Commissions on Geographic Education and on Gender and Geography have been consistent advocates for their members' specialties. The IGU's overall effectiveness would be enhanced if their examples were emulated by other commissions and task forces for which advocacy is appropriate. In some respects, the establishment of a commission or task force devoted to a topic or problem is, in and of itself, implicit advocacy. In today's world and the future, more explicit and intentional publicizing of the result and benefits of specific lines of geographical inquiry and practice should be contemplated.

The IGU does engage in a critical and highly beneficial form of advocacy largely out of the sight of its members by nominating qualified geographers to serve on and in the committees, commissions, projects, and programs that manage international science. That advocacy is done largely by the IGU's Presidents and Secretaries-General through the umbrella international organizations of which the IGU is a member, particularly the former International Council for Science (ICSU) and the former International Social Science Council (ISSC) and now into the unified International Council Science (ISC). Though admitted not necessarily a topic for a centennial volume, the quiet advocacy accomplished by arranging the selection of geographers for influential positions in international science is one of the keys to the IGU's effectiveness.

Validating. Scholarly societies validate their constituencies' research, teaching, and service by publishing their members' research and by conferring honors and awards upon members and non-members for outstanding accomplishments. The IGU's publication efforts have historically been centered on the outputs of its commissions and task forces, published by those originating bodies, rather than as an executive committee's responsibility. Publications emerging from the secretariat have been informal and intermittent throughout much of the IGU's first century, and when undertaken, largely focused on the IGU's internal audience rather than directed to non-IGU audiences. The recently concluded publication arrangements with Springer for an Advances in Geographical and Environmental Sciences series, a Perspectives on Geographical Marginality series, and with Edward Elgar Publishing for The International Geographical Union Series on Contemporary Geographies, are welcome steps toward a more robust program of disseminating research results under the IGU imprimatur.

The validation of accomplishments by the IGU with honors and awards was formalized only during the last decade. The 1972 conception of the IGU *Lauréat d'honneur* was hurried and not completely thought out; an internal debate continues about whether the *Lauréat* should be primarily an award for service to the IGU, whether it should be conferred for substantive research and practice accomplishments, or both. Before 2012, decisions on honors and awards were made in informal and sometimes tense discussions within the IGU Executive Committee, without a call for nominations and formal documentation. The IGU's Planet and Humanity Medal was more mature in conception and has more successfully met its designed purposes. Recognizing and lauding significant scientific accomplishments is a comparatively low-cost means of raising a society's profile within its global and cognate disciplines. Thus, the inauguration of two new IGU awards this year represents major progress in the IGU's validation. The first IGU Early Career Awards in Geography, and the first IGU Awards for Distinguished Geographical Practice will be awarded at the July 2022 Congress in Paris.

I assign no grades to the IGU on these report card headings despite its format. Some of the functions cited above were, for good reasons, not addressed in this volume, and some are of limited applicability for an international society, formulated as they were in the context of national organizations. In the IGU context, a heading of *Coordination* could well have been added to D'Arms' list. Scholarly research and teaching have historically been largely solitary endeavors, especially in the humanities. The value and necessity of coordinating scholarly work among teams of individuals or across political boundaries have been limited in the past, but it is a major *raison d'être* in the IGU, one that it performs well indeed owing to the provisions written into its statutes.

18.4 Toward 2122

In addition to the suggestions regarding the IGU's further development noted above, several points seem to be worth further examination or re-examination given changing national sentiments and social and intellectual mores.

More than half of the IGU's first century after World War Two was devoted to weaning itself from its Euro-Anglo origins, a process that continues. Positively, there's been a persistent quest to increase the IGU's country membership, notwithstanding the reluctance to allow Germany into the Union following World War One and the foot-dragging in admitting Germany and members of the Eastern and Asian Blocs during the Cold War (Chaps. 3 and 4). The British exit from the European Union and growing nationalism and calls for autarky seem to be at least partially at odds with the kinds of international partnerships the IGU and similar organizations have sought to create and strengthen. As suggested in Chap. 2, traditional concepts of internationalism may be less useful than they have been in the past.

In addition to changing attitudes within the global science community, technological choices drive many aspects of scientific practice and create new options for multi- and trans-disciplinary collaboration. The utility of digital dissemination of research findings reflects the similar power the same technologies wield in producing those findings (Chap. 8). Because language is so intimately linked to knowledge acquisition and education, the relationships between language and geographical education (Chap. 10) deserve special and continuing attention. The widespread shifts from in-person learning and instruction to distance education occasioned by the COVID pandemics of the last two years (with successive waves like to continue) will force accelerated innovation in education for as long as the epi- and pandemics persist. In the admittedly macabre tradition of never letting a good crisis go to waste, the IGU and its member countries should make serious efforts to use the emergency funding often available for alternatives to traditional instruction to seek new and effective ways of teaching geography. The IGU Commission on Geographical Education seems in fact to lie at the nexus of several salient needs and themes identified in *A Geographical Century*: geography itself, global awareness and understanding (Chap. 14), sustainability (Chap. 15), language (Chap. 9), feminist perspectives (Chap. 12), and technology (Chap. 8).

18.5 Envoi

Any international organization that commenced operations 100 years ago has endured stresses and shocks and has sometimes overlooked opportunities it should have grasped. Having observed the IGU's leadership for 45 of those years as a Congress and Conference participant, a component of its leadership as a Commission Steering Committee member, and then a Vice President, Secretary-General, and President, I believe that the organization has served its members and its discipline faithfully and well. As documented throughout *A Geographical Century*, the IGU is up to date, alert for opportunities, and well embedded in the world's international scientific community—including its premier scientific consortium: the International Science Council.

We have no guarantee that the years and decades ahead will be any less stressful and dangerous than those the IGU has endured. On the contrary, conditions, and events in late 2021 suggest that anticipating and preparing for even greater disruptions and turbulence would be prudent. Whatever ensues, these accounts, assessments, and descriptions of the IGU's responses to the challenges its leaders and members have encountered since 1922, and the authors' suggestions for the immediate future, bolster my conclusion that the IGU is in good condition and well prepared to succeed in its missions during its next century.

Ronald F. Abler is Emeritus Professor of Geography at The Pennsylvania State University. He was active in the IGU starting with the 1976 IGU Congress in Moscow, USSR, was a charter member of the IGU Commission on the Geography of Communications, IGU Vice President (1996–2000, 2003–2008), Secretary General (2000–2003), President (2008–2012), and Past President (2012–2016). He currently chairs the IGU Honors and Awards Committee.

Printed by Printforce, the Netherlands